The European Families of the Diptera

of the **Diptera**

Identification, diagnosis, biology

The European Families of the Diptera

Identification, diagnosis, biology

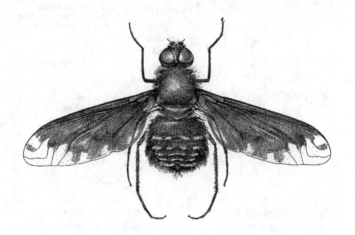

Pjotr Oosterbroek

KNNV Publishing

KNNV
vereniging
voor veldbiologie

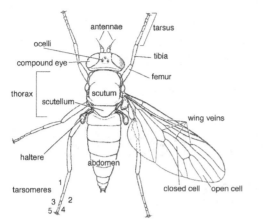

habitus brachycerous fly

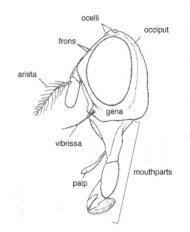

head calyptrate fly

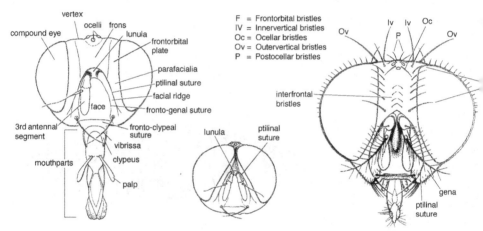

F = Frontorbital bristles
IV = Innervertical bristles
Oc = Ocellar bristles
Ov = Outervertical bristles
P = Postocellar bristles

head calyptrate fly;
eyes separated
(dichoptic)

eyes in contact
with each other
(holoptic)

head calyptrate fly

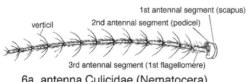

6a antenna Culicidae (Nematocera)

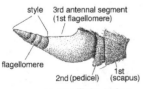

antenna Tabanidae
(lower Brachycera)

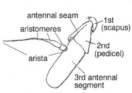

antenna calyptrate fly
(higher Brachycera)

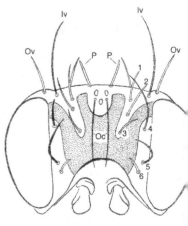

head acalyptrate fly
1-6 = F-bristles

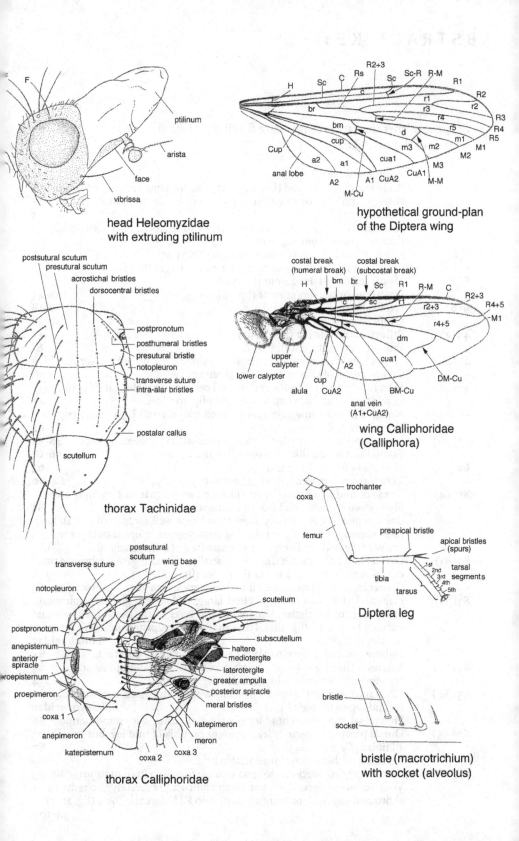

head Heleomyzidae with extruding ptilinum

F, ptilinum, arista, face, vibrissa

hypothetical ground-plan of the Diptera wing

H, Sc, C, Rs, R2+3, Sc, Sc-R, R-M, R1, R2, r1, r2, R3, br, r3, R4, r4, R5, c, bm, d, r5, m1, M1, cup, Cup, m3, m2, M2, a2, a1, cua1, M3, anal lobe, CuA1, A2, A1, CuA2, M-M, M-Cu

thorax Tachinidae

postsutural scutum, presutural scutum, acrostichal bristles, dorsocentral bristles, postpronotum, posthumeral bristles, presutural bristle, notopleuron, transverse suture, intra-alar bristles, postalar callus, scutellum

wing Calliphoridae (Calliphora)

costal break (humeral break), costal break (subcostal break), H, bm, br, Sc, R1, R-M, C, R2+3, c, sc, r1, r2+3, R4+5, M1, r4+5, dm, upper calypter, lower calypter, A2, cua1, alula, cup, CuA2, BM-Cu, DM-Cu, anal vein (A1+CuA2)

Diptera leg

trochanter, coxa, femur, preapical bristle, apical bristles (spurs), tibia, tarsus, 1st, 2nd, 3rd, 4th, 5th, tarsal segments

thorax Calliphoridae

postsutural scutum, transverse suture, wing base, notopleuron, scutellum, postpronotum, subscutellum, anepisternum, haltere, mediotergite, anterior spiracle, laterotergite, proepisternum, greater ampulla, proepimeron, posterior spiracle, meral bristles, coxa 1, katepimeron, anepimeron, meron, katepisternum, coxa 2, coxa 3

bristle (macrotrichium) with socket (alveolus)

bristle, socket

ABSTRACT KEY

Purpose of this abstract is to facilitate finding a particular
section of the main key.

1	Body and wing normal.	2
–	Body strongly flattened (fig. 441-443) and/or wing reduced or absent.	191
2(1)	Antenna filiform or resembling a series of beads, with more than 6 segments (fig. 17-19)	3
–	Antenna with less than 10 segments; third segment nearly always larger than the subsequent segments (fig. 20-21).	44
3(2)	2 anal veins reaching the wing margin (fig. 23-35).	4-7
–	Only 1 (or no) anal veinreaching the wing margin (fig. 24).	8
8(3)	Discal cell or cell dm absent (fig. 39-83).	9
–	Discal cell or cell dm present (fig. 38).	38-43
9(8)	Ocelli absent.	10-18
	Ocelli present (fig. 58, 73).	19-37
44(2)	Ptilinal suture absent (fig. 92-97).	45
–	Ptilinal suture present (fig. 91).	81
45(44)	Thorax humpbacked and abdomen usually globose (fig. 98-99).	**Acroceridae**
–	Combination of characters not as given above.	46
46(45)	Vein R4+5 forked into R4 and R5 and cell cup very elongate(fig. 105, 113), or wing venation aberrant as shown in fig. 107-109.	47
–	Not at the same time vein R4+5 forked into R4 and R5 and cell cup very elongate (e.g. fig. 101-102, 133).	62
47(46)	Empodium pulvilliform (last tarsal segments with 3 cushions; fig. 103).	**48-55**
–	Empodium bristle-like or absent (fig. 104).	**56-61**
62(46)	Vein R4+5 forked (fig. 132-134).	63
–	Vein R4+5 not forked (all wing figures beyond 134).	**64-80**
81(44)	Coxae of mid legs not far apart. Tarsal claws normal, not strongly curved. Fly not ectoparasitic on birds or mammals.	82
–	Coxae of mid legs, and usually of the other legs as well, far apart (fig. 174a). Tarsal claws strongly curved (fig. 174b) or last tarsal segment widened and eyes reduced or absent (fig. 175). Fly living as an ectoparasite on birds or mammals.	169
82(81)	Greater ampulla present (fig. 176). Usually 1 or more strong vibrissae present, as well as incurved lower F-bristles (fig. 178).	170
–	Greater ampulla inconspicuous; if well developed, then vibrissae absent.	83
83(82)	Haltere dark brown to black. Head large, wide and high (hemispherical); only 1 pair of F-bristles, situated at the level of the ocellar triangle and curving backward (fig. 182). Hind margin of anepisternum with a row of bristles (fig. 183).	**Lonchaeidae**
–	Haltere not dark brown to black or other characters different.	84
84(83)	Vibrissae (fig. 184-189) or vibrissa-like bristles (fig. 186) present.	85
–	No vibrissae or vibrissa-like bristles (fig. 187-188).	134
85(84)	First tarsal segment of hind leg much swollen, and usually shorter than the second segment (fig. 184, 185	**Sphaeroceridae**
–	First tarsal segment of hind leg not swollen, usually long and slender.	86
86(85)	Dorsal preapical bristle at least present on tibia of mid leg, but usually on all tibiae (fig. 190-204).	87
–	All tibiae without dorsal preapical bristle.	101
87(86)	Vein Sc complete, reaching the costa separate from vein R1 (fig. 190-193).	**88-93**
–	Vein Sc incomplete, does not reach the costa separately: Sc reduced or shortened (fig. 191), or merging with vein R1 before the costa (fig. 215b).	**94-100**

101(86)	Vein R2+3 ends in the costa about halfway up the wing; vein M1 apically pallid, not reaching the wing margin (fig. 227). **Xenasteiidae**
–	Vein R2+3 is strikingly short (fig. 228) or ends in the costa distinctly beyond the middle (fig. 230-235), or if vein R2+3 ends in the costa about halfway up the wing, then vein M1 reaching the wing margin (fig. 240). **102**
102(101)	Wing base narrow, alula and anal vein small or absent (fig. 228-237). **103-105**
–	Wing base not narrow, alula well developed. **106**
106(102)	Gena with 1 or more strong upcurved bristle(s) (fig. 238a). **Canacidae**
–	Gena without strong upcurved bristles. **107**
107(106)	Vein Sc abruptly bent forward toward the costa at nearly 90°; cell cup closed by a geniculate vein CuA2 and with an acute apical end (fig. 239a). **Tephritidae**
–	Vein Sc and cell cup not as given above. **108**
108(107)	Vein Sc complete, usually reaching the costa separate from vein R1 (fig 243-257). **109-119**
–	Vein Sc absent, or short and not reaching the costa, or merging with vein R1 distinctly before the costa (fig. 240, fig. 258-262, 267-298), or reduced to a fold that may or may not reach the costa (fig. 263-264). **120**
120(108)	Crossvein BM-Cu present (cells bm and dm separate) (fig. 260-272b). **121-127**
–	Crossvein BM-Cu absent (cells bm and dm fused) (fig. 258). **128-133**
134(84)	At the same time cell cup elongate (approaching the hind margin of the wing) and vein CuA2 (i.e., the vein closing cell cup) more or less straight; cell r4+5 narrowing toward the wing tip, or closed (fig. 299- 300). **Conopidae**
–	Cell cup short or absent, its apex not near the hind margin of the wing (fig. 302-347), or if long, then closed by a geniculate vein CuA2 (fig. 301). **135**
135(134)	Dorsal preapical bristle present at least on tibia of mid leg but usually on tibiae of all legs (fig. 309). **136-146**
–	Tibiae without dorsal preapical bristle. **147**
147(135)	Arista absent; third antennal segment very large (fig. 322). **Cryptochetidae**
–	Arista present. **148**
148(147)	Ocelli absent. **Pyrgotidae**
–	Ocelli present. **149**
149(148)	Vein Sc abruptly bent forward toward the costa at nearly 90°, going toward the costa as a transparent line or trace (fig. 326-330). **150**
–	Vein Sc not abruptly bent forward. **151**
151(149)	Vein Sc complete, its apical part not pallid, reaching the costa separately from vein R1 (fig. 333-357). **152-163**
–	Vein Sc absent, shortened, merging with R1 before the costa, or its apical part pallid and vague (fig. 332, 358-369). **164-168**
170(82)	Mouth opening small, its diameter up to 1/4 to 1/8 of the width of the head; mouthparts minute, rudimentary (fig. 375-382). **171-174**
–	Mouth opening large, its diameter at least 1/4 of the width of the head; mouthparts well developed, well visible (fig. 4, 7-8, 393-394). **175**
175(170)	Meron without one or more rows of bristles near its hind margin (fig. 384). **176-180**
–	Meron with bristles near the hind margin, arranged in one (fig. 386) or several (fig. 387) rows, or in a small cluster (fig. 408). **181-190**
191(1)	Antenna usually long, filiform or resembling a string of beads, with 6 or (usually) more segments; usually all segments or those beyond the second more or less similar (fig. 426, 437). **192-205**
–	Antenna not filiform or resembling a string of beads, with less than 6 segments; usually the first 2 segments short, the third segment large and the subsequent segments making up an arista or style. **206-220**

CONTENTS

INTRODUCTION

Some 25 years ago, a key to the European families of the Diptera was published, in Dutch (Oosterbroek 1981). The latest checklist of the Diptera of The Netherlands (Beuk 2002) was reason to completely revise and update this 1981 key, extended for each family with a short chapter on characters, biology and identification references (Oosterbroek et al. 2005). Of this book, also published in Dutch, an English version was prepared. Subsequently, this was substantially improved by the valuable corrections, comments and additions received from over 30 mainly European dipterists, leading to the present key for and overview of the families of the Diptera occurring in Europe, from Iceland and Fennoscandia in the north, the Ural in the east, and Macaronesia and the Caucasus in the south. Other recently published keys are Matile 1993-1995 (in French), Papp & Schumann 2000 (with a limited number of included figures) and Nartshuk 2003 (in Russian).

Identification of Diptera is often considered to be difficult. Experience shows that in particular the smaller species can present the student with various problems. This is partially due to the question whether the vein which forms the front margin of the wing (the costa) shows one or two breaks, or none at all. This character is used in many identification keys because of its reliability to distinguish between (groups of) families. It is nevertheless also sometimes a rather difficult character. Therefore, the key included here has been designed to employ this character as little as possible. This also applies to some other characters that can be difficult to ascertain, such as the bristles next to the posterior spiracle in the Sepsidae (fig. 196, 197). Another aspect of the Diptera is the overwhelming diversity of characters. In view of the ambition to cover all European genera (some 2.320) and species (about 19.200), a fair number of families key out more than once.

An obvious difference with respect to the key published 25 years ago (Oosterbroek 1981), is the number of families, which rose from 107 to 132. This increase is due to the fact that several families have been split up, with these 'new' families being internationally accepted. Some examples: the former Tipulidae have been split up into 4 families; Bibionidae into 3; Mycetophilidae into 5; Empididae into 4 or 5. Moreover, substantial progress has been made in the study of the Diptera. Since 1981, the Manual of the Nearctic Diptera (J.F. McAlpine et al. 1981-1989), the Catalogue of Palaearctic Diptera (Soós & Papp (eds) 1984-1993) as well as the Contributions to a Manual of Palaearctic Diptera (Papp & Darvas 1997-2000) have been published. Our understanding of the systematics also has been much improved (reviews in Yeates & Wiegmann 1999, 2005), as has our knowledge of distribution and faunistics (see the list of checklists of the separate countries).

The Nematocera (mosquitoes, midges, gnats, etc.) and Brachycera (true flies) together constitute the insect order Diptera, a word that derives from the Greek di = two and pteron = wing. A common character of all Diptera is the reduction of the second set of wings (the posterior pair) to a pair of halteres or wing knobs. This reduction has not impaired the flying capacity of these insects. In fact, many Diptera are on a par with other good fliers like dragonflies, some bees and wasps with respect to their manoeuvrability and speed. The transfer of the flying action to just the anterior pair of wings has led to a strong development of the middle part of the thorax, named the mesothorax. This character affords identification of the several Dipterous species that have lost their wings in the course of evolution, in one sex or both. These flightless species include parasites that have their body adopted in such a way that they would not otherwise be easily recognised as belonging to the Diptera.

The Diptera display a huge array of shapes and show a great variety in their biology. The number of species and that of individuals is overwhelming. At nearly all locations, the Diptera are the most numerous insects. For example, the number of Diptera species known from the Netherlands (4,967 after Beuk 2002) is more than twice that of the Lepidoptera (2,336, Kuchlein & De Vos 1999). It is also larger than that of the Coleoptera (4,041, Vorst & Huijbregts 2002). These figures, however, do not mean that in general the Dipterous fauna is well known. Several families have been relatively well studied but there is a considerable terra incognita comprising, for example, families rich in species like the Limoniidae, Ceratopogonidae, Mycetophilidae, Phoridae, Lauxaniidae, Agromyzidae, Ephydridae, Drosophilidae, Chloropidae, Anthomyiidae and many of the smaller families as well.

The study of the Diptera does not have to be limited to inventories of the fauna. In the field of biology much work is still to be done. The available information often derives from a small number of observations in only a few species. In particular, the vast diversity with regard to ecology and behaviour make the Diptera an interesting topic for research suitable for the amateur. For example, rearing phytophagous Diptera can in many cases be done without specialist training or means. One need not be an expert to discover new facts.

Diptera belong to the holometabolous insects, i.e., each individual leads two 'separate' lives: first as a larva, then as an adult. As a rule, the larval stages are considered necessary to build up and store reserves to enable the adult stage to exist. This final stage, the adult dipteron, is mainly directed at propagation. It is worthwhile to study the larvae, as even today very little or even nothing at all is known about this stage of life in a number of Dipterous families. Readers interested in this subject will find more information in Ferrar 1987; Matile 1995; Smith 1989; Stubbs & Chandler 1978; Courtney et al. 2000.

Collecting and preserving in a group as diverse as the Diptera requires various techniques. It is beyond the scope of the present book to deal with this. References on this subject are Gullan & Cranston 2005; Matile 1993; Schauff 2002; Smithers 1982; Stubbs & Chandler 1978; Upton 1991.

CLASSIFICATION

The order Diptera includes two major groups, viz., the Nematocera (mosquitoes, midges, gnats, etc.) and the Brachycera (true flies). Both groups are divided into a number of subgroups (see fig. 1 and Classification table below). There is some general agreement on the divisions mentioned here. The relationships between the various groups (fig. 1), however, are still a matter of debate. This also applies to the question of which groups are to be accorded family status.

In the table below, groupings of families for which no current name is available are indicated as Lower Muscomorpha, Lower Eremoneura, Lower Cyclorrhapha. The terms Higher and Lower are used in the literature on systematics to indicate the position of a group within the overall order of all groups.

The name Nematocera means 'threadlike antennae' and refers to the antennae in this group that are often long and usually comprise many segments. Conversely, the Brachycera have short antennae, most often consisting of three segments plus an arista. The maximum number of segments in the Brachycera is 10, with the exception of the Rachiceridae in which this number reaches about 40 due to secondary division of the segments.

The European Nematocera are split up into seven infraorders and the European Brachycera into four infraorders. One of the latter, the infraorder Muscomorpha, includes the vast majority of families and the large group Cyclorrhapha. In the Cyclorrhapha the larvae pupate inside a usually barrel-shaped puparium made up of the skin of the last larval instar. Adult Cyclorrhaphous flies leave this puparium through a more or less circular orifice. In case of all Brachycera that have been positioned earlier in the systematic order, the adults emerge through a longitudinal or a T-shaped slit, giving rise to the name of Orthorrapha that is often used to denote these Diptera.

In the Cyclorrhapha, the largest group are the Schizophora, comprising all Acalyptrate and Calyptrate fly families, all of which possess a ptilinum (see Terminology). In the Calyptratae the lower calypters are generally well developed, in the Acalyptrate flies this structure is small (see the identification key, no. 82). The latter group includes many families. In the Calyptrate families Hippoboscidae, Nycteribiidae and Streblidae, all parasites of mammals and birds, the larvae develop fully inside the female womb and pupate immediately after deposition. These families are sometimes grouped under the name Pupipara. Another group within the Calyptrate flies are the bots and warbleflies, including the three families of Gasterophilidae, Hypodermatidae and Oestridae. Their larval stages live as parasites in mammals. The adult flies show largely reduced mouthparts and are unable to feed.

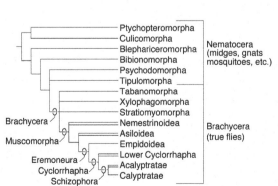

1 Outline of the relationships between the higher categories of the Diptera; after Yeates & Wiegmann 1999

CLASSIFICATION TABLE

Names of higher categories have specific endings, indicating their rank; some of the endings used in zoology are: Infraorder: -morpha; Superfamily: -oidea; Family: -idae; Subfamily: -inae; Tribe: -ini.
Behind the family name the number of genera and species in Europe is given.

SUBORDER INFRAORDER	SUPERFAMILY	FAMILY(132)
Nematocera (Midges and the like)		*32 families*
Tipulomorpha	Tipuloidea	Tipulidae: 10, 470
		Limoniidae: 70, 560
		Pediciidae: 5, 60
		Cylindrotomidae: 4, 6
	Trichoceroidea	Trichoceridae: 2, 50
Blephariceromorpha	Blephariceroidea	Blephariceridae: 5, 38
Axymyiomorpha	Axymyioidea	Axymyiidae: 1, 1
Bibionomorpha	Pachyneuroidea	Pachyneuridae: 1, 1
	Bibionoidea	Bibionidae: 2, 47
		Pleciidae: 1, 2
		Hesperinidae: 1, 1
	Sciaroidea	Mycetophilidae: 70, 945
		Bolitophilidae: 1, 36
		Diadocidiidae: 1, 5
		Ditomyiidae: 2, 2
		Keroplatidae: 16, 110
		Sciaridae: 28, 620
		Cecidomyiidae: 280, 1640
Psychodomorpha	Psychodoidea	Psychodidae: 40, 500
	Anisopodoidea	Anisopodidae: 1, 10
		Mycetobiidae: 2, 4
	Scatopsoidea	Scatopsidae: 22, 100
		Synneuridae: 1, 1
		Canthyloscelidae: 1, 2
Ptychopteromorpha	Ptychopteroidea	Ptychopteridae: 1, 15
Culicomorpha	Culicoidea	Dixidae: 2, 32
		Chaoboridae: 3, 9
		Culicidae: 8, 105
	Chironomoidea	Thaumaleidae: 3, 75
		Simuliidae: 8, 230
		Ceratopogonidae: 30, 590
		Chironomidae: 180, 1190
Brachycera (True Flies)		*100 families*
Xylophagomorpha	Xylophagoidea	Xylophagidae: 1, 5
		Coenomyiidae: 1, 1
		Rachiceridae: 1, 1
Stratiomyomorpha	Stratiomyoidea	Xylomyidae: 2, 8
		Stratiomyidae: 27, 140

9

Tabanomorpha	Tabanoidea	Athericidae: 4, 10
		Rhagionidae: 6, 85
		Tabanidae: 13, 220
		Vermileonidae: 2, 9
Muscomorpha		
Lower Muscomorpha	Nemestrinoidea	Nemestrinidae: 6, 13
		Acroceridae: 8, 35
	Asiloidea	Bombyliidae: 52, 340
		Mythicomyiidae: 7, 30
		Hilarimorphidae: 1, 2
		Therevidae: 17, 100
		Scenopinidae: 2, 16
		Mydidae: 4, 7
		Asilidae: 77, 540
Eremoneura		
Lower Eremoneura	Empidoidea	Atelestidae: 3, 4
		Hybotidae: 30, 440
		Empididae: 23, 810
		Brachystomatidae: 5, 13
		Microphoridae: 4, 16
		Dolichopodidae: 60, 775
(Cyclorrhapha)		
Lower Cyclorrhapha	Platypezoidea	Platypezidae: 12, 45
		Opetiidae: 1, 1
		Phoridae: 35, 605
	Lonchopteroidea	Lonchopteridae: 2, 13
	Syrphoidea	Syrphidae: 90, 830
		Pipunculidae: 14, 200
Schizophora		
Acalyptratae	Nerioidea	Micropezidae: 5, 22
		Pseudopomyzidae: 1, 1
	Diopsoidea	Tanypezidae: 1, 1
		Strongylophthalmyiidae:
		Psilidae: 6, 50
		Megamerinidae: 1, 1
		Diopsidae: 1, 1
	Conopoidea	Conopidae: 14, 85
	Tephritoidea	Lonchaeidae: 8, 100
		Pallopteridae: 5, 23
		Piophilidae: 16, 30
		Ulidiidae: 17, 105
		Platystomatidae: 2, 20
		Tephritidae: 70, 270
		Pyrgotidae: 1, 1
	Lauxanioidea	Lauxaniidae: 18, 160
		Chamaemyiidae: 9, 110
		Cremifaniidae: 1, 2
	Sciomyzoidea	Coelopidae: 2, 3
		Dryomyzidae: 2, 4
		Helcomyzidae: 1, 2
		Heterocheilidae: 1, 1
		Sciomyzidae: 24, 140

		Phaeomyiidae: 1, 3
		Sepsidae: 9, 50
	Opomyzoidea	Clusiidae: 5, 15
		Acartophthalmidae: 1, 3
		Odiniidae: 3, 14
		Agromyzidae: 23, 910
		Opomyzidae: 3, 33
		Anthomyzidae: 9, 30
		Aulacigastridae: 1, 4
		Stenomicridae: 1, 3
		Periscelididae: 1, 4
		Xenasteiidae: 2, 2
		Asteiidae: 4, 18
	Carnoidea	Milichiidae: 9, 45
		Carnidae: 3, 40
		Cryptochetidae: 1, 3
		Braulidae: 1, 3
		Tethinidae: 3, 35
		Canacidae: 2, 4
		Chloropidae: 65, 395
	Sphaeroceroidea	Heleomyzidae: 23, 150
		Trixoscelididae: 1, 25
		Chiropteromyzidae: 2, 2
		Cnemospathidae: 1, 1
		Chyromyidae: 3, 60
		Sphaeroceridae: 40, 260
	Ephydroidea	Drosophilidae: 17, 120
		Campichoetidae: 2, 7
		Diastatidae: 1, 9
		Curtonotidae: 1, 1
		Camillidae: 1, 8
		Ephydridae: 60, 335
	Superfamily unknown	Nannodastiidae: 1, 3
Calyptratae	Hippoboscoidea	Hippoboscidae: 11, 30
		Nycteribiidae: 4, 15
		Streblidae: 1, 1
	Muscoidea	Scathophagidae: 38, 160
		Anthomyiidae: 35, 480
		Fanniidae: 3, 82
		Muscidae: 45, 575
	Oestroidea	Calliphoridae: 22, 115
		Rhinophoridae: 11, 45
		Sarcophagidae: 30, 310
		Tachinidae: 280, 880
		Oestridae: 4, 8
		Gasterophilidae: 1, 6
		Hypodermatidae: 4, 8

(Higher Classification according to Yeates & Wiegmann 1999)

TERMINOLOGY

People can be overwhelmed by the multitude of terms used to describe the characters necessary for insect identification. The terms often vary between groups and are often used only by specialists working on a given group. In this book only those terms are used that have been internationally accepted by Diptera workers. They are mainly concerned with the head and its various bristles, the thorax, wings, wing venation and legs. Terms that occur frequently and those used in the text are enumerated here by order of body parts and in alphabetical order. For a more extensive explanation and a fuller survey of terms, the reader is referred to Ziegler 2003 and the introductory chapters of the Manuals of Nearctic and Palaearctic Diptera, respectively by J.F. McAlpine et al. (1981-1989) and Papp & Darvas (1997-2000).

GENERAL TERMS

Anthropogenous or Synanthropic. Mainly occurring in habitats created by man.

Apterous. Without wings (fig. 440).

Brachypterous. Having short wings (fig. 448).

Brackish. Having a salinity between that of fresh water and seawater.

Bristle. See Macrotrichium (fig. 16).

Carnivorous. Feeding on meat (see also Predator).

Cerci (singular: Cercus). Used in this book to refer to the dorsal valves of the ovipositor in some Nematocera (fig. 31). See Sexual dimorphism.

Commensalism (Commensal). The relation between two kind of organisms (commensals) in which one obtains food, protection or other benefits from the other without damaging or benefiting it.

Coprophagous. Feeding on excrement.

Cosmopolitan. Distributed world-wide rather than regional, usually common in most parts of the world and under varied ecological conditions. See also Ubiquitous.

Facultative. Optional; able to survive and/or propagate itself under a variety of circumstances.

Female. See Sexual dimorphism.

Filiform. Having the shape of a thread or filament.

Forensic entomology. A branch of applied entomology largely dedicated to crime-solving and coroner's inquests. Used, for example, to determine the date of death of a deceased person.

Genitalia, male (hypopygium). See Sexual dimorphism.

Habitat. Ecological range; the preferred environment of a plant or animal species (e.g. marsh, forest, alpine meadows, etc.).

Hairs. See Microtrichium.

Halophilous (Halobiont). Flourishing in a saline or salty habitat; such an organism is called a halobiont.

Hemispherical. Resembling a hemisphere, the half of a sphere.

Hypopygium. Male genitalia. See Sexual dimorphism.

Infuscated. Darkened with a tinge, e.g. the wing membrane can be entirely darkened by a brownish tinge.

Inquiline. An animal that lives in the nest or abode of some other species.

Joint. See segment.

Larviparous. The eggs hatch inside the female which deposits larvae instead of eggs.

Macrotrichium. This is a general term for a bristle or seta. This can be recognised by the reinforcement ring at the base, named the alveolus (fig. 16). Oversized macrotrichia are named spurs or spines, e. g., the apical spurs or spines on the tibiae (fig. 75). Normal macrotrichia are called bristles or setae, small ones are called setulae.

Male. See Sexual dimorphism.

Microtrichium. This term is reserved for elements without an alveolus (fig. 16). Usually they are hair-like; hence, they are often called hairs which is, strictly speaking, erroneous since real hairs occur only in mammals.

Mining. Living in, and eating its way through, plant tissue.

Mycophagous. Feeding on mushrooms, (bracket) fungi, or mycelium.

Myiasis. Infection or disease caused by Diptera larvae.

Necrophagous. Feeding on dead animals.

Obligatory. Restricted to a specific way of life or mode of propagation.

Oviscape. Non retractable basal segment of the ovipositor. See also Sexual dimorphism.

Oviparous. Laying eggs.

Ovipositor. See Sexual dimorphism.

Parasite. Feeding in (endoparasite) or on (ectoparasite) another, usually much larger animal, in an intimate (subtle) manner, for the entire or a relatively long portion of the parasite's life cycle, causing some damage but consuming a small proportion of the host, not killing it.

Parasitoid. Like a parasite feeding in or on the host in an intimate manner for a relatively long period, but unlike a parasite and like a predator in predictably killing the host and consuming most or all of its tissues. In some cases, several hosts are required for the development of a single parasitoid, e.g. in some Sciomyzidae.

Parthenogenetic. Capable of reproduction that involves development of a female (or rarely a male) gamete without fertilisation.

Phytophagous. Feeding on plant materials.

Pilose. Covered with hair, especially of soft texture.

Pilosity. The state of being pilose.

Plumose. Provided with rays of long hairs or hair-like bristles.

Predator (Predatory, Predacious). Attacking prey and feeding upon living animal tissue.

Pruinose. Covered with whitish dust or bloom.

Pruinosity. The state of being pruinose.

Pubescent. Covered with fine soft short hairs.

Pupiparous. The larvae develop inside the female and pupate immediately after having been deposited.

Raptorial. Living on prey; adapted to seize prey.

Reniform. Resembling a kidney in shape.

Saline. Salty to a degree just below or above that of seawater.

Saprophagous. Feeding on decaying organic matter.

Segment. In all Diptera, many body parts consist of a series of externally sclerotized members, the segments. They are separated by joints. Segments are numbered going from the body outward, starting from the head.

Seta(e). See Macrotrichium (fig. 16).

Setula(e). See Macrotrichium.

Sexual dimorphism. Although the differences between male and female are generally limited, there are many cases in the Diptera where there are secondary sexual characters and some where the sexes look quite different (e.g. Platypezidae, Anthomyiidae). Usually males are smaller, show better development of the antennae and a more conspicuous colour pattern. In several families the males have larger eyes which meet each other at a point or along a line (holoptic). Differences in wing shape or venation sometimes occur.

The male can be recognised by the presence of the genitalia or hypopygium. This consists of a set of intricate structures the male needs to hold the posterior end of the female abdomen during copulation. The hypopygium can be largely visible, which is the case in many Nematocera and Brachycera, but quite often it is swung down and protected by a covering plate (e.g. the epandrium) or it can be sunk into the abdomen proper to a large extent.

The female generally has an ovipositor. In the 'primitive' families this is not really a tube, but a set of paired dorsal and ventral appendages (cerci and hypogynial valvae, together named ovipositor). In some of the higher Diptera a tube has been formed through fusion of several elements. In some instances, e.g. in the Tephritoidea, part of this tube, the oviscape, is always visible as an elongate structure (fig. 331, 350, 351, 353, 357).

Spine. See Macrotrichium.

Spur. See Macrotrichium.

Synanthropic (Anthropogenous). Mainly occurring in habitats created by man.

Ubiquitous. Occurring everywhere, or in many places in different ecosystems, but not necessarily throughout the world (= Cosmopolitan).

ORIENTATION AND SHAPE

Note that combinations of orientation occur: Posteroventral means posterior and ventral at the same time. These combinations are often necessary to describe a location more exactly.

Acute. Ending in, or with a sharp angle.

Anterior. At the head end or directed toward the head end.

Anterodorsal. Position intermediate between anterior and dorsal.

Anteroventral. Position intermediate between anterior and ventral.

Apicad (Distad). Toward the apex. Opposite of basad.

Apical (Distal). At or near the top; in the direction of the top, the apex (fig. 2a).

Basad (Proximad). Toward the base. Opposite of apicad.

Basal (Proximal). Near the base, the beginning; in the direction of the base (fig. 2a).

Caudad. Backward, toward the rear. Opposite of rostrad.

Caudal. At or near the rear (fig. 2a).

Cephalic. Forward or near, at or on the head (fig. 2a).

Concave. Hollow, curving inward.

Continuous. Not interrupted (costa).

Converging bristles. Bristles that are directed toward each other, meet or cross each other; e.g. fig. 200: P-bristles.

Convex. Bulging, curving outward.

Distad. Another term for apicad.

Distal. Another term for apical (fig. 2a).

Diverging bristles. Bristles that are directed away from each other; e.g. fig. 9: P-bristles.

Dorsal. On the upper side; viewed from above.

Dorsoventral. Down from the top.

Dorsoventrally compressed or flattened. Conspicuously wider than high. Opposite of laterally compressed.

Elongate. Oblong, (much) longer than wide.

Equidistant. At the same distance.

Exclinate bristles. Bristles curving outward; e.g. fig. 8: F-bristles.

Geniculate. With a kink or sharp curve; a geniculate element consists of two (or more, if it is twice geniculate) parts with different orientation.

Inclinate bristles. Bristles curving inward, mediad; e.g. fig. 9: bristles 4-6.

Interrupted. With one or two breaks (costa).

Laterad. From the centre line toward the side. Opposite of mediad.

Lateral. Of or on the side, viewed from the side.

Laterally compressed. Conspicuously higher than wide. Opposite of dorsoventrally compressed.

Lateroclinate bristles. Bristles directed outward; e.g. fig. 8: F-bristles.

Mediad. From the side(s) toward the middle; toward the plane of symmetry. Opposite of laterad.

Medial. In the middle, in the plane of symmetry (fig. 2a).

Parallel. With the same orientation, equidistant at all points.

Posterior. At the hind end or directed toward the hind end.

Posterodorsal. Position intermediate between posterior and dorsal.

Posteroventral. Position intermediate between posterior and ventral.

Preapical. Near, but not at the top or apex. An example is the dorsal preapical bristle of the tibia (fig. 15). This is inserted on the dorsal side of the tibia, near but distinctly before the end (apex).

Proclinate bristles. Bristles curving forward, rostrad; e.g. fig. 245: lowermost F-bristles.
Proximad. Another term for basad.
Proximal. Another term for basal (fig. 2a).
Pulvilliform. Of the same shape as a pulvillum (fig. 103). See Leg: Tarsus.
Reclinate bristles. Bristles curving backward, posteriorly; e.g. fig. 200: F-bristles.
Rostrad. Toward the front, forward. Opposite of caudad.
Terminal. At the end, at the top.
Ventral. On, of, toward the lower side; viewed from below.

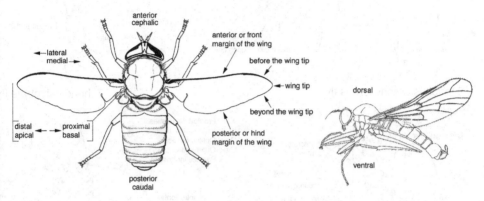

2a Tabanidae (Tabanus)   2b Brachystomatidae (Brachystoma)

2 Terms used in pointing out the orientation

HEAD

Antenna (fig. 3-7). In naming the various parts of the antenna (fig. 6 a-c), a distinction is made between the two basal segments (first = scape, second = pedicel) and all other elements together (= the flagellum). The elements of the flagellum are named flagellomeres and therefore the third antennal segment is also the first flagellomere.

In the Nematocera the flagellum consists of several (in general about 14) usually simple and uniform segments (flagellomeres). They often show a basal ring of long hairs, named the verticils (fig. 6a). In the lower Brachycera the antennal segments can be uniform as well, similar to what is found in the Nematocera. Usually, however, the flagellomeres (some 8 or less) are not that similar. The first flagellomere is often the largest, the consecutive ones are conical and together they make up a style (fig. 6b) or, as in the higher Brachycera, they constitute the arista which can be similar to a bristle and which appears as an elongation of the first flagellomere. Separate segments of the arista are named aristomeres (fig. 6c). The arista can be bare (fig. 6c), pubescent (fig. 10), or plumose (fig. 7, 8).

Arista. See Antenna.

Aristomere. See Antenna.

Bristles of the head (fig. 8, 9). The place of insertion and the orientation of the bristles on the head are of crucial importance to the identification of acalyptrate flies. The vertical bristles are found on the upper margin of the head, the vertex. These bristles are divided into the P-bristles (post-ocellar bristles) immediately behind the ocellar triangle (fig. 8, 9: P); the Iv-bristles (innervertical bristles, fig. 8, 9: Iv) and Ov-bristles (outervertical bristles, fig. 8, 9: Ov). In between the eyes, usually along the eye margin, are the lower frontorbital bristles (fig. 9, bristles 4-6). More dorsally, the upper frontorbital bristles are located (fig. 9, bristles 1-3). In this book, all frontorbital bristles are included under the term F-bristles (fig. 8: F).

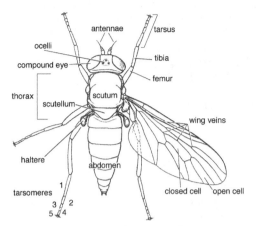

3 habitus brachycerous fly

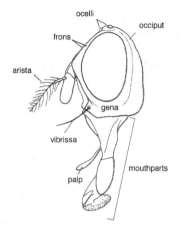

7 head calyptrate fly

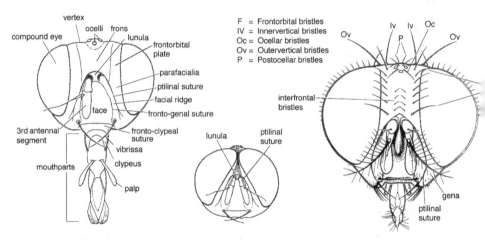

F = Frontorbital bristles
IV = Innervertical bristles
Oc = Ocellar bristles
Ov = Outervertical bristles
P = Postocellar bristles

4 head calyptrate fly;
   eyes separated
   (dichoptic)

5 eyes in contact
   with each other
   (holoptic)

8 head calyptrate fly

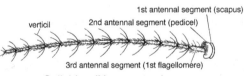

6a antenna Culicidae (Nematocera)

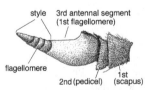

6b antenna Tabanidae
    (lower Brachycera)

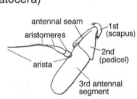

6c antenna calyptrate fly
    (higher Brachycera)

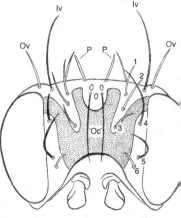

9 head acalyptrate fly
   1-6 = F-bristles

The Oc-bristles (ocellar bristles) are in the dorsomedial region of the head, usually inside the ocellar triangle, in rare cases just outside it. The area between the ocellar triangle, the plates on which the F-bristles are inserted and the ptilinal suture or bases of the antennae is named the frontal vitta or interfrontal area which may, or may not, show characteristic interfrontal bristles or setulae (fig. 8). Below the eyes, where the genae meet the lower margin of the face, the vibrissae are inserted.

Apart from the presence of bristles their orientation is an important character. The bristles can be curving away from each other (diverging), be parallel (equidistant), curving toward or even touching or crossing each other (converging). For the F-bristles it is important to know if they are curving outward (latero- or exclinate), backward (reclinate), inward (inclinate), or forward (proclinate).

Clypeus. A distinct part of the head between the fronto-clypeal suture and the mouthparts (fig. 4). Usually the clypeus is small and inconspicuous, but it can be large and obvious (94, 95, 311, 314), for example in families showing well developed piercing or sucking mouthparts.

Dichoptic. See Eye.

Eye bridge. See Eye.

Eye. An insect eye is a so-called compound eye, made up of a multitude of small lenses, the facets or ommatidia. In the Diptera, the eyes usually occupy a major part of the head. They are usually separate (dichoptic, fig. 3, 4) but sometimes meet each other at a point or along a line (holoptic, fig. 5). In most cases the holoptic condition is limited to the males, but in some families both sexes can be holoptic. Whole series of facets can be larger in size than those elsewhere, this difference in size can be so extensive and/or prominent that it is possible to discriminate between two differently faceted parts of the eye (e.g. an upper and a lower part, as in fig. 55). In some Nematocera the eyes are joined through a narrow bridge, named the eye bridge, above the antennae. It consists of several rows of facets (fig. 58, 73). Many Diptera possess hairs in between the facets, named ommatrichia, which can be so long or numerous as to produce a velvety aspect.

F-bristles (Frontorbital bristles) (fig. 8: F; fig. 9: bristles 1-6). See Bristles of the head.

Face (Prefrons) (fig. 4). The plate dorsally delimited by the antennal sockets, ventrally by the fronto-clypeal suture and laterally by the fronto-genal suture or, if it is absent, by the compound eyes. The face may be small in blood sucking (Culicidae, Ceratopogonidae, Tabanidae) or raptorial Diptera (Empididae, Hybotidae, fig. 96), but is well-developed in most Diptera, often with grooves in which the antennae are situated, separated by a median ridge, the facial carina. Other grooves that can be present are the fronto-genal sutures, separating the medial part of the face from the facial ridge, and the ptilinal suture, separating the parafacialia from the remainder of the face (fig. 4, 91).

Facets. See Eye.

Facial carina. See Face.

Flagellomere. See Antenna.

Flagellum. See Antenna.

Frons (fig. 4, 7, 8). Part of the head bordered by the eyes, the vertex and the lunula or antennal bases. The frons encompasses the frontorbital plates, the ocellar triangle and the interfrontal area in between.

Frontal. Pertaining to the frons. Not to be confused with anterior or rostral.

Frontal vitta. Interfrontal area. See Frons and Bristles of the head.

Frontorbital plates (fig. 4). The sclerotized plates along the eyes on which the frontorbital bristles are inserted.

Frontorbital bristles (fig. 8: F; fig. 9: bristles 1-6). See Bristles of the head.

Gena (fig. 7, 8). The part of the head below the lower eye margin. In the Brachycera it often possesses strong bristles of which the anterior one is the vibrissa.

Holoptic. The eyes are touching in their dorsal region; sometimes also in their ventral region (fig. 5).

Innervertical bristles (fig. 8, 9: Iv). See Bristles of the head.

Interfrontal area (fig. 8). See Bristles of the head.

Interfrontal bristles (fig. 8). See Bristles of the head.

Iv-bristles (Innervertical bristles) (fig. 8, 9: Iv). See Bristles of the head.

Lunula. A somewhat triangular plate above the antennal insertion, just underneath the ptilinal suture (fig. 4, 5, 91). In almost all Schizophora it is well developed. A plate resembling a lunula also occurs in some groups that hold a lower position in the system, e.g. the Syrphidae.

Mouthparts (Proboscis) (fig. 4, 7). In the Diptera, functional mouthparts are always sucking or piercing, never biting or chewing. However, bloodsucking flies are usually referred to in English as "biting". Laterally the mouthparts are provided with palps. These consist of five segments in most Nematocera, two in the lower Brachycera and one in most Muscomorpha.

Nasus. Prolongation at the anterior end of the rostrum (fig. 22).

Oc-bristles (Ocellar bristles) (fig. 8, 9: Oc). See Bristles of the head.

Occiput. The entire back of the head (fig. 7).

Ocellar bristles (fig. 8, 9: Oc). See Bristles of the head.

Ocellar triangle. The plate on which the ocelli are found. It is nearly always larger than the triangle made up by the ocelli themselves.

Ocelli (singular: ocellus). Single lenses in the dorsal area of the head. If present there are nearly always three of them, arranged in a triangle (fig. 3, 4, 7).

Ocular. Pertaining to the eye(s), e.g. ocular margin.

Ommatidia (singular: ommatidium). The facets of the insect eye.

Ommatrichia (singular: ommatrichium). The hairs between the facets of the insect eye.

Oral angle (fig. 187, 188). See Vibrissal angle.

Oral margin (Peristoma) (fig. 308, 311, 314). The lower margin of the face.

Outervertical bristles (fig. 8, 9: Ov). See Bristles of the head.

Ov-bristles (Outervertical bristles) (fig. 8, 9: Ov). See Bristles of the head.

P-bristles (Postocellar bristles) (fig. 8, 9: P). See Bristles of the head.

Palp. See Mouthparts.

Parafacialia. In the Schizophora the plate along the eye between the frontorbital plate and the gena, separated from the remainder of the face by the ptilinal suture (fig. 4, 91).

Peristoma. Another term for oral margin (fig. 308, 311, 314).

Postocellar bristles (fig. 8, 9: P). See Bristles of the head.

Prefrons. See Face.

Proboscis (fig. 4, 7). See Mouthparts.

Pseudopostocellar bristles. In some cases no real P-bristles are present, but other bristles, placed more laterally or caudally are present.

Pseudovibrissae. Vibrissa-like bristles that are not exactly inserted on the vibrissal angle but more caudally, ventrally or laterally.

Ptilinal fissure. Other term for Ptilinal suture.

Ptilinal suture. A suture, fissure, or groove, present in the Schizophora, usually in a horseshoe shape, running above the lunula down on either side of the antennae along the inner margin of the parafacialia (the plates bordering the eyes) (fig. 4, 5, 8, 91). See also Ptilinum.

Ptilinum (fig. 10). In Schizophora an inflatable bag in the head that is used when the emerging fly leaves the puparium. The ptilinum emerges above the antennae and thereby pushes open the puparium valve. In this position the face is swung into an anteroventral position. Subsequently, the bag is retracted and the face regains its normal position, separated from the other elements of the head by the ptilinal suture.

PVT-bristles (Postvertical bristles). Other term for P-bristles.

Rostrum. In particular in some Nematocera the lower part of the head can be transformed into a long snout, the rostrum that may carry a nasus on its anterior end (fig. 22).

Style (Stylus). See Antenna.

Vertex. The top of the head, behind the ocellar triangle. It is bounded by the eyes, the frons and the occiput (fig. 4).

Verticils. Hair-like setae, usually in rings, at the base of the flagellar segments (fig. 6a).

Vibrissa. A usually strong bristle at the vibrissal angle (fig. 4, 7).

Vibrissal angle (Oral angle) (fig. 187, 188). The point below the eye where the lower margin of the gena meets the lower margin of the face (oral margin) and where the vibrissa is inserted (fig. 4, 7).

THORAX

Ac-bristles (Acrostichal bristles) (fig. 11). See Scutal bristles.

Acrostichal bristles (fig. 11). See Scutal bristles.

Air inlet. See Spiracle.

Ampulla. See Greater ampulla.

Anatergite. Part of the laterotergite (fig. 12). See Main parts.

Anepimeron (fig. 12). See Main parts.

Anepisternum (fig. 12). See Main parts.

Dc-bristles (Dorsocentral bristles) (fig. 11). See Scutal bristles.

Dorsocentral bristles (fig. 11). See Scutal bristles.

Greater ampulla (fig. 12). A round bulge on the lateral side of the thorax just below the front end of the wing attachment. This bulge occurs in all calyptrate families, but is also found in some other families, e.g. Periscelididae, many Syrphidae and some Psilidae, Sciomyzidae and Tephritidae.

Haltere (plural: Halteres) (fig. 3, 12, 41). A remnant of the hind wing that has been transformed into an organ of balance. Consisting of a stem (peduncle) and a knob.

Humeral callus. Another term for postpronotum (fig. 11, 12).

Humerus. Another term for postpronotum (fig. 11, 12).

Hypopleuron. Another term for meron (fig. 12).

Intra-alar bristles (fig. 11). See Scutal bristles.

Katatergite. Part of the laterotergite (fig. 12). See Main parts.

Katepimeron (fig. 12). See Main parts.

Katepisternum (fig. 12). See Main parts.

Laterotergite (fig. 12). See Main parts.

Main parts (fig. 11, 12). The main parts of the dorsal side of the thorax are the pronotum (usually reduced), scutum, scutellum and mediotergite. The transverse suture divides the scutum into the presutural and postsutural parts. The subscutellum is found in between the mediotergite and scutellum. It can be strongly convex and pillow-shaped (fig. 374, 399). Laterally of the scutum some minor elements are found. They generally have bristles which are important for identification purposes. In an anterior to posterior order, these elements are: postpronotum (fig. 11, 12), notopleuron (fig. 11, 12) and postalar callus (fig. 11).

The lateral side of the thorax is made up of, in the order of anterior to posterior and dorsal to ventral (fig. 12): pronotum (usually reduced), propleuron consisting of proepisternum and proepimeron, anepisternum, katepisternum, anepimeron, laterotergite (which may comprise two parts: anatergite and katatergite), katepimeron and meron.

Mediotergite (fig. 12, 25, 41). See Main parts.

Meral bristles. Bristles on the meron, e.g. as present in certain Calyptrate families (fig. 386, 387).

Meron (fig. 12). See Main parts.

Mesonotum. Term for scutum + scutellum (fig. 3, 11, 12).

Mesopleuron. Another term for anepisternum (fig. 12).

Metasternum. Sometimes used as the name for the plate between and in front of the coxae of the hind legs.

Notopleuron (fig. 12). See Main parts.

Peduncle. See Haltere.

Pleurotergite. Another term for laterotergite (fig. 12).

Postalar callus (fig. 11). See Main parts.

Postpronotal lobe. Another term for postpronotum (fig. 11)

Postpronotum (fig. 11). See Main parts.

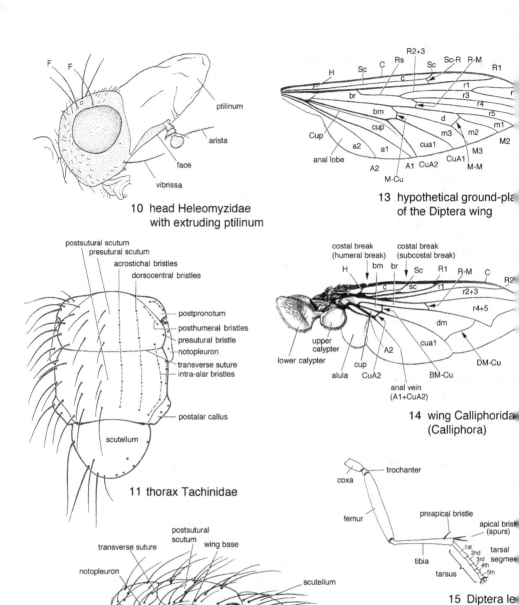

10 head Heleomyzidae with extruding ptilinum

F, F, ptilinum, arista, face, vibrissa

13 hypothetical ground-plan of the Diptera wing

H, Sc, C, Rs, R2+3, Sc-R, R-M, R1, r1, r3, r, r4, r5, m1, br, c, Sc, bm, d, m3, m2, M2, cup, Cup, cua1, M3, a2, a1, M-M, anal lobe, A2, A1, CuA2, CuA1, M-Cu

11 thorax Tachinidae

postsutural scutum, presutural scutum, acrostichal bristles, dorsocentral bristles, postpronotum, posthumeral bristles, presutural bristle, notopleuron, transverse suture, intra-alar bristles, postalar callus, scutellum

14 wing Calliphoridae (Calliphora)

costal break (humeral break), costal break (subcostal break), H, bm, br, Sc, R1, R-M, C, R2, c, sc, r1, r2+3, r4+5, dm, cua1, upper calypter, lower calypter, A2, DM-Cu, alula, CuA2, BM-Cu, cup, anal vein (A1+CuA2)

15 Diptera leg

coxa, trochanter, femur, preapical bristle, apical bristle (spurs), tibia, tarsus, 1st, 2nd, 3rd, 4th, 5th, tarsal segment

12 thorax Calliphoridae

transverse suture, postsutural scutum, wing base, notopleuron, postpronotum, anepisternum, anterior spiracle, proepisternum, proepimeron, coxa 1, anepimeron, katepisternum, coxa 2, coxa 3, meron, katepimeron, meral bristles, posterior spiracle, greater ampulla, laterotergite, mediotergite, haltere, subscutellum, scutellum

16 bristle (macrotrichium) with socket (alveolus)

bristle, socket

Postscutellum. Another term for subscutellum (fig. 12).

Postsutural scutum (fig. 11, 12). See Main parts.

Precoxal bridge (fig. 133, 134, 170). A precoxal bridge is present when the plate on the ventral side of the thorax between the neck and the fore legs (the prosternum) is confluent with the lateral plates on which it borders (the episterna).

Prescutum. Another term for presutural scutum (fig. 11).

Presutural scutum (fig. 11). See Main parts.

Proepimeron (fig. 12). See Main parts.

Proepisternum (fig. 12). See Main parts.

Pronotum. See Main parts.

Propleuron. See Main parts.

Prosternum. Plate on the ventral side of the thorax between the neck and the fore legs (fig. 133, 134, 171). See Precoxal bridge.

Pteropleuron. Another term for anepimeron (fig. 12).

Scutal bristles. Several longitudinal rows of bristles are found on the scutum, in particular in the higher Diptera. They are important for identification, especially on the generic and specific level. Going laterad from the middle the following rows of bristles occur (fig. 11): acrostichal bristles, dorsocentral bristles and intra-alar bristles. In all series the distinction of insertion in front versus behind the transverse suture is made. The term 'acrostichals 1+3' means that one acrostichal bristle is found in front and three are situated behind the transverse suture.

Scutellum. See Main parts.

Scutum (fig. 3, 11, 12). See Main parts.

Spiracle (fig. 12). The air inlet. The lateral side of the thorax has two spiracles. The anterior spiracle is behind the pronotum or proepisternum, the posterior spiracle is between the haltere and the meron and can be partially or entirely covered. The way in which the spiracle is covered is a character used to distinguish between some of the Calyptrate families (fig. 402-404).

Sternopleuron. Another term for katepisternum (fig. 12).

Stigma. Another term for spiracle (fig. 12).

Subscutellum (fig. 12). See Main parts.

Thoracic suture. Another term for transverse suture (fig. 11, 12).

Transverse suture (fig. 11, 12). The transverse suture extends on the dorsal side of the thorax, anterior to the wing insertions, from one side to the other. In several families of the Nematocera this suture is V-shape (fig. 25); in many groups the mid part of the suture is less distinct or even absent.

WING

For orientation purposes, the wings are considered to be fully extended sideways.

Abbreviations of veins and crossveins are given in capitals (e.g. vein R), those of the wing cells in lower case (e.g. cell d).

Wing cells are named after the vein on their anterior side.

A wing cell is closed when it is nowhere bounded by the wing margin (fig. 14: cell r4+5 is open, cell dm is closed).

A-vein(s) (Anal vein(s)) (fig. 13: A1, A2; fig. 14: A1+CuA2). See Longitudinal veins.

Alula (fig. 14). A lobe-like extension of the basal hind part of the wing, situated in between the anal lobe (where that is present) and the upper calypter, and separated from them by notches or narrower parts of the wing.

Anal angle. The angle of the basal part of the hind margin of the wing (fig. 79).

Anal cell. Another term for cell cup.

Anal lobe. The lobe at the basal hind part of the wing (fig. 13, 79).

Anal vein (fig. 13: A1, A2; fig. 14: A1+CuA2). See Longitudinal veins.

Basal cells (fig. 13, 14). Cells br, bm and cup, situated in the basal part of the wing.

Bifurcation of a vein. See Fork.

C-vein (Costa) (fig. 13, 14: C). See Longitudinal veins.

Calypter (Squama) (fig. 14). The two (upper and lower, i.e., relatively apical and basal) lobes at the basal hind margin of the wing next to the thorax.

Cell bm. One of the basal wing cells (fig. 13, 14), abbreviated from base of the Media.

Cell br. One of the basal wing cells (fig. 13, 14), abbreviated from base of the Radius.

Cell cup (Anal cell) (fig. 13, 14). One of the basal wing cells, name not related to the words "cup" or "cup shaped" but abbreviated from posterior branch of the Cubitus.

Cell dm (Discal medial cell) (fig. 14). See Discal cell.

Cell gb (Greater basal cell). The cell resulting from fusion of cells br and bm (fig. 72, 76-82).

Costa (fig. 13, 14: C). See Longitudinal veins.

Costal break (fig. 14: arrows). The costa can be interrupted once, twice or not at all. This is an important character, in particular in the identification of Acalyptrate families. This character, however, is not always clear. Hence it has been used as little as possible in the key. It is a good idea to look at the wing from an oblique position and to light it from below instead of from above.

Crossvein BM-Cu (fig. 14). Crossvein found in the higher Diptera in between cells bm and dm.

Crossvein DM-Cu (fig. 14). Closing vein of cell dm.

Crossvein H (Humeral vein) (fig. 13, 14). Crossvein in between the costa and subcosta in the basal part of the wing.

Crossvein M-Cu (fig. 13). Crossvein found in the more primitive Diptera, in between cell bm and the most basal/posterior m cell.

Crossvein M-M (fig. 13). Crossvein found in the more primitive Diptera, closing cell d posteriorly.

Crossvein R-M (fig. 13, 14). Crossvein in between cell br and the most basal/posterior r cell.

Cu-vein (Cubitus). See Longitudinal veins.

Cubitus (Cu). See Longitudinal veins.

Discal cell (cell d); discal-medial cell (cell dm) (fig. 13: d; fig. 14: dm). The discal cell or discal-medial cell is a closed cell in the central or posterior part of the wing. A discal cell is present when the cell is nowhere bordered by vein CuA1; if the cell is bordered by this vein, then it is named discal-medial cell. The outermost closing vein of the discal cell is crossvein M-M (fig. 13), in case of the discal-medial cell it is crossvein DM-Cu (fig. 14). The definition of the discal cell can be applied to most Nematocera. However, in the Nematocera as well as in the Lower Brachycera (the Lower Muscomorpha in particular) there are several families with genera for which the discal cell definition obtains, alongside other genera where that of the discal-medial cell applies.

In this book the term discal cell is used for all Nematocera and Brachycera excluding the Eremoneura, whereas the term cell dm is used for all families of the Eremoneura (i.e., all families from Atelestidae to, and including, the Hypodermatidae in the classification table given above).

Discal-medial cell (cell dm) (fig. 14: dm). See Discal cell.

Fork. Point of bifurcation of a vein, e.g. fig. 13: vein R2+3 bifurcates into R2 and R3 at the fork.

Greater basal cell. See Cell gb.

Humeral break. An interruption of the costa just beyond crossvein H (fig. 14).

Humeral vein. See Crossvein H.

Longitudinal veins. The six longitudinal veins in the wing are, from the front margin to the hind margin: Costa (fig. 13, 14: C) which occupies the front (anterior) margin; in some primitive Diptera it occupies the entire wing margin, including the hind (posterior) margin. Subcosta (fig. 13, 14: Sc) is a weaker vein between costa and R1. Radius (fig. 13, 14: R) in its primitive layout shows five branches reaching the margin, four of which have a common stem, the radial sector (vein Rs). Media (fig. 13, 14: M) in its primitive layout has three branches reaching the margin. Cubitus, in its primitive layout has two branches, the anterior of which ramifies into CuA1 and CuA2. Anal vein, in the more primitive Diptera this is vein A1, in the higher Diptera this is vein A1+CuA2. In the Tipulomorpha, a second anal vein is found, named A2; this vein

is absent, or less developed, in most other Diptera. The spurious vein (vena spuria) is not connected to any other vein and crosses R-M at a right angle (fig. 135).

Lower Calypter (fig. 14). A basal appendage at the hind margin of the wing, close to the thorax and next to the upper calypter.

M-vein(s) (Media) (fig. 13, 14: M). See Longitudinal veins.

Media (fig. 13, 14: M). See Longitudinal veins.

Petiolate (Petiole). Having a section of a vein (the petiole) between a closed cell and the wing
margin (fig. 109, 389).

Pterostigma. See Stigma.

R-vein(s) (Radius) (fig. 13, 14: R). See Longitudinal veins.

Radial sector (Rs-vein). Basal part or stem of the R veins in between vein R1 and the other R-
veins (fig. 13: Rs).

Radius (fig. 13, 14: R). See Longitudinal veins.

Rs-vein (fig. 13: Rs). See Radial sector.

Sc-vein (Subcosta) (fig. 13, 14: Sc). See Longitudinal veins.

Spurious vein (Vena spuria). (fig. 135). See Longitudinal veins.

Squama. Another term for calypter (fig. 14).

Stem. Part of the vein before the ramification or fork.

Stigma (Pterostigma). Characteristic coloured area of the wing surface alongside the costa
near the end of vein(s) Sc and/or R1 (fig. 71, 165-168).

Subcosta (fig. 13, 14: Sc). See Longitudinal veins.

Subcostal break. Interruption of the costa, just before the point where the subcosta or vein R1
meets or (if reduced or merged) should meet the costa (fig. 14).

Upper Calypter (fig. 14). A basal appendage at the hind margin of the wing, close to the thorax and next to the Lower Calypter.

Vena spuria (Spurious vein) (fig. 135). See Longitudinal veins.

Wing tip. In case of fully extended wing, the point furthest away from the body.

LEG

The leg consists of the following major parts (fig. 15), going outward from the thorax: coxa
(plural: coxae), trochanter, femur (plural: femora), tibia (plural: tibiae), tarsus (plural: tarsi),
claw, empodium and pulvilli when developed.

Basitarsus. See Tarsus.

Claw. See Tarsus.

Distitarsus. See Tarsus.

Empodium. See Tarsus.

Metatarsus. See Tarsus.

Pulvilli. See Tarsus.

Tarsomeres. See Tarsus.

Tarsus. The Tarsus is made up of five segments (tarsomeres or tarsal segments) (fig. 3, 15); the
first (basal) segment is also named metatarsus or basitarsus; the last (apical) one is also
called distitarsus. The fifth tarsomere includes a small terminal sclerite (acropod or post-
tarsus) bearing the two claws, and, when developed, three appendages: two pulvilli with
an empodium in between, the latter can be bristle-like or of the same shape as the pulvilli
(pulvilliform) (fig. 103, 104).

Tibia. The tibia may, in some cases, have apical and/or preapical bristles, spurs, or spine(s)
(fig. 15, 63, 75, 185).

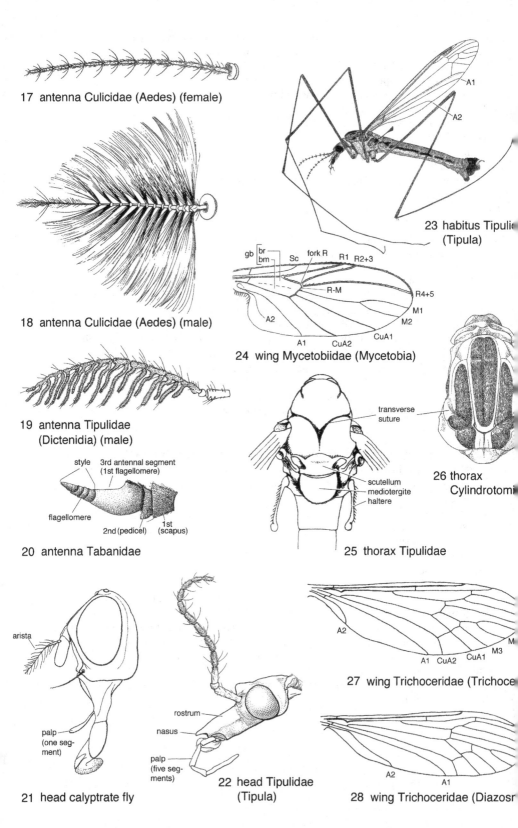

17 antenna Culicidae (Aedes) (female)

18 antenna Culicidae (Aedes) (male)

19 antenna Tipulidae (Dictenidia) (male)

20 antenna Tabanidae

style 3rd antennal segment (1st flagellomere)

flagellomere

2nd (pedicel) 1st (scapus)

21 head calyptrate fly

arista

palp (one segment)

22 head Tipulidae (Tipula)

rostrum

nasus

palp (five segments)

23 habitus Tipuli (Tipula)

A1

A2

24 wing Mycetobiidae (Mycetobia)

gb br bm Sc fork R R1 R2+3

R-M

R4+5

M1

M2

A2

A1 CuA2 CuA1

25 thorax Tipulidae

transverse suture

scutellum
mediotergite
haltere

26 thorax Cylindrotom

27 wing Trichoceridae (Trichoce

A2

M

M3

A1 CuA2 CuA1

28 wing Trichoceridae (Diazosr

A2

A1

2 4

This key is a so-called dichotomous key: every couplet contains two alternatives. It is recommended to read both alternatives before coming to a decision and to have a look at the family descriptions for additional diagnostic characters. In order to be able to identify specimens it is of great importance that they are in good condition. This is especially true for the head, wings, legs and the bristles of the head and legs (bristles which are broken off still reveal former presence by their socket, the alveolus, which stays behind, fig. 16). For specimens larger than 6 mm, a hand lens with 20x magnification will generally be sufficient. For smaller specimens a stereo microscope up to 40x magnification is required.

>>>	Indicates a note about (additional) characters, classification, or something to be aware of.
( )	The numbers between brackets refer to previous couplets so the path back through the key can be retraced easily.

1	Body not strongly dorsoventrally flattened. Wing well developed, not snapped off near the base, not strongly reduced or absent. **2**
–	Body strongly dorsoventrally flattened (fig. 441-443) and/or wing strongly reduced, absent, or snapped off near the base (fig. 427-451). **191**
2(1)	Antenna filiform or resembling a series of beads, with more than 6 segments; all segments or those beyond the second more or less similar (fig. 17); antennal segments sometimes with long verticils (fig. 18) or with appendages (fig. 19, 85). Palp in Nematocera often with 3-5 segments (fig. 22, 29, 48). **Nematocera** (3-38) and **Brachycera with a primitive antenna** (39-43) - **3**
–	Antenna not filiform or resembling a series of beads, with less than 10 segments; the third segment nearly always larger than the subsequent segments which, in that case, strongly differ from the third segment, or together constitute an (un)segmented continuation: a style (fig. 20) or arista (fig. 21). Palp usually with 1-2 segments (fig. 21). **Brachycera - 44**
3(2)	Wing with 2 anal veins (veins A1 and A2) reaching the wing margin (fig. 23, wing figures 27-35). Upper side of thorax with a V-shaped transverse suture (fig. 25; suture in Cylindrotomidae less distinct laterally, fig. 26). **4**
–	Only 1 (or no) anal vein (vein A1) reaching the wing margin (fig. 24). Transverse suture absent, unclear, interrupted or, if completely present and V-shaped, haltere with a basal appendage (fig. 41). **8**
4(3)	Ocelli present. Anal vein A2 short and curved (fig. 27), longest in *Diazosma* (fig. 28). **Trichoceridae**
–	Ocelli absent (in very few cases, 2 rudimentary ocelli present, fig. 29). Anal vein A2 almost always at least half as long as vein A1 (fig. 23, 30-35). **5**

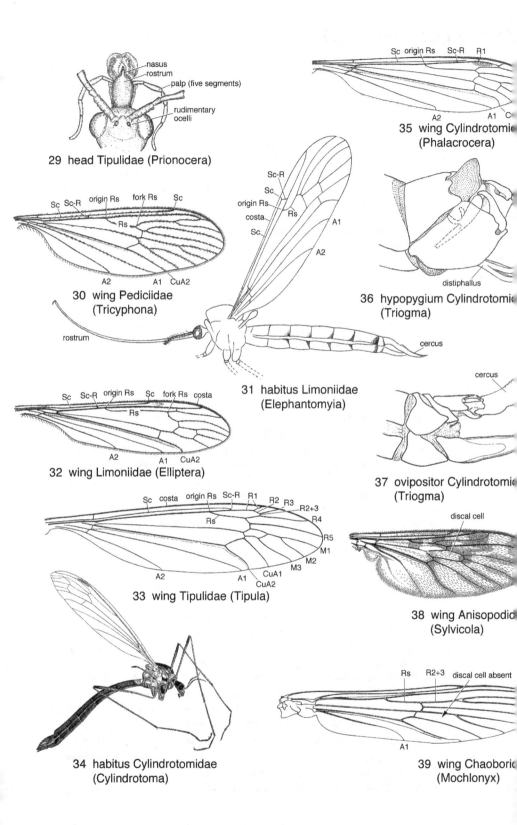

29 head Tipulidae (Prionocera)

35 wing Cylindrotomii (Phalacrocera)

30 wing Pediciidae (Tricyphona)

31 habitus Limoniidae (Elephantomyia)

32 wing Limoniidae (Elliptera)

33 wing Tipulidae (Tipula)

34 habitus Cylindrotomidae (Cylindrotoma)

36 hypopygium Cylindrotomii (Triogma)

37 ovipositor Cylindrotomii (Triogma)

38 wing Anisopodid (Sylvicola)

39 wing Chaoboric (Mochlonyx)

5(4)	Eye pubescent, with ommatrichia in between the facets. Crossvein Sc-R located before the origin of vein Rs; Sc long, past the fork of Rs (fig. 30). **Pediciidae**
–	Eye bare, without ommatrichia. Crossvein Sc-R usually located past the origin of vein Rs (fig. 31, 33, 35), if located before the origin of vein Rs, then vein Sc not distinctly past the fork of Rs (fig. 32). **6**
6(5)	Last (5th) segment of the palp elongate, distinctly longer than the preceding segments; rostrum usually well developed, often with a nasus (fig. 22, 29). Vein Sc usually does not end in the costa (fig. 33). Antenna usually with 13 segments or, in a few cases, more. **Tipulidae**
–	Last palpal segment short, if slightly elongate then combination of characters not as given above. Rostrum usually short and without nasus (fig. 34), rostrum sometimes elongate (fig. 31). Vein Sc usually ends in the costa (fig. 31, 32; but not so in fig. 35). Antenna usually with 14 or more segments, in a few cases less than 13. **7**
7(6)	Apical part of vein CuA2 strongly curving toward the wing margin (fig. 35). V-shaped transverse suture on upper side of thorax less distinct at the sides (fig. 26). Male: apical part of the distiphallus consists of 2 or 3 separate small tubes (well visible from outside) (fig. 36). Female: upper valves of the ovipositor (cerci) short and wide (fig. 37). **Cylindrotomidae**
–	Apical part of vein CuA2 usually not strongly curving toward the wing margin (fig. 32). V-shaped transverse suture on upper side of thorax distinct throughout (fig. 25). Male: apical part of the distiphallus with no more than 2 or 3 (usually 0) separate openings which are hardly or not at all visible from outside. Female: upper valves of the ovipositor (cerci) usually long, slender and acute (fig. 31). **Limoniidae**
8(3)	Discal cell or cell dm absent (wing figures 39-83). **9**
–	Discal cell or cell dm present (fig. 38). **38**
9(8)	Ocelli absent. **10**
	Ocelli present (fig. 58, 73). **19**
10(9)	Nine or more longitudinal veins or longitudinal vein branches reaching the wing margin, including the anal vein (A1) (wing figures 39-45). **11**
–	No more than eight longitudinal veins or longitudinal vein branches reaching the wing margin (wing figures 47-53). **15**

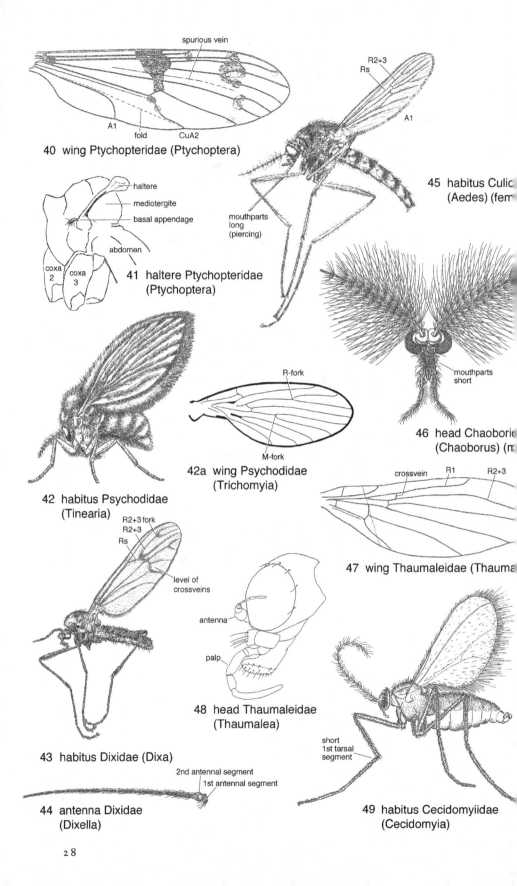

40 wing Ptychopteridae (Ptychoptera)

spurious vein

A1
fold   CuA2

R2+3
Rs

A1

41 haltere Ptychopteridae (Ptychoptera)

haltere
mediotergite
basal appendage
abdomen
coxa 2   coxa 3

mouthparts long (piercing)

45 habitus Culic (Aedes) (fem

mouthparts short

46 head Chaobori (Chaoborus) (m

42 habitus Psychodidae (Tinearia)

42a wing Psychodidae (Trichomyia)

R-fork
M-fork

crossvein   R1   R2+3

47 wing Thaumaleidae (Thauma

R2+3 fork
R2+3
Rs
level of crossveins

antenna
palp

48 head Thaumaleidae (Thaumalea)

43 habitus Dixidae (Dixa)

short 1st tarsal segment

44 antenna Dixidae (Dixella)

2nd antennal segment
1st antennal segment

49 habitus Cecidomyiidae (Cecidomyia)

28

11(10) Haltere with basal appendage (fig. 41). First antennal segment larger than second. Upper side of thorax with U- or V-shaped transverse suture. Wing venation as in fig. 40, with a spurious vein between the R and M veins.   **Ptychopteridae**

(In Europe 1 genus, *Ptychoptera*, with about 15 species.)

– Haltere without basal appendage. First antennal segment usually ring-shaped, second segment largest and spherical (fig. 6a, 44, 46). Combination of other characters not as given above.   **12**

12(11) Crossveins near the wing base or crossveins (partly) absent (fig. 42, 42a); wing often strongly hairy and acute (fig. 42), sometimes rounded (fig. 42a); costa surrounding the wing completely [not clear from the drawing but it should be from the insect].   **Psychodidae**

– Crossveins not limited to near the base; wing not acute (fig. 43, 45); costa surrounding the wing completely or not.   **13**

13(12) Vein R2+3 not in line with vein Rs, but curves toward the fork; wing with pubescence on the veins (fig. 43). Antennal segments without long verticils near their bases (fig. 44).   **Dixidae**

– Vein R2+3 in line with vein Rs (fig. 39, 45); wing on the veins and especially along the hind margin conspicuously hairy or set with scales. Antennal segments with long basal verticils (fig. 45, 46).   **14**

14(13) Scales present on the wing margin and veins, the legs and usually on the abdomen. Mouthparts conspicuously elongate (fig. 45), female bloodsucking.   **Culicidae**

>>> [In specimens preserved in liquid the scales are easily lost.]

– Scales virtually exclusively on the wing margin, the scales on the veins narrow or resembling hair-like bristles; the rest of the wing with pubescence only. Mouthparts not conspicuously elongate (fig. 46), female not bloodsucking.   **Chaoboridae**

15(10) Wing venation as in fig. 47: crossveins near the wing base and with a crossvein between veins R1 and R2+3; vein R2+3 curving; costa surrounding the wing completely [not clear from the drawing but it should be from the insect]. Eyes holoptic in both sexes. Antenna as in fig. 48: first and second antennal segments widened, the other segments narrower.
                    Thaumaleidae

– Combination of characters not as given above.   **16**

16(15) Legs with first tarsal segment much shorter than second, or with at most 4 tarsal segments (fig. 49). Only a limited number of longitudinal veins (usually 2 to 4, at most 6) reach or go in the direction of the wing margin. Margin generally with a break past vein R4+5 (fig. 57). Usually small, delicate insects with conspicuously long antenna.
                    Cecidomyiidae (except Lestremiinae)

– First tarsal segment not conspicuously shorter than the second and tarsus with 5 segments. Usually 6 or more longitudinal veins reach or go in the direction of the wing margin.   **17**

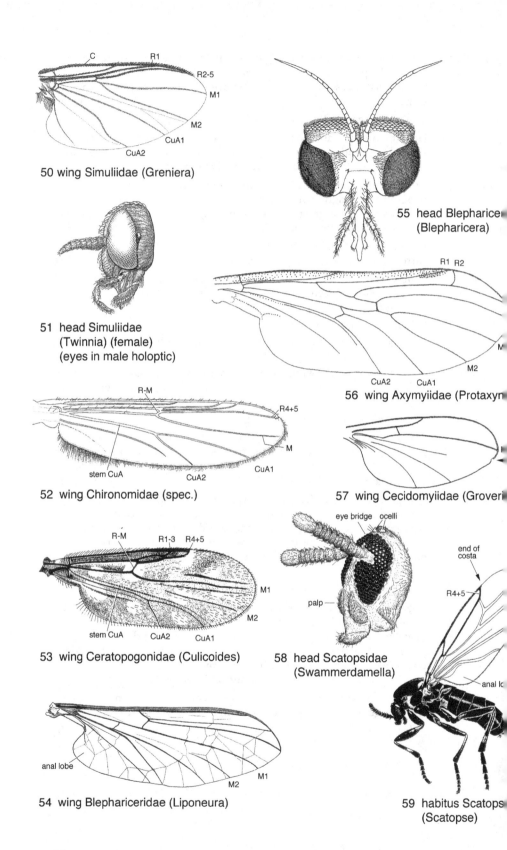

50 wing Simuliidae (Greniera)

55 head Blepharice
(Blepharicera)

51 head Simuliidae
(Twinnia) (female)
(eyes in male holoptic)

56 wing Axymyiidae (Protaxyn

52 wing Chironomidae (spec.)

57 wing Cecidomyiidae (Grover

53 wing Ceratopogonidae (Culicoides)

58 head Scatopsidae
(Swammerdamella)

54 wing Blephariceridae (Liponeura)

59 habitus Scatops
(Scatopse)

17(16)	Veins CuA1 and CuA2 are separate from each other right from the wing base; wing wide; costa and R-veins more conspicuous than the other veins; vein CuA2 usually curving like an elongate, inverted S (fig. 50); vein Rs not forked (fig. 50) or forked in R2+3 and R4+5. Antenna short, wide, tapering toward the apex, with 11 (rarely 9) segments (fig. 51). **Simuliidae**
–	Venation and antenna different. Veins CuA1 and CuA2 have a long joint stem starting from the wing base (fig. 52, 53; stem maybe transparent or vestigial, especially in small species). **18**
18(17)	Vein M continuing singly, not forked; R-veins not shortened with R4+5 ending close to the wing tip (fig. 52). No piercing mouthparts. **Chironomidae**
–	Vein M forked into M1 and M2 and/or R-veins shortened; usually both characters present together (fig. 53); vein M2 often reduced or poorly visible near the fork; if reduced throughout, then R-veins shortened; crossvein R-M absent in *Leptoconops*. Female with piercing mouthparts. **Ceratopogonidae**
19(9)	Wing with a grid of secondary veins and a conspicuous anal lobe; 1 or 2 of the M-veins reach the wing margin; if M2 present then not connected to M1 (fig. 54). Eye often show two separate, differently faceted parts (fig. 55). Legs long. **Blephariceridae**
–	Wing without secondary veins and combination of other characters not as given above. **20**
20(19)	Tibiae of mid and hind legs without apical bristles or spurs (fig. 59, 62). **21**
–	Tibiae of mid and hind legs with apical bristles or spurs (fig. 63, 75, 78). **25**
21(20)	Wing with 4 R-veins of which R2 is short and ends near R1 (fig. 56). **Axymyiidae**
	(In Europe 1 species, *Mesaxymyia kerteszi* (Duda).)
–	No more than 3 R-veins (fig. 57, 59-61). **22**
22(21)	Costa surrounding the wing completely [not clear from the drawing but it should be from the insect] and usually interrupted just beyond vein R4+5; only a small number of veins (6 or less) reaching the wing margin (fig. 57). Small, delicate insects usually with strikingly long antenna. **Cecidomyiidae: Lestremiinae**
–	Costa not surrounding all of the wing but ending at or before the wing tip (fig. 59). Combination of other characters not as given above. **23**
23(22)	Palp with 1 segment only (fig. 58). Wing short, wide and with anal lobe; costa ends well before the wing tip, nearly always near the point where R4+5 reaches the wing margin (fig. 59). **Scatopsidae**

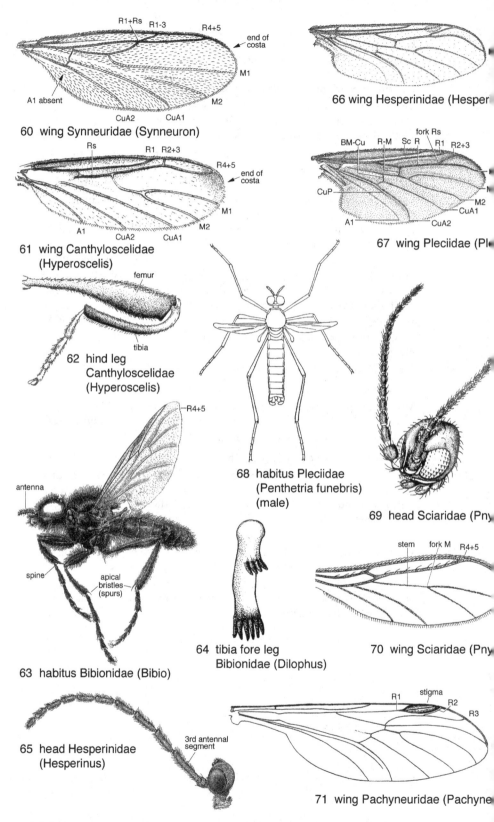

60 wing Synneuridae (Synneuron)

R1+Rs R1-3 R4+5 end of costa M1 M2 A1 absent CuA2 CuA1

61 wing Canthyloscelidae (Hyperoscelis)

Rs R1 R2+3 R4+5 end of costa M1 M2 A1 CuA2 CuA1

62 hind leg Canthyloscelidae (Hyperoscelis)

femur tibia

63 habitus Bibionidae (Bibio)

antenna spine apical bristles (spurs) R4+5

64 tibia fore leg Bibionidae (Dilophus)

65 head Hesperinidae (Hesperinus)

3rd antennal segment

66 wing Hesperinidae (Hesper

67 wing Pleciidae (Pl

fork Rs BM-Cu R-M Sc R R1 R2+3 CuP M2 CuA1 A1 CuA2

68 habitus Pleciidae (Penthetria funebris) (male)

69 head Sciaridae (Pny

70 wing Sciaridae (Pny

stem fork M R4+5

71 wing Pachyneuridae (Pachyne

R1 stigma R2 R3

–	Palp with 2 or more segments. Wing narrower, without anal lobe; costa ends closer to the wing tip (fig. 60, 61). **24**
24(23)	Anal vein (A1) absent or present as a basal remnant only; veins R1 and Rs contiguous over a short stretch; M-fork incomplete, i.e., vein M2 is not connected to vein M1 (fig. 60). Femur and tibia of hind leg slender. Antenna with 12 segments. **Synneuridae**

(In Europe 1 species: *Synneuron annulipes* Lundström.)

–	Anal vein (A1) distinctly visible; veins R1 and Rs separate; M-fork usually complete (fig. 61). Hind leg with apical half of femur swollen and tibia curved (fig. 62). Antenna with 16 segments. **Canthyloscelidae**

(In Europe 1 genus, *Hyperoscelis*, with 2 species.)

25(20)	Antenna inserted at lower part of the head, usually just above the lower margin of the face (fig. 63, 65, 69). **26**
–	Antenna inserted halfway up the eye, or higher (fig. 73, 78). **29**
26(25)	Third antennal segment conspicuously elongate, as long as or even longer than the two following segments and as long as (female) or 1,5 to 2 times as long as the head (male) (fig. 65). Wing venation as in fig. 66, more or less similar to that of the Pleciidae (fig. 67), but with vein R4+5 more strongly curved. **Hesperinidae**

(In Europe 1 species, *Hesperinus imbecillus* (Loew).)

–	Third antennal segment and wing venation not as above (fig. 63, 67-70). **27**
27(26)	Vein R forked into R2+3 and R4+5 with vein R4+5 continuing more or less straight beyond the fork (fig. 67) (in *Penthetria funebris* Meigen, the wing is darkened and, in the male, shorter and narrower, fig. 68). **Pleciidae**

(In Europe 1 genus, *Penthetria*, with 2 species.)

–	Vein R not forked (vein R2+3 absent) (fig. 63, 70). **28**
28(27)	Tibia of fore leg often swollen, in *Bibio* with a long, stout spine and an apical projection (fig. 63); in *Dilophus* with several rows of short, strong spines (fig. 64). **Bibionidae**
–	Tibia of fore leg simple, without spines or projections. Wing with a limited number of longitudinal veins; stem of vein M is about as long as part beyond the M fork (fig. 70). **Sciaridae**
>>>	[Most Sciaridae have the antenna inserted halfway up the eye with an eye bridge above the antennae (fig. 73; couplet 31); in some species the antenna is inserted much lower and/or the eye bridge is absent (fig. 69).]
29(25)	4 R-veins reaching the wing margin (fig. 71). **Pachyneuridae**

(In Europe 1 species, *Pachyneura fasciata* Zetterstedt.)

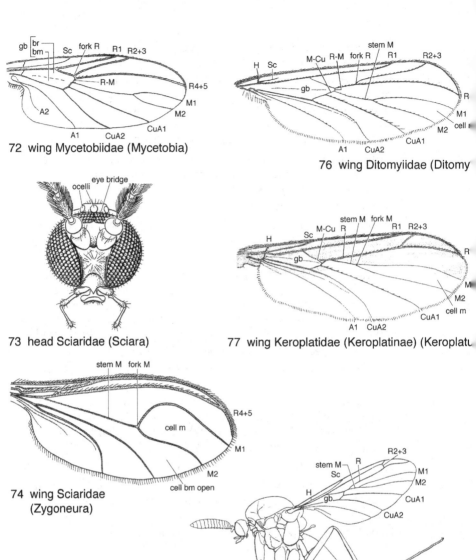

72 wing Mycetobiidae (Mycetobia)

73 head Sciaridae (Sciara)

76 wing Ditomyiidae (Ditomy

77 wing Keroplatidae (Keroplatinae) (Keroplatu

74 wing Sciaridae
(Zygoneura)

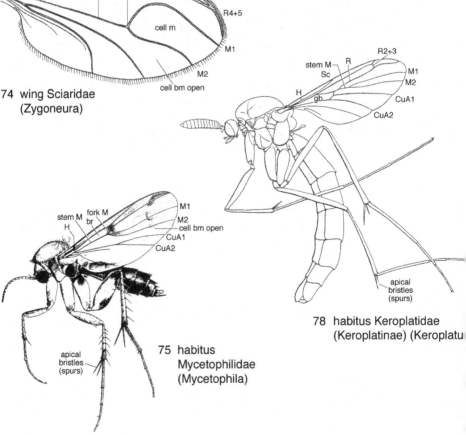

75 habitus
Mycetophilidae
(Mycetophila)

78 habitus Keroplatidae
(Keroplatinae) (Keroplatu

| – | 1 to 3 R-veins reaching the wing margin (wing figures 72-83). | **30** |

30(29) Wing with a large basal cell gb (fig. 72: cells br and bm fused into a single cell) from which 6 veins originate; R-fork before or at the same level as crossvein R-M (fig. 72). Coxae elongate. **Mycetobiidae**

– Cell bm not closed, i.e., open to the wing margin (fig. 74, 75) or, if cell gb is present, then at most 5 veins originate from this cell (fig. 76-82: cell gb); R-fork, if present, beyond crossvein R-M (fig. 76-81). Coxae elongate or not. **31**

31(30) Eyes with a narrow eye bridge above the antennae (fig. 73). Stem of vein M more or less of the same length as M1 and M2; cell m frequently bell-shaped (fig. 74). Coxae not conspicuously elongate (fig. 75, 78). **Sciaridae**

– Eyes separate. Stem of vein M shorter than M1 and M2; cell m nearly always narrow, elongate, not bell-shaped (fig. 75-83). Coxae usually conspicuously elongate. **32**

32(31) Vein CuA1 and stem of M not connected or connected at the level of crossvein H; vein A1 not reaching wing margin (fig. 75). **Mycetophilidae**

– Vein CuA1 connected to stem of M through crossvein M-Cu (fig. 76-83); this connection is always much more apical than the location of crossvein H; vein A1 in most cases reaching wing margin, at least as a distinct fold (fig. 76-82). **33**

33(32) Vein R2+3 more than half as long as R4+5; origin of vein Sc strong, the rest present only as a fold and ending free (fig. 76). **Ditomyiidae**

– Vein R2+3 less than half as long as R4+5 (fig. 77-81) or absent (fig. 82, 83); vein Sc nearly always long, ending in costa or in vein R1 (fig. 77-81). **34**

34(33) Stem of vein M originates from vein Rs or R (fig. 77-79). **35**

– Stem of vein M and Rs are connected by crossvein R-M (fig. 81-83). **36**

35(34) Antenna relatively short, at most as long as head and thorax together (fig. 78). Anal lobe not enlarged (fig. 77, 78).
Keroplatidae, subfamily Keroplatinae

– Antenna more than two thirds body length (fig. 80). Anal lobe large, with a (nearly) right anal angle (fig. 79, 80).
Keroplatidae, subfamily Macrocerinae

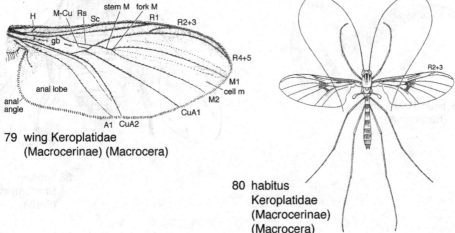

79 wing Keroplatidae
(Macrocerinae) (Macrocera)

80 habitus
Keroplatidae
(Macrocerinae)
(Macrocera)

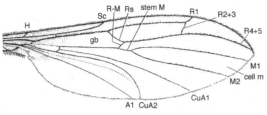

81 wing Bolitophilidae (Bolitophila)

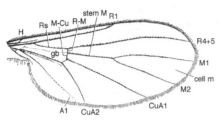

82 wing Diadocidiidae (Diadocidia)

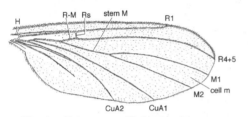

83 wing Sciarosoma (Sciaroidea)

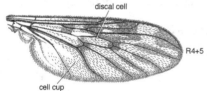

84 wing Anisopodidae (Sylvicola)

85 head Rachiceridae
(Rachicerus)

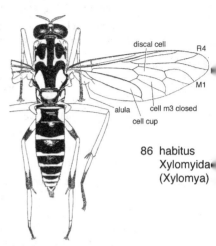

86 habitus
Xylomyida
(Xylomya)

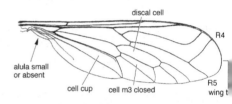

87 wing Vermileonidae (Vermitigris

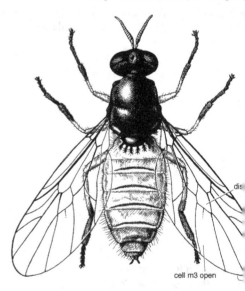

88 habitus
Stratiomyidae
(Beris)

| 36(34) | Vein R2+3 present (fig. 81). | **Bolitophilidae** |

36(34) Vein R2+3 present (fig. 81). **Bolitophilidae**
(In Europe 1 genus, *Bolitophila*, with about 36 species.)

– Vein R2+3 absent (fig. 82, 83). **37**

37(36) Vein R4+5 in alignment with vein Rs; crossveins R-M and M-Cu in line which each other (fig. 82). **Diadocidiidae**
(In Europe 1 genus, *Diadocidia*, with 5 species.)

– Vein Rs more or less vertical with respect to vein R4+5; crossvein R-M more in line with vein R4+5 (fig. 83). *Heterotricha* and *Sciarosoma*

>>> [The genera *Heterotricha* (with only 1 European species, *H. takkae* Chandler, known from Switzerland, Italy and Greece), and *Sciarosoma* (also with only 1 European species, *S. borealis* Chandler, known from Fennoscandia, Northwestern Russia, Germany and the Czech Republic), are 2 primitive Sciaroidea genera not currently assigned to a family; for details see Chandler 2002, Kallweit & Jaschhof 2004, Jaschhof et al. 2006.]

38(8) Vein R4+5 not forked; cell cup widely open toward wing margin (fig. 84). Antenna with over 14 segments. Palp with 4 segments. **Anisopodidae**
(In Europe 1 genus, *Sylvicola*, with about 10 species.)

– Vein R4+5 nearly always forked; if not so, then antenna with less than 14 segments; cell cup strongly narrowing toward or closed before wing margin (fig. 86-90). Palp with 1 to 3 segments. **39**

39(38) More than 11 antennal segments present (fig. 85) (the only European species has over 30 antennal segments). Wing venation more or less as in fig. 86: cell m3 closed, vein R5 ends near the wing tip. **Rachiceridae**
(In Europe 1 species, *Rachicerus tristis* (Loew).)

– No more than 11 antennal segments. **40**

40(39) Cell m3, the cell below the discal cell, closed (veins M3 and CuA1 meet before, fig. 86, or at the wing margin, fig. 87). **41**

– Cell m3 open (fig. 88-90). **42**

41(40) Vein R5 ending near the wing tip; alula normally developed (fig. 86); costa does not surround all of the wing but reaches at most up to vein M1.
Xylomyidae

– Vein R5 ending distinctly beyond the wing tip; alula small or absent (fig. 87); costa surrounding the wing completely [not clear from the drawing but it should be from the insect]. **Vermileonidae**

42(40) R-veins in the front part of the wing with vein R5 almost always reaching the wing margin before the wing tip; discal cell diamond shaped and usually strikingly small (fig. 88); costa does not surround all of the wing but continues up to just beyond the wing tip. **Stratiomyidae**

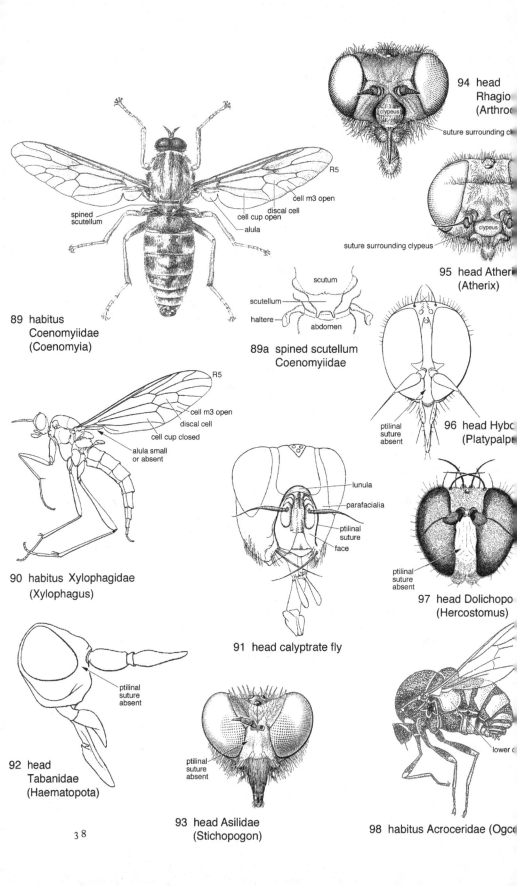

94 head Rhagio (Arthro...

suture surrounding cl...

suture surrounding clypeus

95 head Ather... (Atherix)

R5

cell m3 open

discal cell

cell cup open

alula

spined scutellum

89 habitus Coenomyiidae (Coenomyia)

scutum

scutellum

haltere

abdomen

89a spined scutellum Coenomyiidae

ptilinal suture absent

96 head Hybo... (Platypalp...)

R5

cell m3 open

discal cell

cell cup closed

alula small or absent

90 habitus Xylophagidae (Xylophagus)

lunula

parafacialia

ptilinal suture

face

ptilinal suture absent

97 head Dolichopo... (Hercostomus)

91 head calyptrate fly

ptilinal suture absent

92 head Tabanidae (Haematopota)

ptilinal suture absent

93 head Asilidae (Stichopogon)

lower...

98 habitus Acroceridae (Ogc...

–	Venation different; vein R5 ending at or distinctly beyond the wing tip; discal cell elongate (fig. 89, 90); costa surrounding the wing completely [not clear from the drawing but it should be from the insect]. **43**
43(42)	Scutellum with 2 spiny outgrowths (fig. 89a). Alula present; cell cup strongly narrowing toward wing margin, but nearly always open (fig. 89). **Coenomyiidae**

(In Europe 1 widespread species, *Coenomyia ferruginea* (Scopoli).)

–	Scutellum without outgrowths. Alula small or absent; cell cup closed (veins A1 and CuA2 meet before or at the wing margin) (fig. 90). **Xylophagidae**

(In Europe 1 genus, *Xylophagus*, with 5 species.)

44(2)	It is possible to skip couplet 44 and to continue with couplet 45, in which case the decision about the ptilinal suture is postponed to couplet 71. All families with a more complete wing venation (e.g. fig. 99, 100) or with a long spurious vein (fig. 135) key out before couplet 71, the large majority of the families with a more simplified wing venation (e.g. fig. 101, 102) key out beyond couplet 71.
–	Ptilinal suture present (fig. 91; a usually horseshoe shaped groove or suture running above the lunula down on either side of the antennae, separating the parafacialia from the remainder of the face). Wing venation with at most 7 veins reaching the wing margin (wing figures 102, 179-420), venation usually simple, vein R4+5 never forked, cell bm always short (fig. 102) or fused with cell dm (fig. 207). **Acalyptrate and Calyptrate families - 81**
–	Ptilinal suture absent (although a closed suture might be apparent above the antennae this does not extend below the antennae (fig. 92-97), if a suture is present then this surrounds the clypeus (fig. 94, 95) and does not extend to above the antennae). Venation sometimes simple (e.g. Dolichopodidae, fig. 162) but usually very much different, as shown in wing figures 98-101, 105-168), often cells br, bm and cup long and therefore the discal cell, if present, in distal part of the wing, vein R4+5 often forked (fig. 100). **45**
45(44)	Thorax humpbacked and abdomen usually globose (fig. 98, 99). Head small, situated (somewhat) below the highly vaulted thorax; eyes holoptic; antenna with 3 segments of which the first can be very small. Wing venation varies from reduced (fig. 98) to complete (fig. 99); lower calypter conspicuous, large, ear-shaped. Empodium pulvilliform (last tarsal segments with 3 cushions; fig. 103). **Acroceridae**
–	Combination of characters not as given above. **46**
46(45)	At the same time vein R4+5 forked into R4 and R5 and cell cup very elongate (this cell closed at (fig. 100) or not far before (fig. 105, 131) the wing margin, or open toward the wing margin (fig. 113, 129)), or wing venation aberrant as shown in fig. 107-109. **47**
–	Not at the same time vein R4+5 forked into R4 and R5 and cell cup very elongate (e.g. fig. 101, 139: R4+5 not forked and cell cup elongate; fig. 102: R4+5 not forked and cell cup short; fig. 133: R4+5 forked and cell cup short; fig. 228: R4+5 not forked and cell cup absent). **62**

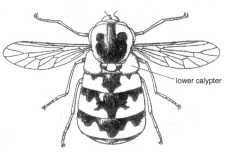

99 habitus Acroceridae (Cyrtus)

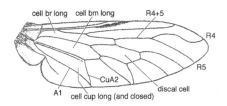

100 wing Therevidae (Thereva)

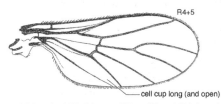

101 wing Mythicomyiidae (Empidideicus)

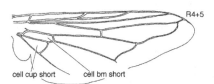

102 wing Oestridae (Oestrus)

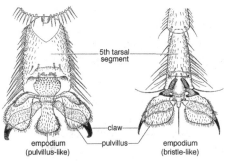

103 5th tarsomere Stratiomyidae (Inopus)

104 5th tarsomere Muscidae (Musca)

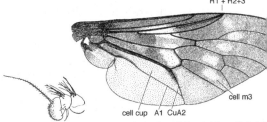

105 wing Athericidae (Ibisia)

106 antenna Athericidae (Atherix)

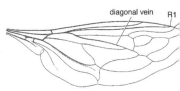

107 wing Nemestrinidae (Fallenia)

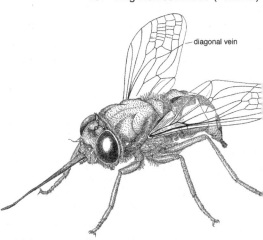

108 habitus Nemestrinidae (Nemestrinus)

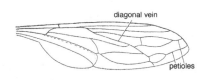

109 wing Nemestrinidae (Trichopsidea)

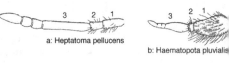

110 a-d antennae Tabanidae

47(46)	Empodium pulvilliform (last tarsal segments with 3 cushions; fig. 103). **48**
>>>	[In *Lampromyia* (Vermileonidae) pulvilli and/or empodium sometimes extremely reduced; such species can be recognised by the very long mouthparts in combination with the wing venation (fig. 114).]
–	Empodium bristle-like or absent (last tarsal segments with at most 2 cushions) (fig. 104). **56**
>>>	[A key for most of the families and genera which key out in couplets 48-55 is given by Nagatomi & Ding Yang 1998.]
48(47)	Veins R1 and R2+3 reach the costa at or almost at the same point (fig. 105). Third antennal segment kidney-shaped (fig. 95, 106). **Athericidae**
–	Veins R1 and R2+3 reach the costa at a clear distance from each other (most of the wing figures 107-119) and/or wing venation very different (fig. 107-109, 116); third antennal segment virtually never kidney-shaped. **49**
49(48)	R- and M-veins more or less converging toward costa and wing tip, M-veins ending before or just beyond the wing tip; apical part of wing with several, usually longitudinally aligned cells that may in turn be divided into smaller cells; a so-called diagonal vein present (fig. 107-109). Head large, usually with elongate mouthparts. Tibiae without apical bristles (fig. 108). **Nemestrinidae**
–	Combination of characters not as given above. **50**
50(49)	Lower calypter conspicuously large; vein R5 reaching the wing margin distinctly beyond the wing tip and fork of veins R4 and R5 usually strikingly diverging (fig. 111). Antenna with large third segment (fig. 110 a-d) and usually a tapering style (fig. 110 d). **Tabanidae**
–	Lower calypter small. Other characters as specified above or not, but rarely in the above combination. **51**
51(50)	Alula small or absent (fig. 112-114). **52**
–	Alula normally developed (fig. 115-119). **53**
52(51)	Cell cup closed (veins A1 and CuA2 meet before or at the wing margin); vein R5 ends near the wing tip (fig. 112). All antennal segments past the second more or less similar (fig. 112 a). **Xylophagidae**

(In Europe 1 genus, *Xylophagus*, with 5 species.)

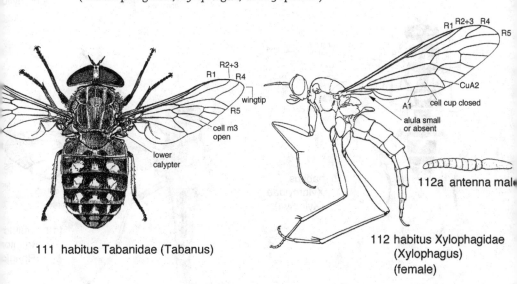

111 habitus Tabanidae (Tabanus)

112 habitus Xylophagidae (Xylophagus) (female)

112a antenna male

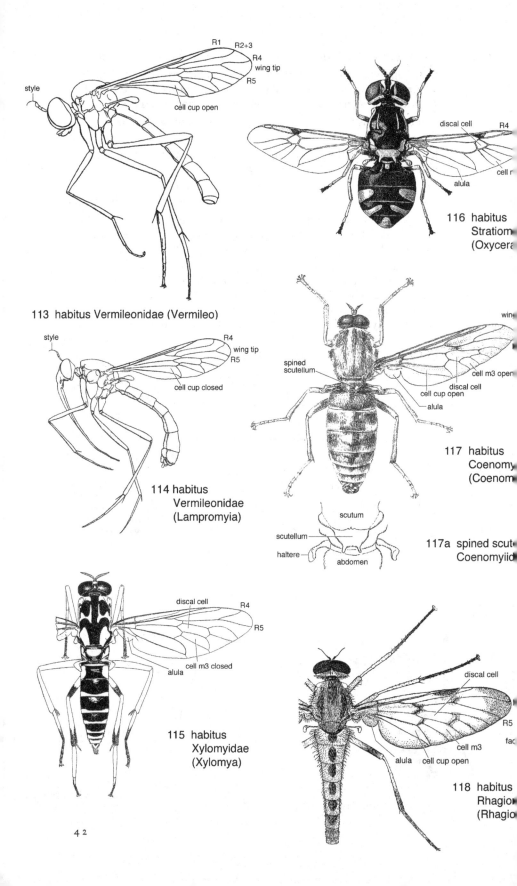

R1
R2+3
R4
wing tip
R5
style
cell cup open

113 habitus Vermileonidae (Vermileo)

discal cell
R4
cell r
alula

116 habitus
Stratiom
(Oxycera

style
R4
wing tip
R5
cell cup closed

114 habitus
Vermileonidae
(Lampromyia)

win
spined
scutellum
cell m3 open
discal cell
cell cup open
alula

117 habitus
Coenomy
(Coenom

scutum
scutellum
haltere
abdomen

117a spined scut
Coenomyiid

discal cell
R4
R5
alula
cell m3 closed

115 habitus
Xylomyidae
(Xylomya)

discal cell
R5
fac
cell m3
alula cell cup open

118 habitus
Rhagio
(Rhagio

| – | Cell cup open (fig. 113), or, if closed, then vein R5 ending distinctly beyond the wing tip (fig. 114). Antennal segments past the second not similar, but forming a style. **Vermileonidae** |

–   Cell cup open (fig. 113), or, if closed, then vein R5 ending distinctly beyond the wing tip (fig. 114). Antennal segments past the second not similar, but forming a style.   **Vermileonidae**

53(51) Cell m3, the cell below the discal cell, is closed (veins M3 and CuA1 meet before or at the wing margin, fig. 115).   **Xylomyidae**

–   Cell m3 open (fig. 116-119).   **54**

54(53) R-veins situated in front part of the wing with R5 virtually always ending well before the wing tip; discal cell diamond-shaped and small, often strikingly small (fig. 116); costa not surrounding all of the wing but reaching at most to just beyond the wing tip.   **Stratiomyidae**

–   Wing venation different; vein R5 ending at or well beyond the wing tip; discal cell elongate (fig. 117-119); costa surrounding the wing completely [not clear from the drawing but it should be from the insect].   **55**

55(54) Scutellum with 2 spiny outgrowths (fig. 117a). Wing venation as in fig. 117: cell cup strongly narrowing toward the wing margin but nearly always open; 4 veins originating from the discal cell reach the wing margin.

Coenomyiidae

(In Europe 1 widespread species, *Coenomyia ferruginea* (Scopoli).)

–   Scutellum without spiny outgrowths. Wing venation with cell cup open (fig. 118) or closed (fig. 119) and usually 3 veins originating from the discal cell reaching the wing margin.   **Rhagionidae**

56(47) Veins R4, R5 and M curving forward so that at least vein R4 ends in vein R1 and with at most 2 but usually just 1 vein reaching the hind margin of the wing (fig. 120). Last antennal segment conspicuously swollen (fig. 121) or aberrant.   **Mydidae**

–   Wing venation and antenna different.   **57**

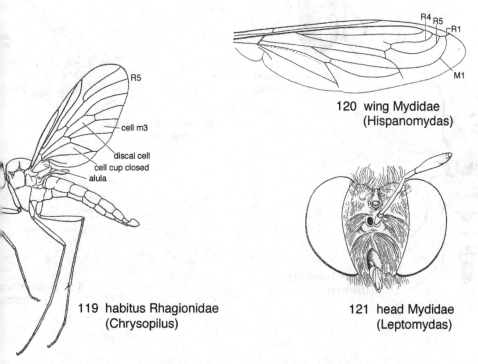

120 wing Mydidae
(Hispanomydas)

119 habitus Rhagionidae
(Chrysopilus)

121 head Mydidae
(Leptomydas)

43

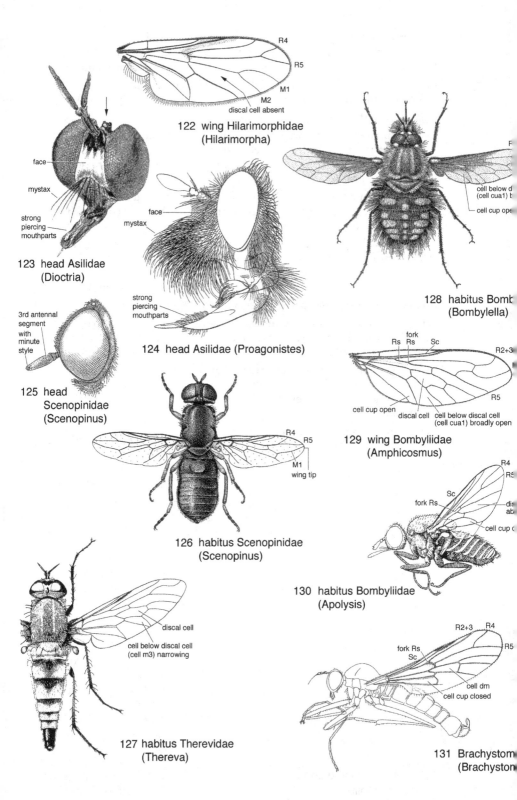

122 wing Hilarimorphidae
(Hilarimorpha)

R4
R5
M1
M2
discal cell absent

face
mystax
strong
piercing
mouthparts

123 head Asilidae
(Dioctria)

face
mystax

strong
piercing
mouthparts

124 head Asilidae (Proagonistes)

3rd antennal
segment
with
minute
style

125 head
Scenopinidae
(Scenopinus)

R4
R5
M1
wing tip

126 habitus Scenopinidae
(Scenopinus)

discal cell
cell below discal cell
(cell m3) narrowing

127 habitus Therevidae
(Thereva)

cell below d
(cell cua1) b
cell cup ope

128 habitus Bomb
(Bombylella)

fork
Rs Rs   Sc   R2+3
R5
cell cup open
discal cell   cell below discal cell
(cell cua1) broadly open

129 wing Bombyliidae
(Amphicosmus)

R4
R5
Sc
fork Rs   dis
ab
cell cup (

130 habitus Bombyliidae
(Apolysis)

R2+3   R4
R5
fork Rs
Sc
cell dm
cell cup closed

131 Brachystom
(Brachyston

57(56)	Veins R4+5 and M1+2 are forked in (almost) similar manner and discal cell absent (fig. 122). **Hilarimorphidae**
	(In Europe 1 genus, *Hilarimorpha*, with 2 species.)
–	Combination of characters not as given above. **58**
58(57)	Upper part of head, between the eyes, with a saddle-shaped depression; ocelli on a separate elevation which, however, does not project above the upper margin of the eyes (fig. 123, arrow); face relatively long and nearly always with long bristles, forming the mystax or beard; powerful, piercing mouthparts present (fig. 123, 124). **Asilidae**
–	Combination of characters not as given above. **59**
59(58)	Antenna with 3 segments of which the third is the largest; this segment with a minute style and/or forked at the apex (fig. 125). Vein M1 curving forward, ending before the wing tip or in vein R5 (fig. 126). **Scenopinidae**
–	Combination of characters not as given above. **60**
60(59)	The cell below the discal cell closed or becoming narrower toward the hind margin of the wing (fig. 127). **Therevidae**
–	The cell below the discal cell open and widening toward the hind margin of the wing (fig. 128, 129), or discal cell absent (fig. 130). **61**
61(60)	Vein Sc ends in the costa beyond the fork of Rs; veins R2+3 and R4 often strongly curving forward; cell cup usually open to the wing margin (fig. 128, 129); if closed, then ending in an acute angle (fig. 130); wing often with conspicuous markings. Mouthparts may or may not constitute a long sucking proboscis. **Bombyliidae**
–	Vein Sc ends in the costa before the fork of Rs; Veins R2+3 and R4 not strongly curved; cell cup closed, ending obtusely; wing without conspicuous markings (fig. 131). **Brachystomatidae**: *Brachystoma*
62(46)	Vein R4+5 forked (fig. 132-134). **63**
–	Vein R4+5 not forked (all wing figures beyond 134). **64**

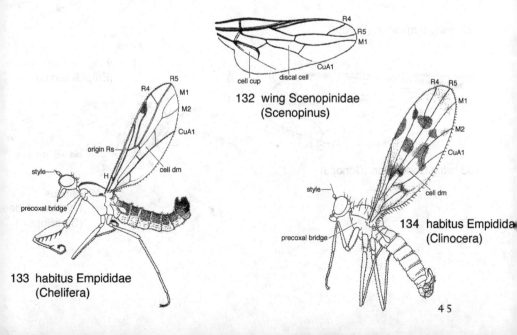

132 wing Scenopinidae
(Scenopinus)

133 habitus Empididae
(Chelifera)

134 habitus Empidida
(Clinocera)

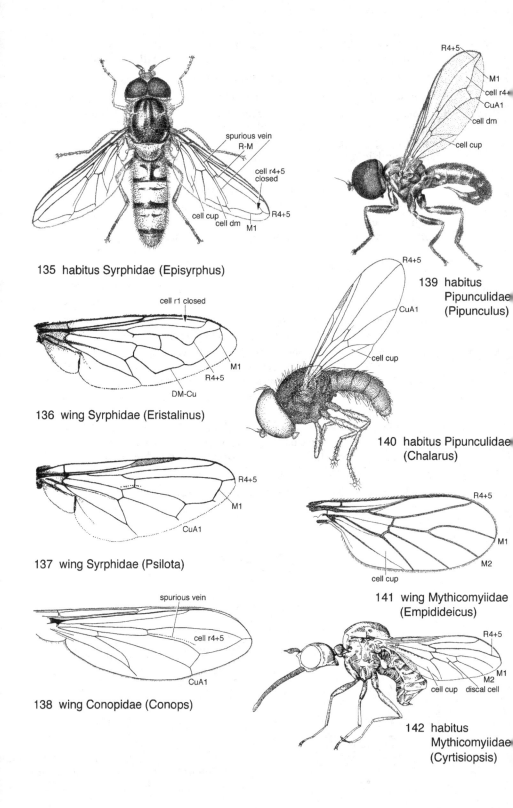

135 habitus Syrphidae (Episyrphus)

136 wing Syrphidae (Eristalinus)

137 wing Syrphidae (Psilota)

138 wing Conopidae (Conops)

139 habitus Pipunculidae (Pipunculus)

140 habitus Pipunculidae (Chalarus)

141 wing Mythicomyiidae (Empidideicus)

142 habitus Mythicomyiidae (Cyrtisiopsis)

63(62) Antenna with 3 segments of which the third is the largest; this segment sometimes forked at the apex and/or with a minute style (fig. 125). 2 veins originating from cell dm run in the direction of the wing margin, M1 and CuA1 (fig. 132), vein M1 ends in vein R5 or in the wing margin close to vein R5. **Scenopinidae**

– Style virtually always well developed (fig. 133, 134). 3 veins originating from cell dm reach the wing margin: M1, M2 and CuA1 (fig. 133, 134).
Empididae

>>> [Including the brachystomatid genera with vein R4+5 forked: *Brachystoma, Gloma, Pseudoheleodromia* and *Trichopeza* .]

64(62) A longitudinal spurious vein (vena spuria) present on both sides of crossvein R-M; vein M1 does not reach the wing margin but curves forward and ends in vein R4+5 before that vein reaches the margin (cell r4+5 closed); cell dm present; cell cup long and ending acutely just before the wing margin (fig. 135). **Syrphidae**

– Wing venation different; vena spuria usually absent, but if present (some Conopidae), then restricted to cell r4+5 (fig. 138). **65**

>>> [In the Syrphidae genera *Eristalinus* and *Psilota* the vena spuria is hardly indicated or absent, and confusion is possible with some Conopidae genera that also possess a closed cell r4+5 (fig. 138, 144). Furthermore, some Conopidae do possess a vena spuria but this is always restricted to cell r4+5 (fig. 138). Discriminating characters are: cell r1 closed and vein R4+5 strongly curved (fig. 136): Syrphidae: *Eristalinus*; vein M1 more or less rectangular with respect to vein R4+5 (fig. 137): Syrphidae: *Psilota*; vein CuA1 reaches the wing margin (fig. 138, 144); vena spuria, if present, restricted to cell r4+5 (fig. 138): Conopidae.]

65(64) Head large, hemispherical to nearly spherical, the eyes occupying almost all of the head. Wing venation as in fig. 139 or 140: cell cup long and ending acutely just before the wing margin; cell r4+5 narrowing toward wing margin but open; sometimes vein M1 and cell dm absent (fig. 140). Mouthparts small. **Pipunculidae**

– Combination of characters not as given above; if cell cup elongate and acute and cell r4+5 narrowing, then mouthparts and/or antenna elongate (fig. 143-144). **66**

66(65) Cell cup open and conspicuously long (fig. 141, 142). **Mythicomyiidae**

– Cell cup different or absent. **67**

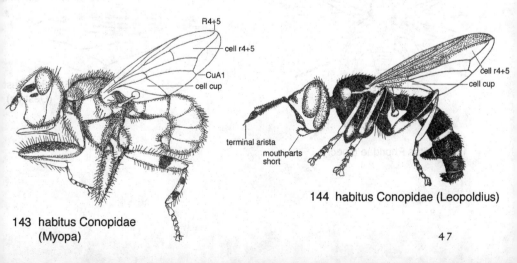

143 habitus Conopidae
(Myopa)

144 habitus Conopidae (Leopoldius)

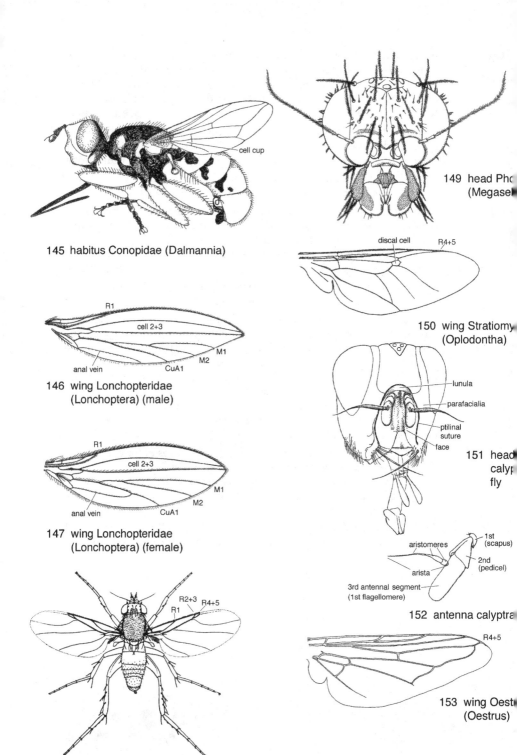

145 habitus Conopidae (Dalmannia)

146 wing Lonchopteridae
(Lonchoptera) (male)

147 wing Lonchopteridae
(Lonchoptera) (female)

148 habitus Phoridae (spec.)

149 head Pho
(Megasel

150 wing Stratiomy
(Oplodontha)

151 head
calyp
fly

152 antenna calyptra

153 wing Oest
(Oestrus)

4 8

67(66)	Cell r4+5 narrowing toward wing tip (fig. 143, 145) or closed (fig. 144); cell cup usually elongate, ending acutely, sometimes short (fig. 145). Second antennal segment elongate. Mouthparts usually long, slender, geniculate at the base, sometimes again geniculate in their mid region (fig. 143, 145). If mouthparts short, then arista terminal (fig. 144). **Conopidae**
–	Combination of characters not as given above. **68**
68(67)	Wing lancet-shaped with characteristic venation: vein R1 short, cell r2+3 narrowing toward wing tip, vein M forked into M1 and M2, crossveins present only in basal part of the wing; longitudinal veins with black bristles on dorsal side (fig. 146, 147). **Lonchopteridae**
–	Wing different. **69**
69(68)	Wing venation characteristic, reduced: the strong R-veins end in the costa, about halfway up the wing; the other veins are weaker and usually cross the wing in an oblique direction, more or less parallel to each other (fig. 148). Head, palp and legs usually with strong, dentated or feathered bristles (fig. 149). Small and hump-backed, usually dark flies. **Phoridae**
–	Combination of characters not as given above. **70**
70(69)	Wing venation as in fig. 150, with a small, diamond-shaped discal cell. Legs without bristles; empodium pulvilliform (fig. 103). **Stratiomyidae**
–	Combination of characters not as given above. **71**
71(70)	Ptilinal suture present (fig. 151; a usually horseshoe shaped groove or suture running above the lunula down on either side of the antennae, separating the parafacialia from the remainder of the face). Arista with a few exceptions situated on dorsal side of the 3rd antennal segment, near the 2nd segment (fig. 152, head figures 182-416), arista rarely terminal or absent. **Acalyptrate and Calyptrate families - 81**
–	Ptilinal suture absent (although a closed suture is often apparent above the antennae this does not extend below the antennae, thus there are no vertical sutures between the eyes below the antennae; fig. 154, 155). Arista frequently terminal, situated at the end of the 3rd antennal segment (fig. 154, 156, 159-161), sometimes situated more dorsally (fig. 155). **72**

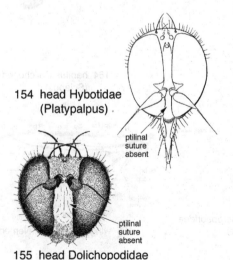

154 head Hybotidae
(Platypalpus) .

ptilinal
suture
absent

5th   4th   3rd ant. segment 2nd 1st
(3rd) (2nd) (1st flagellomere)

156 antenna Empididae (Empis)

ptilinal
suture
absent

155 head Dolichopodidae
(Hercostomus)

49

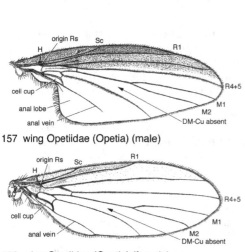

157 wing Opetiidae (Opetia) (male)

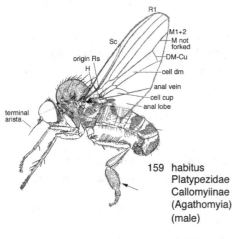

158 wing Opetiidae (Opetia) (female)

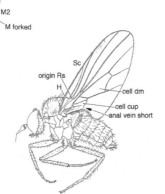

159 habitus
Platypezidae
Callomyiinae
(Agathomyia)
(male)

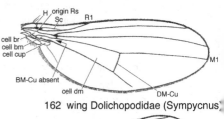

162 wing Dolichopodidae (Sympycnus)

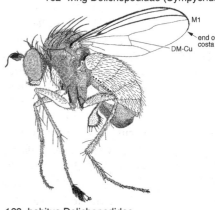

163 habitus Dolichopodidae
(Dolichopus)

164 habitus Dolichopodidae
(Sciapus)

160 habitus Platypezidae
Platypezinae
(Polyporivora) (female)

161 habitus Microphoridae
(Microphor)

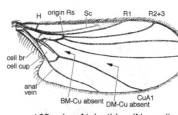

165 wing Atelestidae (Nemedina)

72(71)	Origin of vein Rs before or opposite crossvein H, or at a distance beyond crossvein H less than the length of that crossvein (fig. 157-164). **73**
–	Origin of vein Rs farther beyond crossvein H, at a distance greater than the length of the crossvein (fig. 165, 168). **78**
73(72)	Wing venation as in fig. 157 or 158: vein Sc complete, ending in the costa; crossvein DM-Cu absent; veins M1 and M2 long; cell cup small, narrowing toward its acute apical angle; in male: anal lobe large and anal vein not reaching wing margin; in female: anal lobe virtually absent and anal vein reaching wing margin. **Opetiidae**
	(In Europe 1 widespread species, *Opetia nigra* Meigen.)
–	Wing venation different. **74**
74(73)	Vein Sc complete, ending in costa; anal vein reaching wing margin; cell cup ending in an acute angle; alula and well developed anal lobe present (fig. 159, 160). **Platypezidae**
>>>	[Most Platypezidae show the following additional characters: terminal arista; vein M forked and crossvein DM-Cu present; first tarsal segment of hind leg in the subfamily Callomyiinae long and cylindrical, usually swollen in male (arrow fig. 159), slender in female; the subsequent tarsal segments cylindrical. In the subfamily Platypezinae the first tarsal segment of hind leg short, laterally compressed, subsequent tarsal segments laterally compressed as well, in particular in the female (arrow fig. 160). Crossvein DM-Cu absent in *Microsania*.]
–	Combination of characters not as given above (not at the same time the following four characters: vein Sc reaching wing margin, anal lobe well developed, cell cup ending in an acute angle, anal vein reaching wing margin). **75**
>>>	[A key for the families which key out in couplets 75-80 is given by Sinclair & Cumming 2006, including unplaced genera and a somewhat different concept of the families Empididae and Dolichopodidae.]
75(74)	Costa surrounding the wing completely [not clear from the drawing but it should be from the insect]; vein Sc ending in the costa; anal vein short or absent; cell dm present; 3 veins originating from this cell reach the wing margin (fig. 161). **Microphoridae**
–	Costa ending near the wing tip (fig. 163, 165, 166b); vein Sc usually ending in R1 (fig. 162) or incomplete and not reaching the costa (fig. 165, 166b); anal vein and cell dm variable. **76**
76(75)	Cells br, bm and cup small; crossvein BM-Cu absent (cells bm and dm are fused) (fig. 162-164); 2 veins from crossvein DM-Cu toward the wing margin, the anterior of which may be curved (fig. 163) or forked into M1 and M2 (fig. 164). Body usually with a greenish metallic lustre, in a few species dull, yellow, brown or black. **Dolichopodidae**

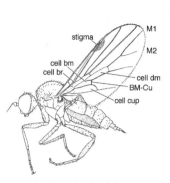

166a habitus Atelestidae (Meghyperus)

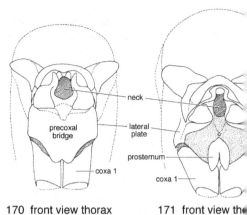

170 front view thorax Empididae (Empis)

171 front view th Hybotidae (H

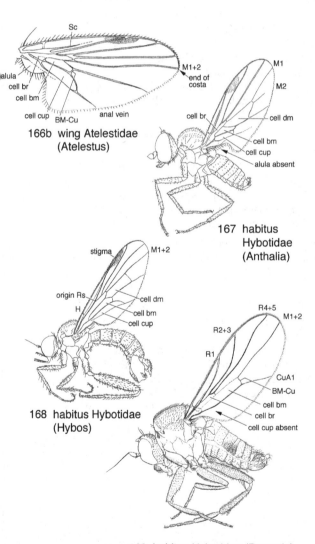

166b wing Atelestidae (Atelestus)

167 habitus Hybotidae (Anthalia)

168 habitus Hybotidae (Hybos)

169 habitus Hybotidae (Drapetis)

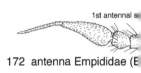

172 antenna Empididae (E

173 antenna Hybotidae (H

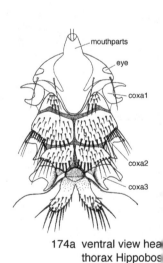

174a ventral view hea thorax Hippobos (Melophagus)

–	Cells br and bm large; cell cup variable but usually large; crossvein BM-Cu present (fig. 166a-167); cell dm may be present (fig. 166a) or not (fig. 166b). Combination of other characters not as given above. **77**
77(76)	Wing with alula; cell cup large and at least as long as cells br and bm, usually longer; either cell dm present and vein M forked into M1 and M2 (fig. 166a); or cell dm absent and vein M not forked (fig. 166b).
	Atelestidae: Atelestinae
–	Wing without alula; cell cup usually shorter than cells br and bm (fig. 167) or cell cup absent. In case of cell cup being as long as or longer than cells br and bm, then cell dm present and vein M not forked (fig. 168).**Hybotidae**
78(72)	Either a precoxal bridge present (prosternal sclerite on the ventral side of the thorax, between the neck and the fore legs, large and fused with the lateral plates bordering it) (fig. 133, 134, 170), and/or the costa surrounding all of the wing. First antennal segment usually distinctly bristled, but in any case with some bristles on its dorsal and/or ventral side (fig. 172).
	Empididae
>>>	[Including the one brachystomatid genus with vein R4+5 not forked, *Heleodromia*.]
–	No precoxal bridge present (prosternum small, separated from lateral plates by a membrane) (fig. 171). Costa ends near the wing tip. First antennal segment without bristles (fig. 173). **79**
79(78)	Crossveins BM-Cu and DM-Cu absent; cell cup long and almost rectangular; anal vein and vein CuA1 arched to wing margin (fig. 165).
	Atelestidae: Nemedininae
	(In Central Europe 1 species, *Nemedina alamirabilis* Chandler.)
–	Combination of wing characters not as given above (fig. 166-169). **80**
80(79)	Wing with alula; cell cup large and at least as long as cells br and bm, usually longer; either cell dm present and vein M forked into M1 and M2 (fig. 166a); or cell dm absent and vein M not forked (fig. 166b).
	Atelestidae: Atelestinae
–	Wing without alula; cell cup usually shorter than cells br and bm (fig. 167) or cell cup absent (fig. 169). In case of cell cup being as long as or longer than cells br and bm, then cell dm present and vein M not forked (fig. 168).
	Hybotidae
81(44/71)	Coxae of mid legs not far apart. Tarsal claws normal, not strongly curved. Fly not ectoparasitic on birds or mammals. **82**
	–Coxae of mid legs, and usually of the other legs as well, far apart (fig. 174a). Tarsal claws strongly curved (fig. 174b) or last tarsal segment widened and eyes reduced or absent (fig. 175). Fly living as an ectoparasite on birds or mammals. **Calyptratae: Hippoboscoidea – 169**

aws

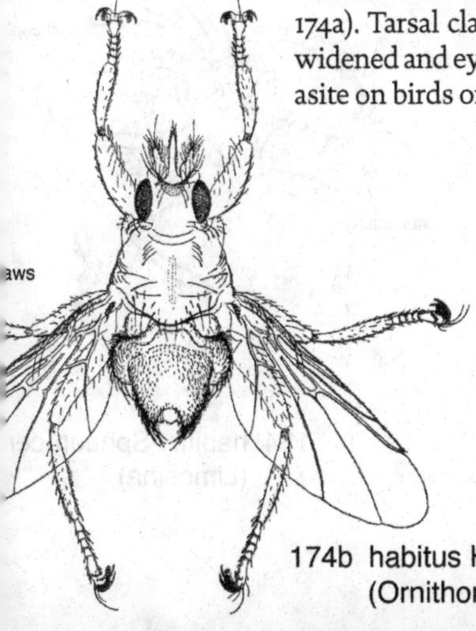

eyes rudimentary
or absent

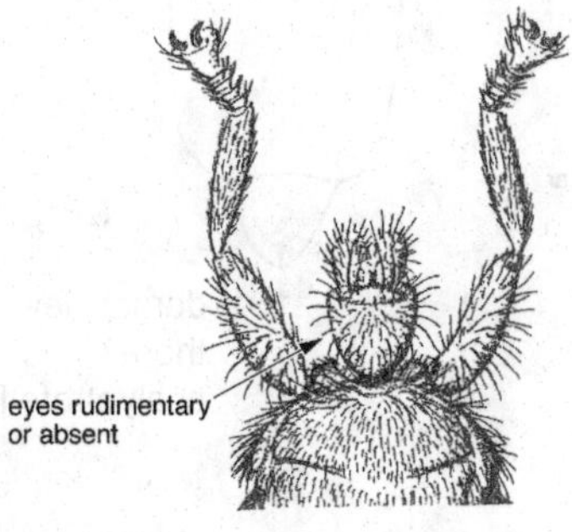

175 head and fore
legs Streblidae
(Brachytarsina)

174b habitus Hippoboscidae
(Ornithomyia)

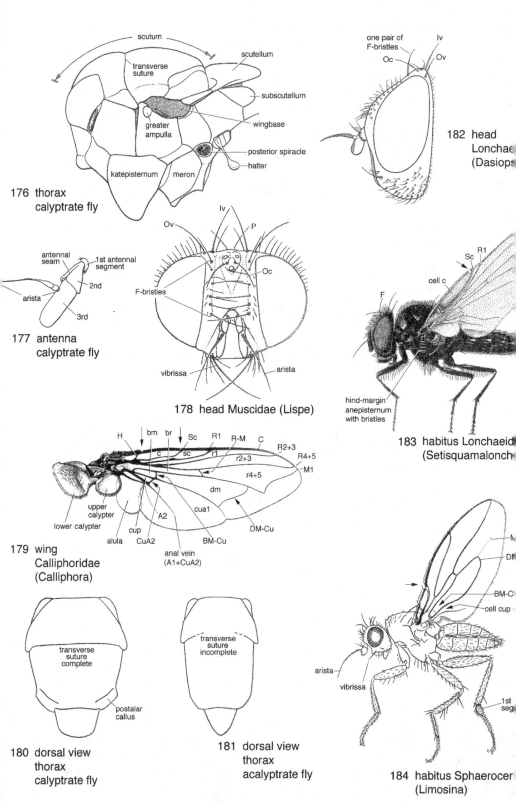

176 thorax calyptrate fly

177 antenna calyptrate fly

antennal seam
1st antennal segment
2nd
arista
3rd

scutum
transverse suture
scutellum
subscutellum
greater ampulla
wingbase
posterior spiracle
halter
katepisternum · meron

one pair of F-bristles
Oc
Iv
Ov

182 head Lonchae (Dasiops

Ov
Iv
P
Oc
F-bristles
vibrissa
arista

178 head Muscidae (Lispe)

F
cell c
Sc
R1
hind-margin
anepisternum with bristles

183 habitus Lonchaeid (Setisquamalonch

H
bm
br
Sc
R1
R-M
C
R2+3
R4+5
c
sc
r1
r2+3
r4+5
M1
dm
upper calypter
lower calypter
cup
alula
CuA2
A2
cua1
anal vein (A1+CuA2)
BM-Cu
DM-Cu

179 wing Calliphoridae (Calliphora)

transverse suture complete
postalar callus

180 dorsal view thorax calyptrate fly

transverse suture incomplete

181 dorsal view thorax acalyptrate fly

M
DI
BM-C
cell cup
arista
vibrissa
1st seg

184 habitus Sphaerocer (Limosina)

82(81)   Greater ampulla present as a major bulge immediately below the anterior part of the wing insertion (fig. 176). Second antennal segment with a distinct seam on its upper side (fig. 177). Usually 1 or more strong vibrissae present, as well as incurved lower F-bristles (fig. 178). Lower calypter usually strongly developed and vein Sc more or less complete, separate from vein R1 (fig. 179, 396). Upper side of thorax usually with a distinct and complete transverse suture (fig. 176, 180; not complete in most Scathophagidae and some Muscidae, Anthomyiidae or Fanniidae); postalar callus usually distinctly separated (fig. 180).

Other Calyptrate families – 170

–   Greater ampulla inconspicuous, if however well developed, then vibrissae absent and other characters different (second antennal segment usually without a distinct seam; lower calypter small or absent; central part of transverse suture usually less distinct or absent (fig. 181); postalar callus not distinctly separated or absent (fig. 181)).   **Acalyptrate families – 83**

83(82)   Haltere dark brown to black. Head large, wide and high (hemispherical); only 1 pair of F-bristles, situated at the level of the ocellar triangle and curving backward (fig. 182). Hind margin of anepisternum with a row of bristles. Wing always without markings. Cell c wide; vein Sc complete, ends in the costa separate from vein R1 (fig. 183).   **Lonchaeidae**

>>>   [Usually compact, 3 to 6 mm long dark flies with or without lustre and the following characters: Lunula large; diverging P-bristles weakly developed. Costa with subcostal break. Tibiae usually without dorsal preapical bristle, sometimes present on mid leg only. Abdomen short, wide, compressed. Female with a lanceolate, incompletely retractable ovipositor.]

–   Haltere not dark brown to black or other characters different. If haltere dark brown, then head not hemispherical and/or more than 1 pair of F-bristles present and/or anepisternum without a row of bristles along its hind margin, and/or wing with a pattern of markings.   **84**

84(83)   Vibrissae (fig. 184, 185, 189) or vibrissa-like bristles (fig. 186) present at or near the vibrissal angle.   **85**

–   No vibrissae or vibrissa-like bristles (fig. 187, 188).   **134**

85(84)   First tarsal segment of hind leg much swollen, and usually shorter than the second segment (fig. 184, 185). Third antennal segment short, arista long. Wing venation ranging from somewhat reduced (fig. 184: crossvein BM-Cu absent, cell cup open, veins beyond crossvein DM-Cu less distinct or largely absent) to more or less complete (fig. 185).   **Sphaeroceridae**

–   First tarsal segment of hind leg not swollen, usually long and slender. Other characters variable.   **86**

186 head Lauxaniidae (Prosopomyia)

187 head Lauxaniidae (Lauxania)

185 habitus Sphaeroceridae (Copromyza)

55

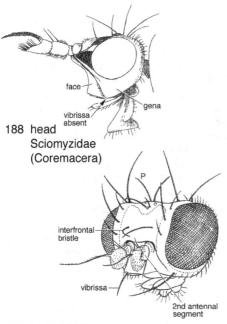

188 head
Sciomyzidae
(Coremacera)

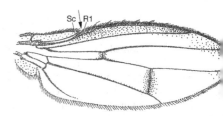

192 wing Trixoscelididae (Trixoscelis)

189 head Clusiidae
(Clusioides)

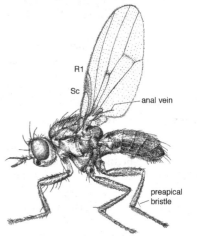

193 habitus Lauxaniidae (Pachycerina)

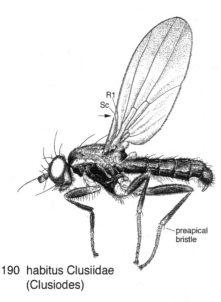

190 habitus Clusiidae
(Clusiodes)

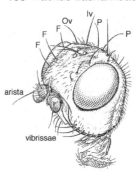

194 head Sepsidae (Orygma)

191 basal part wing Camillidae (Camilla)

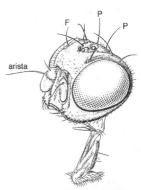

195 head Sepsidae (Saltella)

86(85)   Dorsal preapical bristle at least present on tibia of mid leg, but usually on all tibiae (fig. 190, 193, 198, 201, 204).   **87**

–   All tibiae without dorsal preapical bristle.   **101**

87(86)   Vein Sc complete, reaching the costa separate from vein R1 (fig. 190, 192, 193).   **88**

–   Vein Sc incomplete, does not reach the costa separately: Sc reduced or shortened (fig. 191), or merging with vein R1 before the costa (fig. 215b).   **94**

88(87)   Second antennal segment triangular at outer side; third antennal segment short; P-bristles diverging or absent; a single pair of interfrontal bristles, curving forward (fig. 189). Costa with a subcostal break. Vein Sc complete and separate from vein R1 (fig. 190). Slender, dark-brown, dark brownish-yellow or yellow flies.   **Clusiidae**

–   Second antennal segment not triangular at outer side. Other characters as specified above or not, but rarely in the above combination.   **89**

89(88)   Veins Sc and R1 distinctly separate and reaching the costa at a considerable distance from each other (fig. 193, 198, 201, 203).   **90**

–   Veins Sc and R1 are parallel and reach the costa close together (fig. 192). **93**

90(89)   Anal vein short, not reaching the wing margin and at the same time costa without strong bristles (fig. 193, 198).   **91**

–   Anal vein long, reaching the wing margin or ending just short of it (fig. 201, 203); costa often with strong bristles (fig. 201); if anal vein not conspicuously long, then costa with strong bristles.   **92**

91(90)   P-bristles diverging or absent; arista bare (fig. 194, 195). Posterior spiracle with 1 or more bristles at the posteroventral margin (fig. 196, 197).   **Sepsidae**

>>>   [Usually glossy-black ant-like flies with relatively few bristles or pubescence, a round head and a basally constricted abdomen; wing often with a dark marking at the wing tip (*Sepsis*, fig. 198); in the male, femur and tibia of fore leg showing projections, spines, teeth, etc. (fig. 244). Exception: *Orygma*: brown, somewhat flattened, robust fly provided with bristles, fore leg simple, head as in fig. 194, 199, in coastal habitats.]

–   P-bristles almost always converging (fig. 200), very rarely parallel. Posterior spiracle without bristles at the posteroventral margin.   **Lauxaniidae**

>>>   [Small to medium-sized (2.5 to 6 mm) robust flies, not ant-like; third antennal segment can be elongate (fig. 187, 193); arista bare to long pubescent; wing usually clear, in a number of species with darkening alongside the veins or with markings all over (fig. 303).]

196 posterior spiracle
Sepsidae (Orygma)

197 posterior spiracle
Sepsidae (Nemopoda)

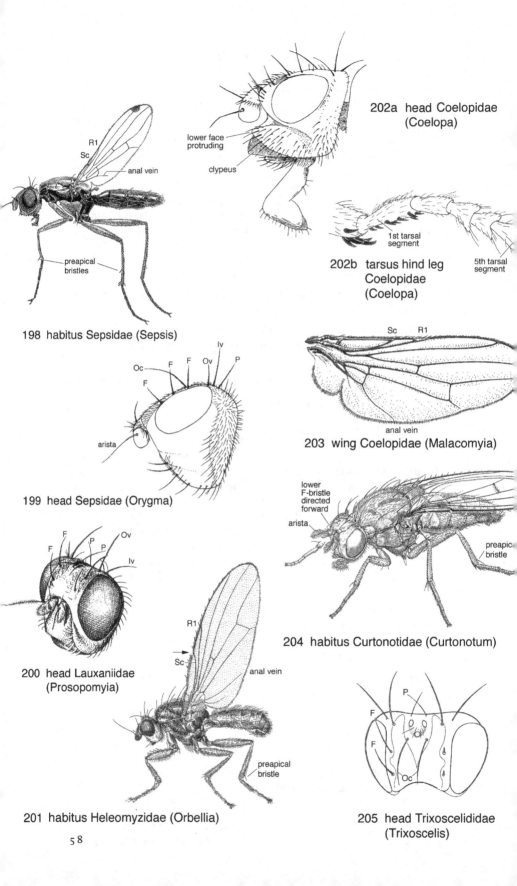

202a head Coelopidae (Coelopa)

R1
Sc
anal vein

lower face protruding
clypeus

1st tarsal segment
5th tarsal segment

202b tarsus hind leg Coelopidae (Coelopa)

preapical bristles

198 habitus Sepsidae (Sepsis)

Iv
Oc F F Ov P
F
arista

Sc R1

anal vein

203 wing Coelopidae (Malacomyia)

199 head Sepsidae (Orygma)

lower F-bristle directed forward
arista
preapical bristle

F P Ov
F P
Iv

204 habitus Curtonotidae (Curtonotum)

R1
Sc
anal vein

200 head Lauxaniidae (Prosopomyia)

preapical bristle

201 habitus Heleomyzidae (Orbellia)

P
F
F
Oc

205 head Trixoscelididae (Trixoscelis)

92(90)	Costa with both larger and smaller bristles (fig. 201). Head not as in fig. 202a (lower part of face not strongly protruding). Scutum not dorsoventrally flattened (fig. 201). Fifth tarsal segment not enlarged and triangular. **Heleomyzidae**
–	Costa with smaller bristles only (fig. 203). Head with lower part of face strongly protruding below circular third antennal segment (fig. 202a). Fifth tarsal segment large, triangular, wider than other tarsal segments; first tarsal segment of mid and hind leg usually with spines on ventral side (fig. 202b). In *Coelopa* scutum somewhat flattened and legs densely set with fine bristles. [Flies of coastal habitats.] **Coelopidae**
93(89)	Oc-bristles inserted on ocellar triangle; arista long plumose; plates on which the F-bristles are inserted are separate from the margins of the eyes; the lowermost F-bristles curving forward (fig. 204). Crossvein BM-Cu absent (cells bm and dm fused). **Curtonotidae**
	(In Europe 1 widespread species, *Curtonotum anus* (Meigen).)
–	Oc-bristles inserted next to the ocellar triangle at the same level as the anterior ocellus or lower; arista usually short pubescent, rarely short plumose; plates on which the F-bristles are inserted bordering the margins of the eyes; F-bristles curving backward (fig. 205). Crossvein BM-Cu present (cells bm and dm separate) (fig. 206). **Trixoscelididae**
	(In Europe 1 genus, *Trixoscelis*, with about 25 species.)
94(87)	Cell cup open; crossvein BM-Cu absent (cells bm and dm fused); anal vein absent (fig. 207). At most tibia of the mid leg with dorsal preapical bristle. **95**
–	Cell cup closed; vein BM-Cu present (cells bm and dm separate, fig. 208), or vein BM-Cu absent (cells bm and dm fused, fig. 219). Anal vein usually present (short in Diastatidae and Campichoetidae, sometimes absent in Drosophilidae). Usually all tibiae with dorsal preapical bristle (fig. 215a, 222), though sometimes thin and delicate in Odiniidae and Chiropteromyzidae. **96**

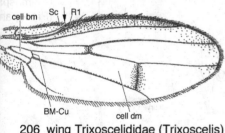

206 wing Trixoscelididae (Trixoscelis)

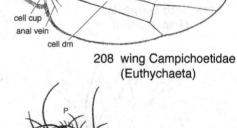

208 wing Campichoetidae
(Euthychaeta)

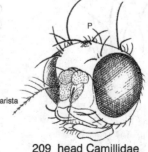

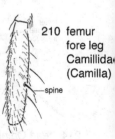

207 wing Camillidae
(Camilla)

209 head Camillidae
(Camilla)

210 femur
fore leg
Camillidae
(Camilla)

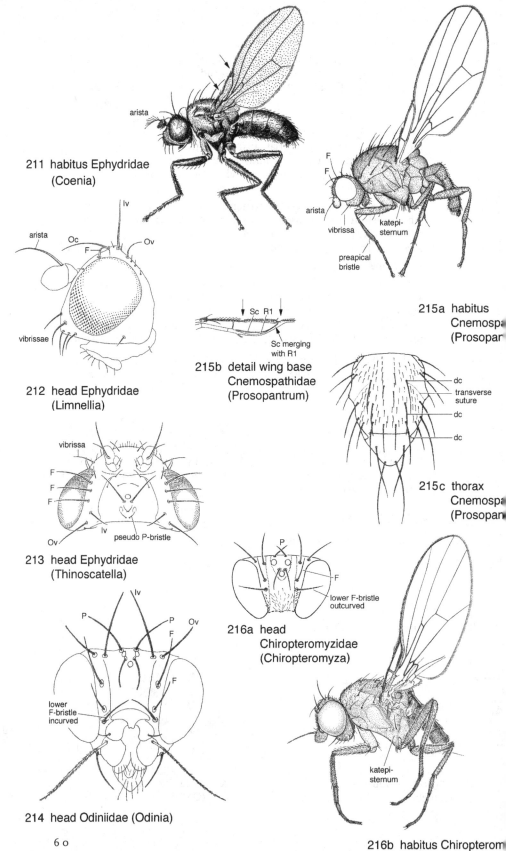

211 habitus Ephydridae (Coenia)

212 head Ephydridae (Limnellia)

213 head Ephydridae (Thinoscatella)

214 head Odiniidae (Odinia)

215a habitus Cnemosp... (Prosopar...

215b detail wing base Cnemospathidae (Prosopantrum)

215c thorax Cnemosp... (Prosopar...

216a head Chiropteromyzidae (Chiropteromyza)

216b habitus Chiropterom... (Chiropteromyza)

95(94) Arista plumose with long rays above and very short rays below; P-bristles converging (fig. 209). Femur of fore leg with a spine on its ventral side, beyond the middle (fig. 210). Wing venation as in fig. 207.   **Camillidae**

(In Europe 1 genus, *Camilla*, with about 8 species.)

– Arista usually with long rays, limited to the upper side (fig. 211), arista sometimes bare (fig. 212) or short pubescent; no real P-bristles, but sometimes diverging pseudopostocellar bristles are present (fig. 213); face usually swollen, convex (fig. 211, 212). Femur of fore leg with or without ventral spines.   **Ephydridae**

96(94) 3 pairs of F-bristles present; the lowest pair curving inward, the upper and middle pairs curving backward; P-bristles diverging (fig. 214). Legs often with alternating brownish and yellowish bands.   **Odiniidae**

– 2 or 3 pairs of F-bristles present: 1 pair curving forward, obliquely forward or outward, 1 or 2 pair(s) curving backward; P-bristles converging, sometimes absent (fig. 215-218, 220-224).   **97**

97(96) 2 pairs of F-bristles present, the lower pair curving outward (fig. 216a); arista bare to pubescent (fig. 215a, 216b).   **98**

– F-bristles not curving outward (fig. 217, 218, 220-224); arista usually plumose (fig. 217, 218) but in some cases pubescent (fig. 223-224).   **99**

98(97) Thorax with 3 pairs of dorsocentral bristles of which the first pair is located near or just before the transverse suture (fig. 215c: dc); katepisternum with 2 strong bristles (fig. 215a). All tibiae with strong dorsal preapical bristle.   **Cnemospathidae**

(In Europe 1 species, *Prosopantrum flavifrons* (Tonnoir & Malloch).)

– Thorax with 4 pairs of dorsocentral bristles, of which 1 pair in front, and 3 pairs behind the transverse suture; katepisternum with 1 strong bristle with a few weak, small bristles in front (fig. 216b). Dorsal preapical bristle present, sometimes smaller on tibiae of fore and hind leg.

Chiropteromyzidae

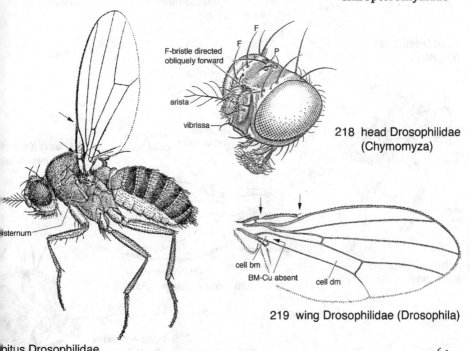

F-bristle directed obliquely forward

arista

vibrissa

218 head Drosophilidae (Chymomyza)

cell bm
BM-Cu absent   cell dm

219 wing Drosophilidae (Drosophila)

sternum

bitus Drosophilidae
osophila)

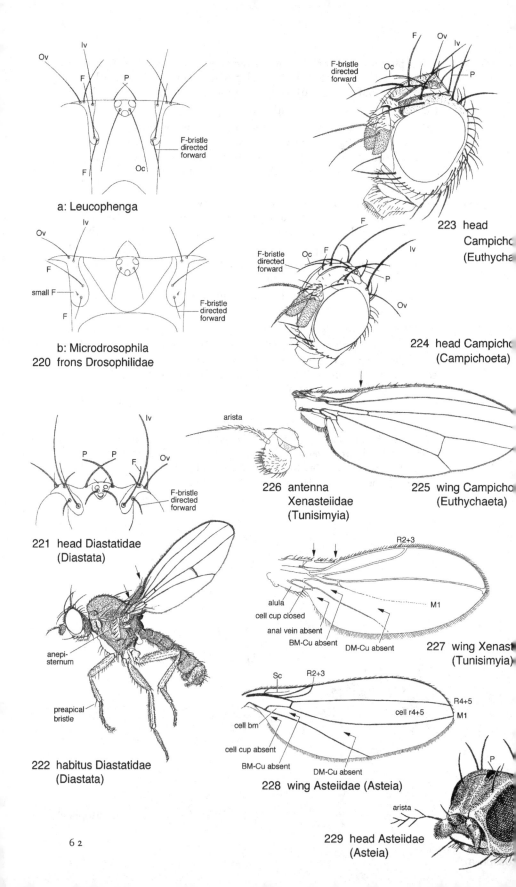

a: Leucophenga

b: Microdrosophila
220 frons Drosophilidae

221 head Diastatidae
(Diastata)

222 habitus Diastatidae
(Diastata)

223 head
Campicho
(Euthycha

224 head Campicho
(Campichoeta)

226 antenna
Xenasteiidae
(Tunisimyia)

225 wing Campicho
(Euthychaeta)

227 wing Xenast
(Tunisimyia)

228 wing Asteiidae (Asteia)

229 head Asteiidae
(Asteia)

99(97)	The F-bristle that curves forward or obliquely forward is inserted on the same level or further away from the eye margin than the nearest bristle that curves backward (fig. 218, 220, 271); this nearest backward directed bristle can be very small (fig. 220b). Anepisternum bare (fig. 217). Wing venation as in fig. 219; crossvein BM-Cu usually absent but sometimes present. **Drosophilidae**
–	The F-bristle that curves forward is inserted closer to the eye margin than the nearest bristle that curves backward (fig. 221-224). Anepisternum bare or with bristles (fig. 222). **100**
100(99)	Arista short to moderately long plumose; 1 strong F-bristle curving forward, 1 strong F-bristle curving backward (fig. 221). Anepisternum with short bristles on upper and posterior part, strongest near hind margin. Costa with both humeral and subcostal breaks (fig. 222). **Diastatidae**
	(In Europe 1 genus, *Diastata*, with about 9 species.)
–	Arista pubescent (fig. 223, 224); either 1 strong F-bristle curving forward and 1 strong F-bristle curving backward (fig. 223) or with yet another, weaker bristle curving backward (fig. 224). Anepisternum bare. Costa with subcostal break only (fig. 225). **Campichoetidae**
101(86)	Vein R2+3 ends in the costa about halfway up the wing; vein M1 apically pallid, not reaching the wing margin (fig. 227). **Xenasteiidae**
>>>	[Very small (1.3 to 1.7 mm) dark coloured flies with a pubescent arista (fig. 226); crossveins BM-Cu and DM-Cu absent; cell cup closed; anal vein absent; alula with long hairs.]
>>>	[Nannodastiidae could key out here if the anterior weak bristles along the lower margin of the genae have been interpreted as vibrissae. Distinguishing characters are: all F-bristles curving backward (in Xenasteiidae lower pair curving inward); crossvein DM-Cu sometimes present (fig. 363), alula absent (fig. 364); see also couplet 167 and the family descriptions.]
–	Vein R2+3 is strikingly short (fig. 228) or ends in the costa distinctly beyond the middle (fig. 230, 234, 235), or if vein R2+3 ends in the costa about halfway up the wing, then vein M1 reaching the wing margin (fig. 240). **102**
102(101)	Wing base narrow, alula and anal vein small or absent, cell cup sometimes absent as well (fig. 228, 230, 234-237); vein Sc incomplete, shortened or merging with vein R1 at some distance from the costa (in *Geomyza*, fig. 236, apical part sometimes visible as a thin line reaching the costa). P-bristles small, hair-like or absent (fig. 229, 232). **103**
–	Wing base not narrow, alula well developed. Other characters as specified above or not, but rarely in the above combination. **106**
103(102)	Cell r4+5 narrowing because veins R4+5 and M1 converging toward the wing tip; crossvein BM-Cu absent (cells bm and dm fused); cell cup open or absent (fig. 228, 230). Oc-bristles present or absent. **104**

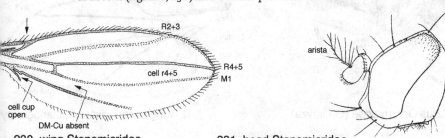

230 wing Stenomicridae
(Stenomicra)

231 head Stenomicridae
(Stenomicra)

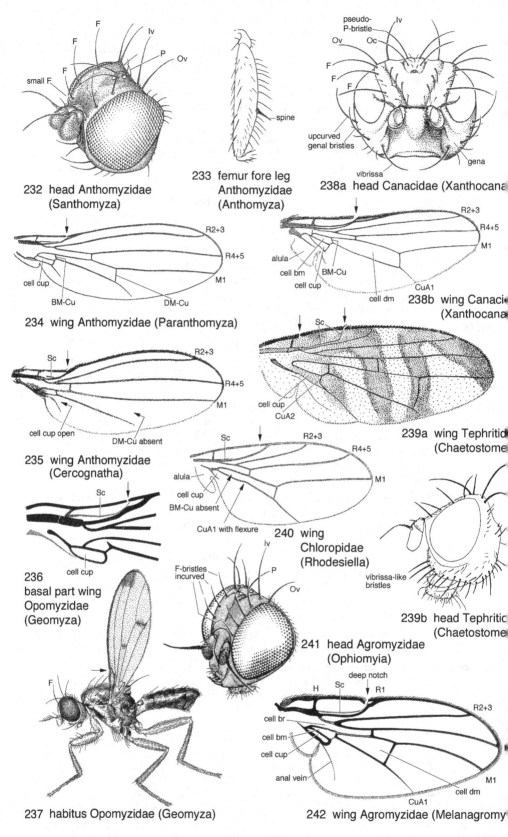

232 head Anthomyzidae (Santhomyza)

233 femur fore leg Anthomyzidae (Anthomyza)

238a head Canacidae (Xanthocana

234 wing Anthomyzidae (Paranthomyza)

238b wing Canaci (Xanthocana

235 wing Anthomyzidae (Cercognatha)

239a wing Tephritic (Chaetostome

236 basal part wing Opomyzidae (Geomyza)

240 wing Chloropidae (Rhodesiella)

239b head Tephritic (Chaetostome

241 head Agromyzidae (Ophiomyia)

237 habitus Opomyzidae (Geomyza)

242 wing Agromyzidae (Melanagromy

–	Cell r4+5 not narrowing; crossvein BM-Cu present (cells bm and dm separate); cell cup closed (fig. 234, 236), rarely open (fig. 235). Oc-bristles present. **105**
104(103)	Vein R2+3 conspicuously short, ending in the basal half of the wing; crossvein DM-Cu absent (fig. 228). Arista sometimes bare or pubescent but usually with a characteristic zig-zag of longer or shorter dorsal and ventral rays (fig. 229) or, in one species, arista absent; ocellar triangle situated at the normal position, in the upper part of the frons. **Asteiidae**
–	Vein R2+3 not strikingly short; crossvein DM-Cu absent (fig. 230) or present. Arista with long rays (fig. 231); ocellar triangle situated more or less in the middle of the frons. **Stenomicridae**
	(In Europe 1 genus, *Stenomicra*, with 3 species.)
105(103)	2 to 3 pairs of F-bristles, often with 1 or more weaker F-bristles in front. P-bristles small, converging, rarely absent (fig. 232, 269). Femur of fore leg in several genera with a short spine on the ventral side, beyond the middle (fig. 233). Wing venation as in fig. 234, 235; cell cup sometimes open, crossvein DM-Cu sometimes absent. **Anthomyzidae**
–	Only 1 pair of F-bristles present (fig. 237); P-bristles absent. Femur of fore leg without short spine on ventral side. Wing venation as in fig. 236, 237. **Opomyzidae**: *Geomyza*
106(102)	Gena with 1 or more strong upcurved bristle(s); the 2 to 5 F-bristles curving outward over the margin of the eye; no real P-bristles, but often replaced by diverging pseudopostocellar bristles (fig. 238a). Vein Sc parallel to, and just before the costa, touching or merging with vein R1; crossvein BM-Cu present (cells bm and dm separate); cell cup closed (fig. 238b). [Flies of coastal habitats.] **Canacidae**
–	Gena without strong upcurved bristles. Combination of other characters not as given above. **107**
>>>	[Of the families keying out here (vibrissae present, preapical bristles absent) outward curving F-bristles can be found in: Piophilidae and Heleomyzidae (*Oldenbergiella* and the Borboropsinae) (anal vein long, usually reaching the wing margin, fig. 254), Tethinidae (lustrous bulge between antenna and vibrissa-like bristle, fig. 255), Carnidae and Milichiidae (lower F-bristles curving inward, fig. 272a-274) and Chloropidae (crossvein BM-Cu absent; cell cup open or absent, fig. 240).]
107(106)	Vein Sc abruptly bent forward toward the costa at nearly 90°; cell cup closed by a geniculate vein CuA2 and with an acute apical end (fig. 239a). **Tephritidae**
>>>	[Tephritidae are without true vibrissae but some genera (e.g. *Chaetostomella*, *Chetostoma*) have strong bristles near the vibrissal angle (fig. 239b); their wing pattern is usually as in fig. 239a. See also couplet 150.]
–	Vein Sc and cell cup not as given above. **108**
108(107)	Vein Sc complete, usually reaching the costa separate from vein R1 (wing figures 243-257a), sometimes Sc and R1 ending close together (fig. 256), or vein Sc merges with R1 just before the costa (e.g. Agromyzinae, fig. 242; Tethinidae, fig. 251; Chamaemyiidae, fig. 257a). Cell cup closed. **109**
–	Vein Sc absent, or short and not reaching the costa, or merging with vein R1 distinctly before the costa (fig. 240, wing figures 258-262, 267-298), or reduced to a fold that may or may not reach the costa (Phytomyzinae, fig. 263, 264). Cell cup closed, open or absent. **120**
>>>	[In most of the species belonging to the 21 families keying out here (vibris-

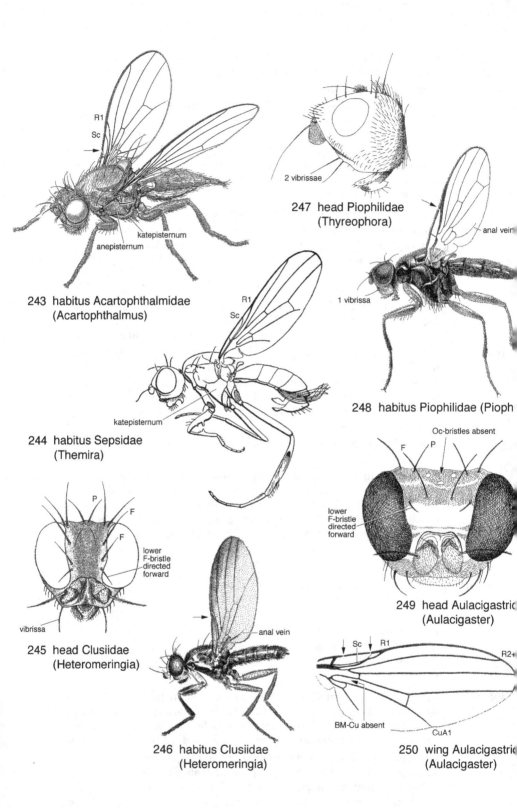

243 habitus Acartophthalmidae (Acartophthalmus)

247 head Piophilidae (Thyreophora)

2 vibrissae

1 vibrissa

anal vein

katepisternum
anepisternum

R1
Sc

244 habitus Sepsidae (Themira)

katepisternum

R1
Sc

248 habitus Piophilidae (Pioph

Oc-bristles absent

F   P

lower F-bristle directed forward

249 head Aulacigastri (Aulacigaster)

245 head Clusiidae (Heteromeringia)

P
F
F

lower F-bristle directed forward

vibrissa

246 habitus Clusiidae (Heteromeringia)

anal vein

250 wing Aulacigastri (Aulacigaster)

Sc   R1

R2+

BM-Cu absent

CuA1

sae and alula present, preapicals absent), vein Sc complete or incomplete is clearly visible. But there are some border cases and therefore some families have been keyed out twice. Nevertheless, in case of doubt it is recommended to follow both alternatives and to consult the family descriptions before coming to a decision.]

109(108) P-bristles diverging (fig. 241, 245), parallel, reduced or absent, in the latter case veins Sc and R1 end in the costa at a considerable distance from each other (fig. 243, 244).   110

– P-bristles converging or absent, in the latter case veins Sc and R1 merge over a short distance ((fig. 250, 257a), have a joint ending in the costa (fig. 251), or end in the costa close to each other (fig. 256).   115

110(109) Vein Sc merging with R1 just before the costa; vein R1 curving toward the costa from this point; costa with subcostal break and with a distinct, deep notch (fig. 242). Usually, at least the lower 2 pairs of the F-bristles curved inward; interfrontal bristles absent (fig. 241). Female with a non-retractable ovipositor, visible as a conspicuous conical structure (fig. 266).
Agromyzidae: Agromyzinae

>>> [A deep notch in the costa is also found in some species of Milichiidae and Carnidae; in both families the interfrontal bristles are present (fig. 273, 274, 287).]

– Veins Sc and R1 different; costa with or without subcostal break but in any case without a distinct notch there (fig. 243, 244). F-bristles and interfrontal bristles variable, usually not similar to what is described above (fig. 243, 244).   111

111(110) Veins Sc and R1 end in the costa at a considerable distance from each other, this distance being at least one-third of the distance between crossvein H and the end of vein Sc (fig. 243, 244).   112

– Veins Sc and R1 end in the costa not far apart, at a distance which is less than one-fourth of the distance between crossvein H and the end of vein Sc (fig. 246, 248).   114

112(111) Wing darkened with numerous small spots (fig. 306), or with a distinct wing pattern.   **Platystomatidae**

– Wing clear or infuscated, but without a pattern of many spots or markings, at most with a dark spot at the wing tip.   113

113(112) 3 pairs of dorsocentral bristles; katepisternum with a strong bristle (fig. 243). Costa with a humeral break (fig. 243). Posterior spiracle without bristles at the posteroventral margin.   **Acartophthalmidae**

(In Europe 1 genus, *Acartophthalmus*, with 3 species.)

– 1 or 2 pair(s) of dorsocentral bristles; katepisternum without a bristle (fig. 244). Costa continuous. Posterior spiracle with 1 or more bristles at the posteroventral margin (fig. 196, 197).   **Sepsidae**

114(111) The lowermost of the 3 pairs of F-bristles curving inward or forward, the other F-bristles curving backward; vibrissal angle with 1 strong vibrissa (fig. 245). Anal vein not reaching the wing margin (fig. 246).   **Clusiidae**

– F-bristles absent, or present in 1-4 pairs, usually curving outward or backward, if the lower 1 or 2 pair(s) curving inward, then the other F-bristles curving outward and/or vibrissal angle with 2 strong vibrissae (fig. 247). Anal vein long (fig. 248), usually reaching the wing margin. **Piophilidae**

>>> [Pallopteridae could key out here if the anterior setulae along the lower

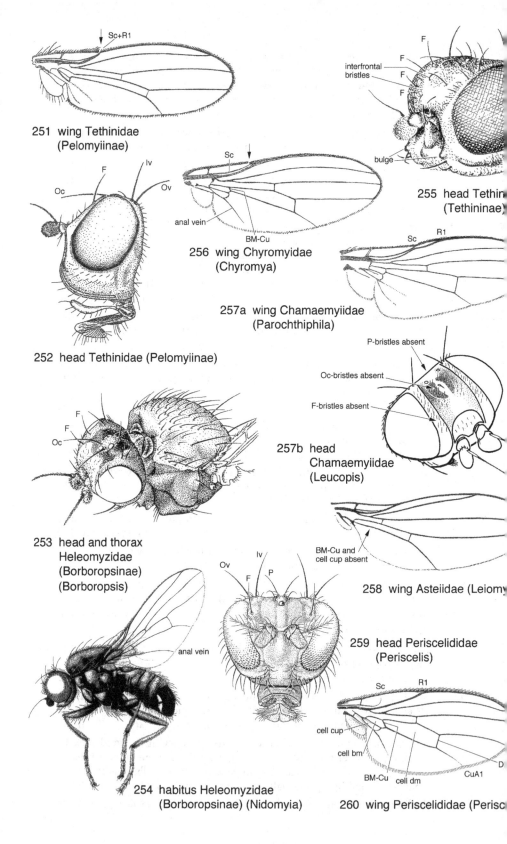

251 wing Tethinidae (Pelomyiinae)

255 head Tethin (Tethininae)

256 wing Chyromyidae (Chyromya)

257a wing Chamaemyiidae (Parochthiphila)

252 head Tethinidae (Pelomyiinae)

257b head Chamaemyiidae (Leucopis)

253 head and thorax Heleomyzidae (Borboropsinae) (Borboropsis)

258 wing Asteiidae (Leiomy

259 head Periscelididae (Periscelis)

254 habitus Heleomyzidae (Borboropsinae) (Nidomyia)

260 wing Periscelididae (Perisc

margin of the genae have been interpreted as vibrissae. Distinguishing characters are that Piophilidae have strongly developed vibrissae and the clypeus small and withdrawn, in Pallopteridae the clypeus is exposed (fig. 311).]

115(109)   2 pairs of F-bristles, lower pair curving forward and somewhat inward, upper pair curving backward, lower pair inserted closer to eye margin than upper pair. Oc-bristles reduced or absent (fig. 249). Veins Sc and R1 reaching the costa separately, but just before they merge for a short distance or touch each other; crossvein BM-Cu absent (cells bm and dm fused) (fig. 250).   **Aulacigastridae**

(In Europe 1 genus, *Aulacigaster*, with 4 species.)

–   F-bristles arranged differently or absent. Other characters as specified above or not, but not in the above combination.   116

116(115)   Crossvein BM-Cu absent (cells bm and dm fused) (fig. 251) or very weakly indicated; vein Sc merging with R1 just before or at the costa. 1 or 2 pair(s) of F-bristles, curving backward (fig. 252).   **Tethinidae: Pelomyiinae**

–   Crossvein BM-Cu distinctly present (cells bm and dm separate) (fig. 254, 256, 257). F-bristles usually different, only in a few species as given above.   117

117(116)   Anal vein long, reaching the wing margin (fig. 254). 1 or 2 pair(s) of F-bristles, curving outward, sometimes curving somewhat obliquely backwards as well (fig. 253).   **Heleomyzidae**

–   Anal vein not reaching the wing margin (fig. 256, 257a). F-bristles absent (fig. 257b), or not curving outward, or more than 2 pairs of F-bristles present (fig. 255).   118

118(117)   A small, lustrous bulge present in between antenna and most anterior vibrissa-like bristle; this bulge slightly larger than an ocellus and more or less of the same colour (fig. 255); interfrontal bristles present; 3 to 5 pairs of F-bristles present.   **Tethinidae: Tethininae**

–   Lustrous bulge absent. Combination of other characters not as given above.   119

119(118)   Usually yellow or pale flies with light-coloured bristles and pubescence; sometimes with brown, grey or black markings on head, thorax and abdomen. Veins Sc and R1 more or less parallel; costa with a subcostal break (fig. 256).   **Chyromyidae**

–   Usually silvery grey flies with dark bristles and pubescence, in some cases with brown, grey or black markings on head, thorax and abdomen, or flies black and lustrous. In most species, veins Sc and R1 approach each other before they separately reach the costa; costa without breaks (fig. 257a). Especially in the large genus *Leucopis* dorsal bristles of the head largely absent (fig. 257b).   **Chamaemyiidae**

120(108)   Crossvein BM-Cu present (cells bm and dm separate); cell cup usually closed (wing figures 260-272b; open in some Periscelididae and Drosophilidae).   121

>>>   [In several species of Milichiidae, cells bm and cup are rather inconspicuous and the clarity of crossvein BM-Cu is variable. Apart from the incurved F-bristles (fig. 273, 274), these species show one or more of the following characters: wing venation as in fig. 277 (basal end of cell dm is acute); body colour black, lustrous; mouthparts conspicuously elongate and folded back (fig. 276), a distinct deep notch at the point where veins Sc and R1 reach the costa (fig. 278).]

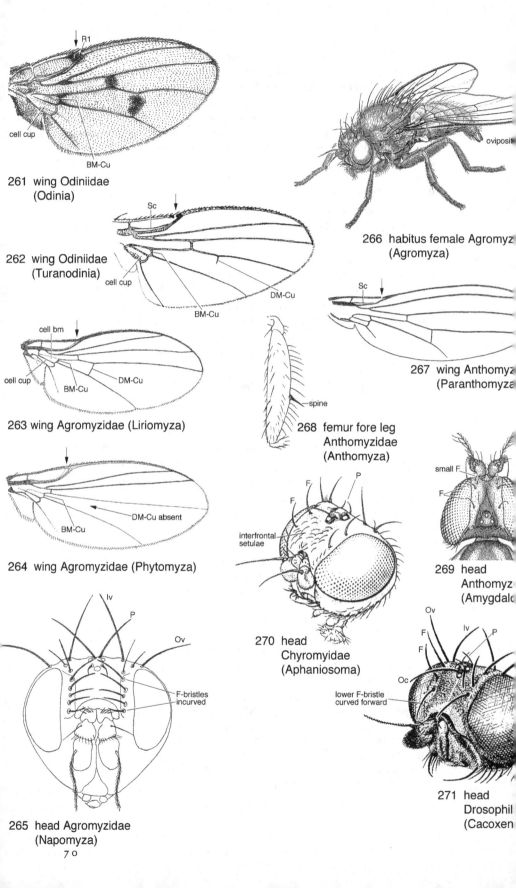

261 wing Odiniidae
(Odinia)

262 wing Odiniidae
(Turanodinia)

263 wing Agromyzidae (Liriomyza)

264 wing Agromyzidae (Phytomyza)

265 head Agromyzidae
(Napomyza)

266 habitus female Agromyz
(Agromyza)

267 wing Anthomyz
(Paranthomyza

268 femur fore leg
Anthomyzidae
(Anthomyza)

269 head
Anthomyz
(Amygdalc

270 head
Chyromyidae
(Aphaniosoma)

271 head
Drosophil
(Cacoxen

70

–	Crossvein BM-Cu absent (cells bm and dm fused); cell cup open or absent (fig. 258), closed in Pseudopomyzidae (fig. 296), closed or open in Drosophilidae. **128**
121(120)	Only 1 pair of F-bristles, curving backward; second antennal segment covering the third segment like a hood; arista with long rays (fig. 259). Cell cup open or closed (fig. 260), crossvein DM-Cu sometimes weakened and in 1 species absent. **Periscelididae**
	(In Europe 1 genus, *Periscelis*, with 4 species.)
–	More than 1 pair of F-bristles present (head figures 265-276). Cell cup closed (fig. 261-264). Other characters variable. **122**
122(121)	Wing venation as in fig. 261 (crossveins and end of vein R1 with dark spots) or fig. 262 (crossvein DM-Cu incomplete). 3 pairs of F-bristles present, of which the lower pair curving inward (fig. 214). **Odiniidae**
–	Combination of characters not as given above. No F-bristles curving inward (fig. 269, 270, 271) or more than 1 pair curving inward (fig. 265, 272a, 273, 274); if only 1 pair of incurved F-bristles is present (some Agromyzidae), then crossvein DM-Cu in basal half of the wing (fig. 263) or absent altogether (fig. 264). **123**
123(122)	P-bristles diverging; at least the lowermost F-bristles curving inward; interfrontal bristles absent, at most some setulae present (fig. 265). Wing not marked; costa with a subcostal break; vein bordering cell bm on the upper side often pallid or absent (fig. 263, 264); crossvein DM-Cu often absent (fig. 264). Female with a non-retractable ovipositor, visible as a conspicuous conical structure (fig. 266). **Agromyzidae: Phytomyzinae**
>>>	[In the genus *Xeniomyza*, with one species in Kazakhstan and another species in Spain, which is one of the smallest European agromyzid species, measuring 0.5 mm, there are no P-bristles. All other characters obtain, including the wing venation characters as in fig. 264.]
–	Combination of characters not as given above. If the lowermost F-bristles are curving inward, then interfrontal bristles present. Female ovipositor retractable, no conspicuous external conical structure present. **124**
124(123)	Costa with subcostal break only (fig. 256, 267). 2 to 6 pairs of F-bristles present (rarely F-bristles absent), usually curving backward or slightly outward as well, in some cases the lower pair curving inward (fig. 269, 270). **125**
–	Costa with humeral and subcostal breaks (fig. 272b, 277, 278). Almost always 1 or 2 pair(s) of lower F-bristles curving forward (fig. 271) or inward (fig. 272a, 273, 274). **126**

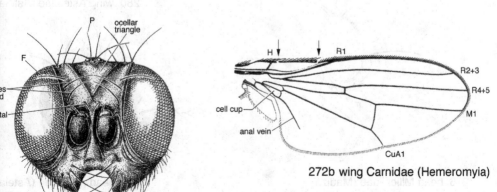

272a head Carnidae (Hemeromyia)

272b wing Carnidae (Hemeromyia)

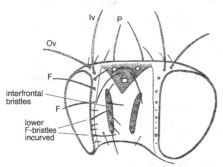

interfrontal bristles

lower F-bristles incurved

273 head Milichiidae (Desmometopa)

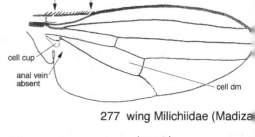

cell cup

anal vein absent

cell dm

277 wing Milichiidae (Madiza

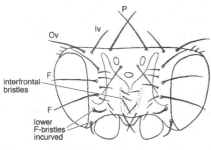

interfrontal bristles

lower F-bristles incurved

274 head Milichiidae (Phyllomyza)

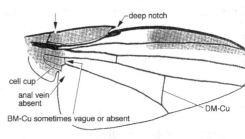

deep notch

cell cup

anal vein absent

BM-Cu sometimes vague or absent

DM-Cu

278 wing Milichiidae (Milichia

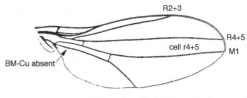

R2+3

R4+5

cell r4+5

M1

BM-Cu absent

279 wing Asteiidae (Leiomyza

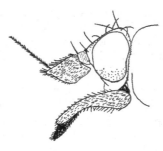

275 head Milichiidae (Phyllomyza)

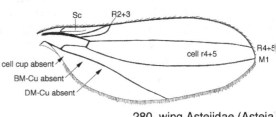

Sc   R2+3

cell cup absent

BM-Cu absent

DM-Cu absent

cell r4+5

R4+5

M1

280 wing Asteiidae (Asteia

276 head Milichiidae (Madiza)

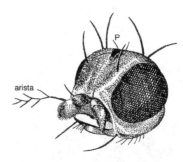

P

arista

281 head Asteiidae (Asteia

125(124)	2 to 3 pairs of F-bristles curving backward, often with 1 or more pairs of weaker bristles in front; interfrontal area bare or with few setulae (fig. 269). Femur of fore leg in several genera with a short spine on the ventral side, beyond the middle (fig. 268). Bristles and pubescence not yellow. Wing venation as in fig. 267. **Anthomyzidae**
–	2 to 6 pairs of F-bristles present (rarely F-bristles absent), usually curving backward or slightly outward as well, in some cases the lower pair curving inward; interfrontal area with many setulae (fig. 270). Femur of fore leg without spine on ventral side. Usually yellow or pallid flies, their bristles and pubescence also of a pale colour, in some cases with brown, grey or black markings on head, thorax and abdomen. Wing venation as in fig. 256, but vein Sc apically pallid. **Chyromyidae**
126(124)	Interfrontal bristles absent, head as in fig 271. **Drosophilidae**: *Cacoxenus*
–	Interfrontal bristles present, head as in figures 272a-276. **127**
127(126)	Only 1 pair of interfrontal bristles, inserted just above the antennae; ocellar triangle large, obvious; P-bristles somewhat diverging to parallel (fig. 272a). Anal vein long (fig. 272b). **Carnidae**: *Hemeromyia*
–	More than a single pair of interfrontal bristles, arranged in rows (fig. 273) or more irregular; ocellar triangle smaller than shown in fig. 272a; P-bristles parallel or converging (fig. 273, 274); third antennal segment and mouthparts in some cases conspicuously large or elongate (fig. 275, 276). Anal vein short or absent (fig. 277, 278). **Milichiidae**
128(120)	Wing with a conspicuous notch where vein R1 ends in the costa; crossvein DM-Cu in the apical part of the wing (fig. 278). The 2 lower pairs of F-bristles curving inward; more than 1 pair of interfrontal bristles, arranged in rows or more irregular; P-bristles parallel or converging (fig. 273, 274). **Milichiidae**
–	Wing and bristles of the head not as described above. If a notch present where vein R1 ends in the costa and 2 lower pairs of F-bristles curving inward, then crossvein DM-Cu either absent (fig. 288) or in the basal half of the wing (fig. 289) and only 1 pair of interfrontal bristles present, inserted just above the antennae (fig. 286, 287). **129**
129(128)	Cell r4+5 ends near the wing tip and is slightly narrowing since veins R4+5 and M1 are somewhat converging; costa continuous (fig. 279, 280); in several genera, vein R2+3 strikingly short, ending in the basal half of the wing (fig. 280). Arista sometimes bare or pubescent but usually with a characteristic zig-zag of longer or shorter dorsal and ventral rays (fig. 281) or, in one species, arista absent. **Asteiidae**

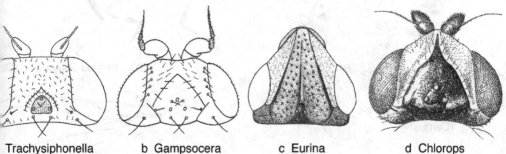

Trachysiphonella   b Gampsocera   c Eurina   d Chlorops

a-d head Chloropidae

73

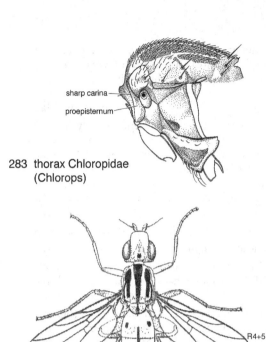

283 thorax Chloropidae (Chlorops)

284 habitus Chloropidae (Meromyza)

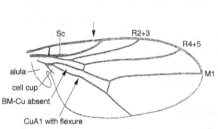

285 wing Chloropidae (Rhodesiella)

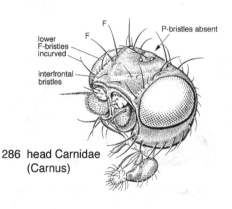

286 head Carnidae (Carnus)

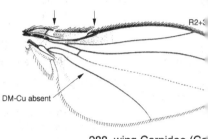

287 head Carnidae (Meo...

288 wing Carnidae (Ca...

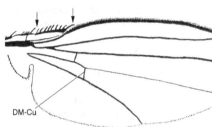

289 wing Carnidae (Meon...

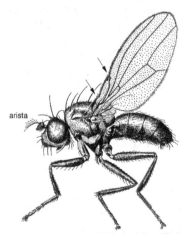

290 habitus Ephydridae (Co...

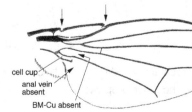

291 wing Ephydridae (Para...

<table>
<tr><td>–</td><td>Cell r4+5 not as described above, if narrowing, then usually because veins R4+5 and M1 are both curving toward the costa; costa nearly always with a subcostal break (continuous in some species of Chloropidae); vein R2+3 not strikingly short (wing figures 284-298). Arista bare, pubescent or with shorter to longer rays, but without a characteristic zig-zag arrangement (fig. 282, 290, 297).</td><td>130</td></tr>
</table>

130(129) Frons nearly always with a strikingly large, often lustrous ocellar triangle (fig. 282 b-d). Proepisternum with a sharp anterior carina (fig. 283). Vein CuA1 usually with a flexure halfway cell bm + dm; often vein R4+5 and sometimes also vein M1 curving toward the costa; costa with a subcostal break (fig. 284, 285; in only a few species the costa is continuous). **Chloropidae**

– Ocellar triangle not conspicuously large or lustrous (except in Carnidae, fig. 286). Proepisternum anteriorly without a sharp carina. Vein CuA1 without flexure; veins R4+5 and M1 rarely (in some Ephydridae) curving toward the costa; costa twice interrupted, with humeral and subcostal breaks (fig. 288-291, 296, 298). **131**

131(130) 4 pairs of F-bristles present, the lower 2 pairs curving inward, the upper 2 pairs curving outward or backward; 1 pair of interfrontal bristles present, inserted just above the antennae (fig. 286, 287). **Carnidae:** *Carnus, Meoneura*

>>> [*Carnus*: P-bristles absent (fig. 286); crossvein DM-Cu absent (fig. 288), wing might be shed off near the base (fig. 449). *Meoneura*: P-bristles parallel (fig. 287); crossvein DM-Cu present (fig. 289).]

– F-bristles and interfrontal bristles not as described above. **132**

132(131) No real P-bristles present, but sometimes parallel or diverging pseudopostocellar bristles present (fig. 293); arista usually with long rays, limited to the upper side (fig. 290), arista sometimes bare (fig. 202) or short pubescent; face usually swollen, convex (fig. 290, 292). Cell cup open, anal vein absent (fig. 291). **Ephydridae**

– P-bristles present, converging; if arista plumose then with long rays on both sides (e.g. fig. 297), but arista may be bare or pubescent only. Face not swollen (fig. 294, 297). Cell cup open or closed; anal vein present or absent (fig. 296, 298). **133**

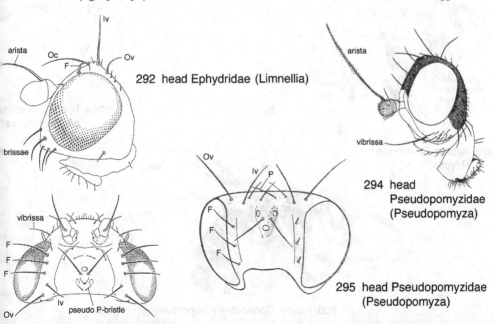

292 head Ephydridae (Limnellia)

294 head Pseudopomyzidae (Pseudopomyza)

295 head Pseudopomyzidae (Pseudopomyza)

3 head Ephydridae (Thinoscatella)

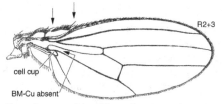

296 wing Pseudopomyzidae (Pseudopomyza)

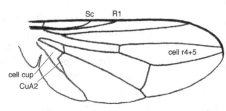

301 wing Ulidiidae (Uli...

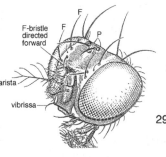

297 head Drosophilidae
(Chymomyza)

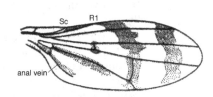

302 wing Sciomyzidae (Coloba...

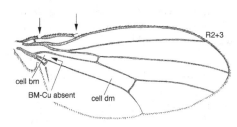

298 wing Drosophilidae (Drosophila)

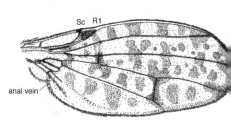

303 wing Lauxaniidae (Eusapromyz...

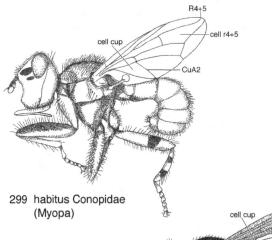

299 habitus Conopidae
(Myopa)

304 head Lauxaniid...
(Sciosapromyz...

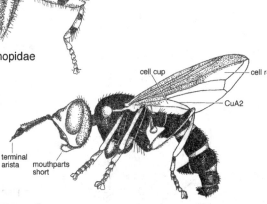

300 habitus Conopidae (Leopoldius)

133(132)	Apart from the converging P-bristles, another pair of converging bristles present; all 3 pairs of F-bristles curving backward; arista pubescent (fig. 294, 295). Wing venation as shown in fig. 296. **Pseudopomyzidae**
	(In Europe 1 widespread species, *Pseudopomyza atrimana* (Meigen).)
–	Only the converging P-bristles present; of the 3 pairs of F-bristles, 1 pair is curving forward or obliquely inward; arista usually with long rays (fig. 297; see for F-bristles and arista also couplet 99 and fig. 271). Wing venation as in fig. 298. **Drosophilidae**
134(84)	At the same time cell cup elongate (approaching the hind margin of the wing) and vein CuA2 (i.e., the vein closing cell cup) more or less straight; cell r4+5 narrowing toward the wing tip, or closed (fig. 299, 300). Mouthparts usually long, slender, geniculate at the base, in some cases twice geniculate (fig. 299); in some cases mouthparts short (fig. 300). **Conopidae**
–	Cell cup short or absent, its apex not near the hind margin of the wing (wing figures 302-347), or if long, then closed by a geniculate vein CuA2 (fig. 301). **135**
>>>	[A geniculate vein CuA2, the vein closing cell cup, only occurs in Ulidiidae (fig. 301), *Salticella* (Sciomyzidae, fig. 307), Pyrgotidae (fig. 324) and Tephritidae (fig. 330).]
135(134)	Dorsal preapical bristle present at least on tibia of mid leg but usually on tibiae of all legs (fig. 309). Vein Sc complete, reaching the costa independently (wing figures 301-320). **136**
–	Tibiae without dorsal preapical bristle. Vein Sc complete or incomplete. **147**
136(135)	Anal vein ending distinctly before the wing margin (fig. 302, 303). **137**
–	Anal vein long, reaching the wing margin (wing figures 306-324) or ending close to it, in some cases very thin or almost fold-like near the end. **139**
137(136)	Wing with characteristic, banded markings (fig. 302). **Sciomyzidae**: *Colobaea*
–	Wing unmarked or with markings different. **138**
138(137)	P-bristles almost always converging (fig. 304), very rarely parallel. Posterior spiracle without bristles at the posteroventral margin (see also couplet 91). **Lauxaniidae**
>>>	[Some specimens of the sciomyzid genus *Pteromicra* may key out here because sometimes their P-bristles are slightly converging and the continuation of the anal vein to the wing margin can be a very thin fold only. *Pteromicra* differs from Lauxaniidae by the following combination of characters: body shiny brownish black or brownish black and yellow, fore legs largely brownish black, other legs mainly yellow, anepisternum without bristles.]
–	P-bristles diverging (fig. 305) or absent. Posterior spiracle with 1 or more bristles at the posteroventral margin (fig. 196, 197) (see also couplet 91). **Sepsidae**

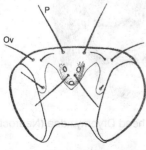

305 head Sepsidae
(Saltella)

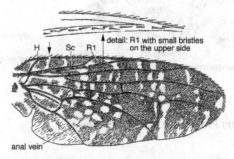

306 wing Platystomatidae (Platystoma)

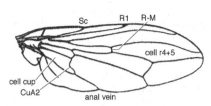

307 wing Sciomyzidae (Salticella)

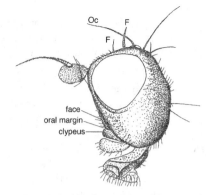

311 head Pallopteridae (Eurygnathomyi

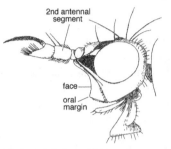

308 head Sciomyzidae
(Coremacera)

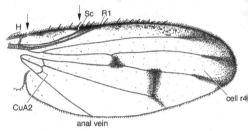

312 wing Pallopteridae (Eurygnathomyi

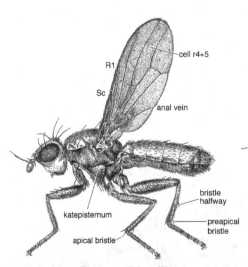

309 habitus Sciomyzidae (Psacadina)

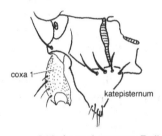

313 katepisternum Pallopterid:
(Eurygnathomyia)

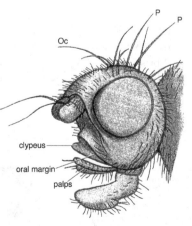

310 habitus Phaeomyiidae (Pelidnoptera)

314 head Dryomyzidae (Neurocten

139(136)	Vein R1 with setulae on its dorsal side (as in fig. 306) and at the same time wing membrane with a patterned (not necessarily as in fig. 306). **Ulidiidae**
–	Combination of characters not as given above; if vein R1 with setulae on its dorsal side, then wing not patterned but clear, shaded around the veins (Dryomyzidae) or entirely infuscated (fig. 310: Phaeomyiidae). **140**
140(139)	Vein CuA2 (the vein closing cell cup) geniculate, causing cell cup to taper to an acute apical angle; cell r4+5 narrowing toward the wing tip (fig. 301, 307). **141**
–	Vein CuA2 straight or convex; cell R4+5 nearly always broad toward the wing tip (fig. 309, 310, 312, 320). **142**
141(140)	Thorax lustrous, greenish blue; abdomen lustrous, blackish brown. Wing without a brownish tinge; cell cup long, narrowing to an acute apical end (fig. 301). Female with a lancet-shaped, incompletely retractable ovipositor (as in fig. 350, 351). **Ulidiidae**
–	Body not conspicuously lustrous; thorax brown with darker longitudinal stripes or spots above and at sides; abdomen largely grey. Wing with brownish tinge, crossvein R-M with dark borders; cell cup less elongate and less acute (fig. 307). Female without a lancet-shaped ovipositor. **Sciomyzidae**: *Salticella*
142(140)	Clypeus small and withdrawn, in lateral view more or less hidden behind the oral margin; second antennal segment often elongate (fig. 308). Katepisternum bare or slightly pubescent, rarely with 1 or 2 bristles near its upper margin (fig. 309, 310). **143**
–	Clypeus well visible, protruding below the oral margin (fig. 311, 314) or protruding from an emargination of this margin (fig. 319, 321); second antennal segment always short. Katepisternum always with more than 1 bristle near its upper margin (usually 3 or more) (fig. 313, 315) or, in case of *Helcomyza*, katepisternum grey and with a dense cover of dark hair-like bristles. **144**
143(142)	Tibiae with apical and dorsal preapical bristles only, no bristles halfway up (fig. 309). Wing often with a pattern, in some species only infuscated or all clear; vein R1 without setulae on its dorsal side. Arista pubescent or with longer rays; second antennal segment often elongate (fig. 308). **Sciomyzidae**
–	Tibia of hind leg, and sometimes tibia of mid leg as well, with apical and dorsal preapical bristles but also with bristles halfway up their length (fig. 310). Wing infuscated, without pattern; vein R1 with setulae on its upper side. Arista pubescent; second antennal segment always short. **Phaeomyiidae** (In Europe 1 genus, *Pelidnoptera*, with 3 species.)
144(142)	Head as in fig. 311: 2 pairs of F-bristles curving backward, the lower one the smallest. Wing as in fig. 312: costa weakened just beyond crossvein H and with a subcostal break; vein CuA2 convex. **Pallopteridae**: *Eurygnathomyia*

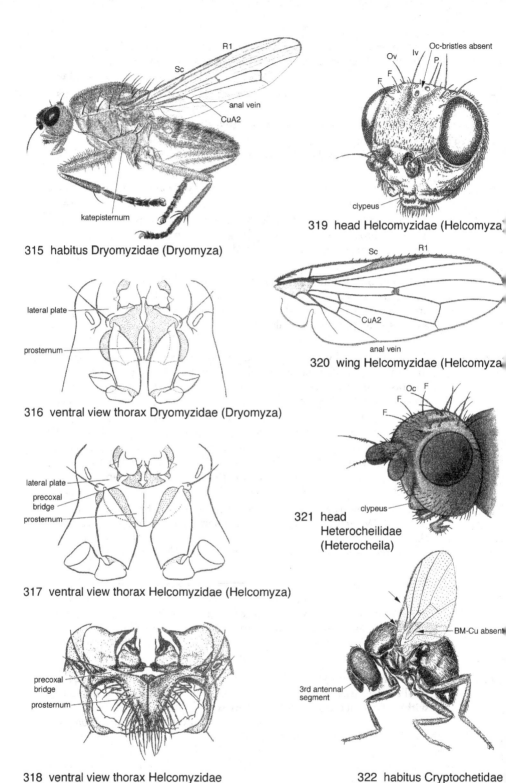

315 habitus Dryomyzidae (Dryomyza)

319 head Helcomyzidae (Helcomyza)

320 wing Helcomyzidae (Helcomyza)

316 ventral view thorax Dryomyzidae (Dryomyza)

321 head Heterocheilidae (Heterocheila)

317 ventral view thorax Helcomyzidae (Helcomyza)

318 ventral view thorax Helcomyzidae (Helcomyza)

322 habitus Cryptochetidae (Cryptochetum)

	Head and wing different; costa continuous; vein CuA2 straight (fig. 315, 320). **145**
145(144)	Antennae inserted close together, not further apart than the length of their second segment (fig. 314). Costa with closely spaced pubescence, without strong short bristles (fig. 315). Underside of thorax without precoxal bridge (prosternum linear to ovoid, separate from the lateral plates by a membrane, fig. 316). Yellow to reddish coloured flies. **Dryomyzidae**
–	Antennal insertions further apart, about twice the length of their second segment (fig. 319, 321). Costa in some cases with strong short bristles (fig. 320). Underside of thorax with a precoxal bridge (prosternum fused with the lateral plates, fig. 317, 318). Flies of coastal habitats, brownish or grey coloured. **146**
146(145)	Body colour mainly grey. Oc-bristles absent or weakly developed; 1-2 pairs of less strong F-bristles present (fig. 319). Wing with strong bristles along costa (fig. 320). Prosternum with long bristles (fig. 318). [Flies of coastal habitats.] **Helcomyzidae**

(In Europe 1 genus, *Helcomyza*, with 2 species.)

–	Body colour mainly brown. Oc- and F-bristles well developed; 3 pairs of F-bristles present (fig. 321). Wing without strong bristles along costa. Prosternum bare. [Flies of coastal habitats.] **Heterocheilidae**

(In Europe, along North Sea and Baltic Sea, 1 species, *Heterocheila buccata* (Fallén).)

147(135)	Arista absent; third antennal segment conspicuously large; ocellar triangle large. Vein Sc complete up to the costa, but weak, hardly visible; crossvein BM-Cu absent (cells bm and dm fused). Small, stout flies with a black or metallic blue lustre (fig. 322). **Cryptochetidae**

(In Europe 1 genus, *Cryptochetum*, with 3 species.)

–	Arista present and combination of other characters not as given above. **148**
148(147)	Ocelli absent; second antennal segment very elongate (fig. 323). Vein Sc ending just before the costa; vein R2+3 nearly always with an appendix (fig. 324). **Pyrgotidae**

(In Central and Southern Europe 1 species, *Adapsilia coarctata* Waga.)

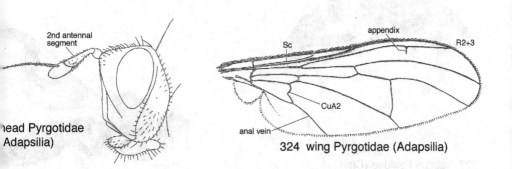

2nd antennal segment

head Pyrgotidae
Adapsilia)

appendix
Sc
R2+3
CuA2
anal vein

324 wing Pyrgotidae (Adapsilia)

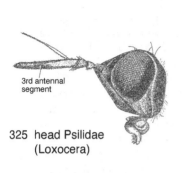

325 head Psilidae
(Loxocera)

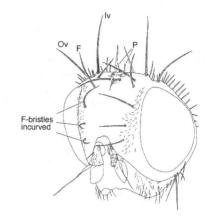

328 head Tephritidae (Or

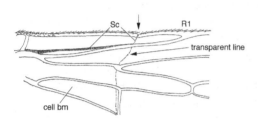

326 wing base
Psilidae (spec.)

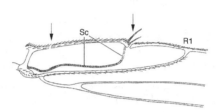

329 wing base Tephritidae (s

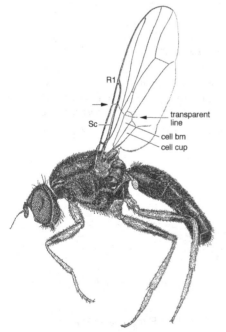

327 habitus Psilidae (Chyliza)

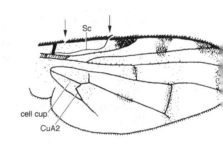

330 wing Tephrit
(Stemonoce

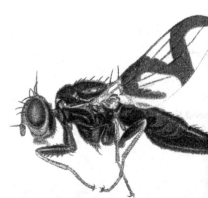

331 habitus Tephritidae (Uroph

–	Ocelli present; second antennal segment usually not elongate (fig. 325, 327). Wing venation different. **149**
149(148)	Vein Sc abruptly bent forward toward the costa at nearly 90°, going toward the costa as a transparent line or trace (fig. 326, 329, 330). **150**
–	Vein Sc not abruptly bent forward. **151**
150(149)	F-bristles, if present, restricted to the upper part of the head, not curving inward; face often receding, third antennal segment in some cases elongate (fig. 325). Wing unmarked; cell cup closed by a straight vein CuA2, not narrowing to an acute end; basal part of wing with a transparent line connecting the (virtual) ending of Sc and the end of cell bm (fig. 326, 327). **Psilidae**
–	Almost always several lower pairs of F-bristles curving inward (fig. 328). Wing often marked; cell cup nearly always closed by a geniculate vein CuA2 and with an acute apical end, vein CuA2 rarely straight or convex; no transparent line in basal part of the wing (fig. 330, 331). **Tephritidae**
151(149)	Vein Sc complete, its apical part not pallid, reaching the costa separately from vein R1 (wing figures 333-357). Note that the ends of veins Sc and R1 may be close together, as in fig. 333, 339). **152**
–	Vein Sc absent, shortened, merging with R1 before the costa, or its apical part pallid and vague (fig. 332, wing figures 358-369). **164**
152(151)	Mouthparts elongate, slender, twice geniculate. Veins Sc and R1 end in the costa close together, halfway up the wing; costa continuous, cell r4+5 narrowing toward wing tip, cell cup closed by straight vein CuA2; anal vein long (fig. 333). **Conopidae**: *Dalmannia*
–	Mouthparts and wing venation not as described above. **153**

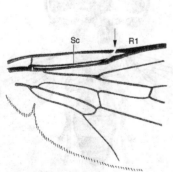

332 wing base Strongylophthalmyiidae
(Strongylopthalmyia)

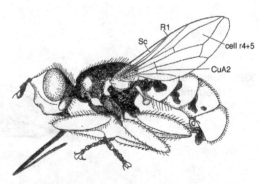

333 habitus Conopidae (Dalmannia)

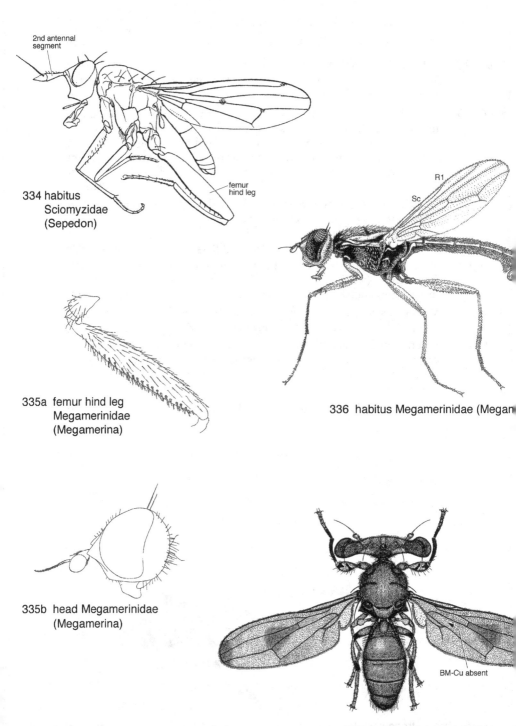

2nd antennal segment

334 habitus
Sciomyzidae
(Sepedon)

femur
hind leg

R1

Sc

335a femur hind leg
Megamerinidae
(Megamerina)

336 habitus Megamerinidae (Megam

335b head Megamerinidae
(Megamerina)

BM-Cu absent

337 habitus Diopsidae (Sphyrace

153(152)	Femur of hind leg swollen, its lower side with 2 rows of spines (fig. 334, 335a). Oc-bristles absent. Habitus, especially head and wing as in figure 334 or 336. **154**
–	Femur of hind leg not swollen, without spines on lower side. Oc-bristles absent or present. Habitus otherwise. **155**
154(153)	Second antennal segment lengthened. Habitus as in fig. 334. **Sciomyzidae**: *Sepedon*
–	Second antennal segment not lengthened (fig. 335b). Habitus as in fig. 336. **Megamerinidae**

(In Europe 1 widespread species, *Megamerina dolium* (Fabricius).

155(153)	Head unusually wide, eyes on short stalks. Crossvein BM-Cu absent (cells bm and dm fused) (fig. 337). **Diopsidae**

(In Central Europe 1 species, *Sphyracephala europaea* Papp & Földvári.)

–	Head not as wide, eyes not on stalks. Crossvein BM-Cu present (cells bm and dm separate) (wing figures 338-357), except in Micropezidae, subfamily Micropezinae (fig. 339, cell r4+5 narrowing). **156**
156(155)	Largely black lustrous, small, ant-like flies with relatively few bristles or pubescence. Head round, abdomen basally constricted. In male, femur and tibia of fore leg often with protrusions, spines, serrate structures, etc. (fig. 338). Arista bare. Wing often with a dark spot at the wing tip (*Sepsis*, fig. 198). Posterior spiracle with 1 or more bristles at the posteroventral margin (fig. 196, 197). **Sepsidae**
–	Combination of characters not as given above. Posterior spiracle without bristles along the posteroventral margin. **157**
>>>	[Black lustrous Ulidiidae, such as some species of *Herina* and *Seioptera*, fig. 350, may key out here. In the Ulidiidae the posterior spiracle has no bristles, the female has an incompletely retractable ovipositor, the apical end of cell cup is acute (fig. 349-351) and/or vein R1 has setulae on its upper side (fig. 354).]

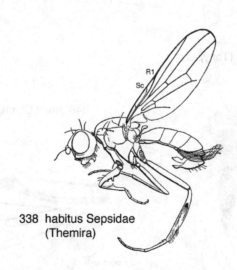

338 habitus Sepsidae
(Themira)

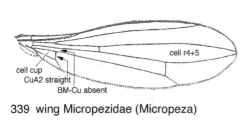

339 wing Micropezidae (Micropeza)

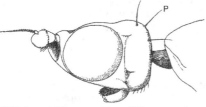

340 head Micropezidae (Micropeza)

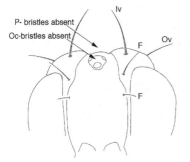

341 head Micropezidae (Calobata)

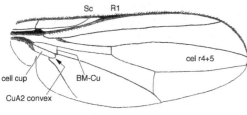

342 wing Tanypezidae (Tanypeza)

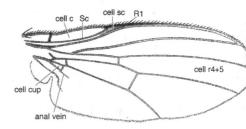

344 wing Chamaemyiidae (Lipoleucopis)

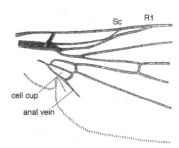

345 wing base Chamaemyiidae (Parochthiphila)

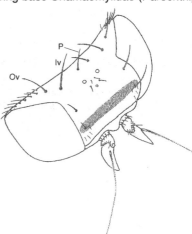

346 head Chamaemyiidae (Parochthiphila)

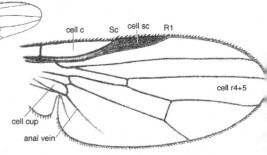

347 wing Cremifaniidae (Cremifania)

157(156) Cell r4+5 narrowing toward wing tip, apical end of cell cup not acute (fig. 339, 342); wing not marked. Legs conspicuously long and slender.   **158**

– Cell r4+5 parallel or widening toward the wing tip (fig. 344); if narrowing toward wing tip, then apical end of cell cup acute and vein CuA2 geniculate (fig. 349) and/or wing with markings. Legs not conspicuously long and slender.   **159**

158(157) Oc-bristles absent; P-bristles diverging (fig. 340) or absent (fig. 341). Vein Sc separate from R1; cell cup closed by a straight vein CuA2; crossvein BM-Cu sometimes absent (fig. 339: subfamily Micropezinae).   **Micropezidae**

– Oc-bristles present; P-bristles diverging. Veins Sc and R1 touch or approach each other over some distance before they separately end in the costa; cell cup closed by a convex vein CuA2; crossvein BM-Cu present (fig. 342). Habitus as in fig. 343.   **Tanypezidae**

(In Europe 1 widespread species, *Tanypeza longimana* Fallén.)

159(157) Silvery grey flies. Female without a lancet-shaped ovipositor. Wing often with darkened cell(s) c and/or sc; anal vein short (fig. 344, 345, 347). P-bristles converging (fig. 346) or absent (fig. 348).   **160**

– Flies of different colour. Female with a lancet-shaped, incompletely retractable ovipositor (fig. 350, 351, 353, 357). Wing often with extensive markings; anal vein long, often reaching the hind margin of the wing. P-bristles diverging or absent (fig. 355, 356).   **161**

160(159) Vein Sc reaching the costa near the end of vein R1 (fig. 344; in most species veins Sc and R1 touch or approach each other before they separately end in the costa, fig. 345). P-bristles converging (fig. 346) or absent. Especially in the large genus *Leucopis* dorsal bristles of the head largely absent (fig. 257b)   **Chamaemyiidae**

– Veins Sc and R1 reaching the costa at some distance of each other (fig. 347). P-bristles absent (fig. 348).   **Cremifaniidae**

(In Europe 1 genus, *Cremifania*, with 2 species.)

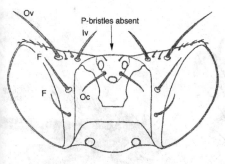

348 head Cremifaniidae
   (Cremifania)

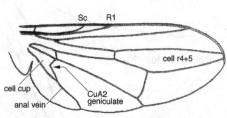

349 wing Ulidiidae
   (Ulidia)

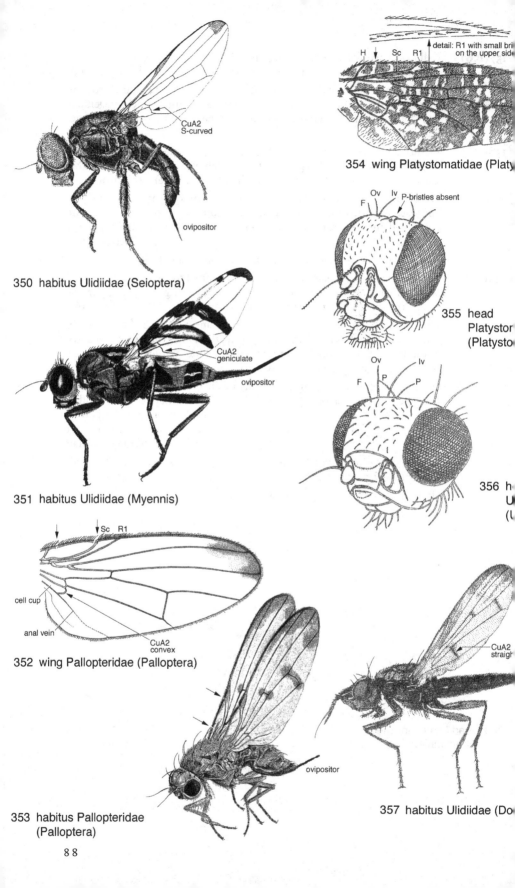

350 habitus Ulidiidae (Seioptera)

CuA2
S-curved

ovipositor

detail: R1 with small bri
on the upper side

H Sc R1

354 wing Platystomatidae (Platy

351 habitus Ulidiidae (Myennis)

CuA2
geniculate

ovipositor

Ov Iv P-bristles absent
F

355 head
Platystor
(Platysto

Ov Iv
F P P

356 h
U
(U

Sc R1

cell cup

anal vein

CuA2
convex

352 wing Pallopteridae (Palloptera)

ovipositor

CuA2
straig

353 habitus Pallopteridae
(Palloptera)

357 habitus Ulidiidae (Do

161(159)	Apical end of cell cup acute; vein CuA2 geniculate or with an S-curve (fig. 349-351). **Ulidiidae**
–	Cell cup different, vein CuA2 convex (fig. 352) or straight (fig. 357). **162**
162(161)	Vein R1 on upper side without setulae (fig. 352, 353). **Pallopteridae**
–	Vein R1 on upper side with setulae (fig. 354). **163**
163(162)	Costa with only a humeral break (just beyond crossvein H) (fig. 354). P-bristles reduced or absent (fig. 355). **Platystomatidae**
–	Costa continuous, or with only a subcostal break, or with both humeral and subcostal breaks. P-bristles usually present, diverging (fig. 356). **Ulidiidae**
164(151)	Vein R2+3 long, nearly reaching R4+5 and wing tip; wing base usually narrow; anal area reduced; often a small cell present where R1 ends in the costa; wing often with markings (wing tip with dark spot, crossveins banded); cell cup and crossvein BM-Cu present (fig. 358, 359). 1 pair of F-bristles curving backward (fig. 360). **Opomyzidae**
>>>	[In case of having overlooked the pale yellow vibrissae in Chyromyidae, this family will key out here. Chyromyidae are mostly yellow flies, with a normal wing base width (fig. 256), no markings on the wing, converging P-bristles and usually more than 1 pair of F-bristles (fig. 270).]

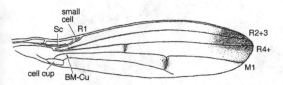

358 wing Opomyzidae (Geomyza)

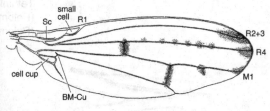

359 wing Opomyzidae (Opomyza)

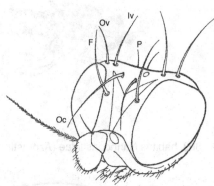

360 head Opomyzidae (Anomalochaeta)

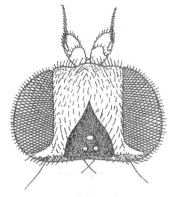

361 head Chloropidae
(Calamoncosis)

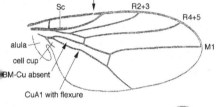

362 wing Chloropidae
(Rhodesiella)

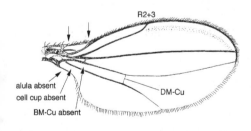

363 habitus Nannodastiidae (Azorastia)

364 wing Nannodastiidae (Azorastia)

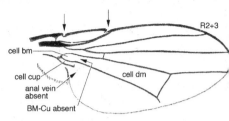

365 wing Ephydridae (Para

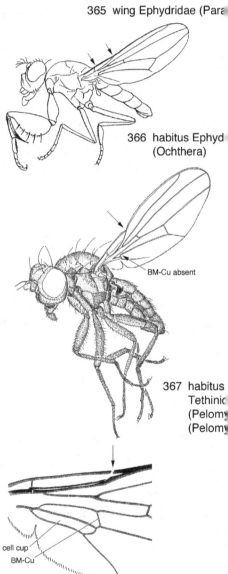

366 habitus Ephyd
(Ochthera)

367 habitus
Tethini
(Pelom
(Pelom

368 wing base
Strongylophthalm
(Strongylophthaln

–	Vein R2+3 not (nearly) reaching R4+5 and wing tip; wing base not narrow; anal area well developed (fig. 362); wing rarely with markings; if wing base somewhat narrow, then vein R2+3 short, ending about halfway up the costa (Nannodastiidae: fig. 363, 364); cell cup and crossvein BM-Cu present or absent. F-bristles variable. **165**
165(164)	Cell cup open (fig. 362, 365) or absent (fig. 364); crossvein BM-Cu absent (cells bm and dm fused) (fig. 362-365). **166**
–	Cell cup present, closed; crossvein BM-Cu present or absent (fig. 367-369). **168**
166(165)	Ocellar triangle conspicuously large (fig. 361, 282). Vein CuA1 usually with a flexure halfway down cell bm + dm (fig. 362); often vein R4+5 and sometimes also vein M1 curving toward the costa; costa with a subcostal break (fig. 362; in only a few species the costa is continuous). Proepisternum with a sharp anterior carina (fig. 283). **Chloropidae**
–	Ocellar triangle not conspicuously large. Vein CuA1 without flexure; costa with both humeral and subcostal breaks (fig. 364-366). Proepisternum without carina. **167**
167(166)	Vein R2+3 short, ending about halfway up the costa; wing base narrow, cell cup and alula absent (fig. 363, 364). Second and third antennal segments with long bristles (fig. 363). [Flies of coastal habitats.] **Nannodastiidae**
	(In Europe 1 genus, *Azorastia*, with 3 species.)
–	Combination of characters not as given above. Wing base rarely narrow, cell cup open or absent, alula usually present; vein r2+3 nearly always long (fig. 365), rarely reaching only halfway up the wing. Arista usually with long rays on the upper side only (fig. 366), sometimes bare or pubescent; face usually swollen, convex. In *Ochthera* fore femur raptorial (fig. 366). **Ephydridae**
168(165)	Crossvein BM-Cu absent (cells bm and dm fused). P-bristles converging. Small, stout flies with greyish, yellowish or brownish pruinosity (fig. 367). **Tethinidae: Pelomyiinae**
–	Crossvein BM-Cu present (cells bm and dm separate) (fig. 368). P-bristles diverging. Long, slender, black flies (fig. 369). **Strongylophthalmyiidae**
	(In Europe 1 genus, *Strongylophthalmyia*, with 2 species).

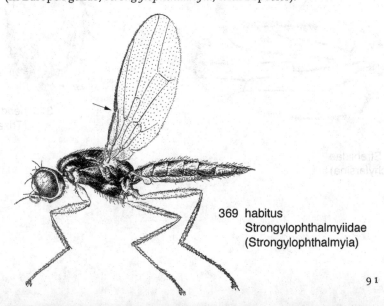

369 habitus
Strongylophthalmyiidae
(Strongylophthalmyia)

370 tarsal claw
Hippoboscidae
(Ornithomya)

eyes rudimentary
or absent

373 head and fore
legs Streblidae
(Brachytarsina)

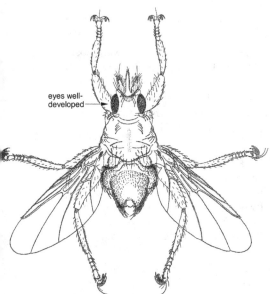

eyes well-
developed

371 habitus Hippoboscidae
(Ornithomya)

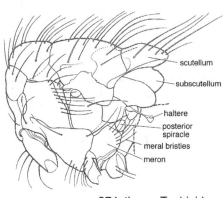

scutellum

subscutellum

haltere

posterior
spiracle

meral bristles

meron

374 thorax Tachinidae
(spec.)

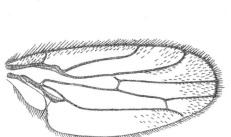

372 wing Streblidae
(Brachytarsina)

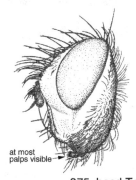

at most
palps visible

375 head Tachinidae
(Trixa)

| 169(81) | Head adpressed to the thorax and eyes well-developed (fig. 371). Tarsal claws long, dentate at the base (fig. 370). Veins in front part of wing strong, remaining veins weak (fig. 371; wing may be present or absent, see couplet 209). Flies ectoparasitic on birds and mammals (not on bats). **Hippoboscidae** |

Hippoboscidae

– Head not adpressed to the thorax. Eyes and ocelli rudimentary or absent. Last tarsal segment broadened (fig. 373). Wing venation as in fig. 372. Flies ectoparasitic on bats. **Streblidae**

(In Southern Europe 1 species, *Brachytarsina flavipennis* Macquart.)

170(82) Mouth opening small, its diameter up to 1/4 to 1/8 of the width of the head; mouthparts minute, rudimentary, in some cases only the swollen tip of the palp visible (fig. 375, 378, 379, 382). **171**

– Mouth opening large, its diameter at least 1/4 of the width of the head; mouthparts well developed, well visible (fig. 4, 7, 8, 393, 394). **175**

171(170) Ocelli absent. **Tachinidae**: *Therobia*

– Ocelli present **172**

172(171) Head (fig. 375), thorax, legs and abdomen with stout black bristles. Meron with a single or double row of bristles along its hind margin near the posterior spiracle; subscutellum convex (swollen, bulging) (fig. 374). **Tachinidae**: *Trixa, Trixiceps*

– Head and usually also legs and remainder of body completely or almost completely devoid of stout black bristles, in most species covered with soft hair-like bristles, giving the flies the aspect of bees or bumblebees (fig. 376-383). Meron largely white or yellow pubescent. Subscutellum variable (from flat to convex). **173**

173(172) Vein M1 more or less straight and ending beyond the wing tip (fig. 376). **Gasterophilidae**

(In Europe 1 genus, *Gasterophilus*, with 6 species.)

376 habitus Gasterophilidae
(Gasterophilus)

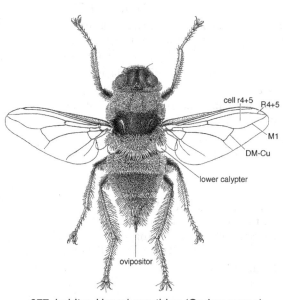

377 habitus Hypodermatidae (Oedemagena)

379 head Oestridae (O

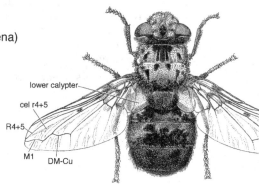

380 habitus Oestridae (Pharyng

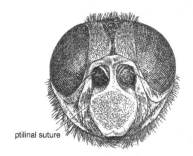

378 head Hypodermatidae
(Crivellia)

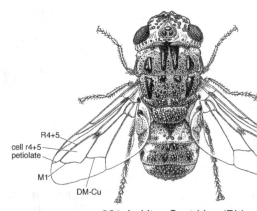

381 habitus Oestridae (Rhino

–   Vein M1 curving and ending near (fig. 377, 380) or in vein R4+5 (fig. 381, 383). **174**

174(173)   Vein M1 beyond crossvein DM-Cu gradually curving toward the costa, or proceeding more or less straight; cell r4+5 open or closed at the wing margin, without a petiole (fig. 377). Ends of the ptilinal suture below the antennae not strongly curved toward each other (fig. 378). Female oviparous and possessing an ovipositor (fig. 377).   **Hypodermatidae**

–   Vein M1 strongly curving or with an angle toward the costa (fig. 380) or ending in vein R4+5 making cell r4+5 petiolate (fig. 381, 383). Ends of the ptilinal suture below the antennae strongly curved toward each other (fig. 379), except in *Cephalopina* (fig. 382) which shows a wing venation as in fig. 383. Female larviparous and without ovipositor.   **Oestridae**

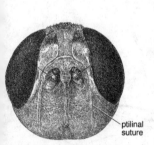

382 head Oestridae (Cephalopina)

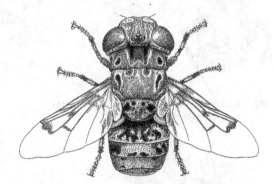

383 habitus Oestridae (Cephalopina)

95

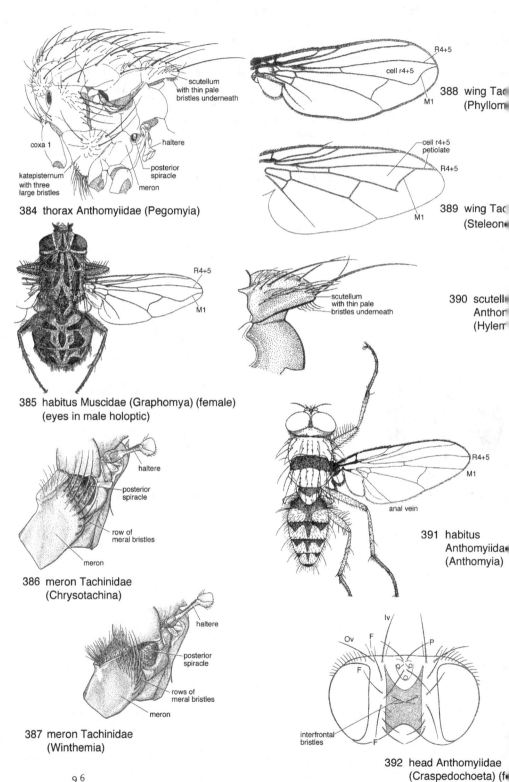

scutellum with thin pale bristles underneath

coxa 1

haltere

katepisternum with three large bristles

posterior spiracle

meron

384 thorax Anthomyiidae (Pegomyia)

R4+5
cell r4+5
M1

388 wing Tac (Phyllom

cell r4+5 petiolate
R4+5
M1

389 wing Tac (Steleon

R4+5
M1

385 habitus Muscidae (Graphomya) (female)
(eyes in male holoptic)

scutellum with thin pale bristles underneath

390 scutell Anthor (Hylem

haltere

posterior spiracle

row of meral bristles

meron

386 meron Tachinidae
(Chrysotachina)

R4+5
M1
anal vein

391 habitus Anthomyiidae (Anthomyia)

haltere

posterior spiracle

rows of meral bristles

meron

387 meron Tachinidae
(Winthemia)

Iv
Ov F P
F
interfrontal bristles
F

392 head Anthomyiidae
(Craspedochoeta) (fe

175(170)	Meron without one or more rows of bristles near its hind margin (fig. 384). Vein M1 usually straight (fig. 391, 394, 395), in Muscidae sometimes curving gently or sharply toward the costa (fig. 396). **176**
–	Meron with bristles near the hind margin, arranged in one (fig. 386) or several (fig. 387) rows, or in a small cluster (fig. 408). Vein M1 usually curving sharply forward (fig. 388, 409), sometimes ending in vein R4+5 making cell r4+5 petiolate (fig. 389), M1 rarely curving gently (fig. 385) or almost straight (fig. 407). **181**
>>>	[A key with good figures for the families and genera which key out in couplets 176-183 is given by Gregor et al. 2002.]
176(175)	Ventral surface of scutellum in the middle with a patch of fine, pale hairs (fig. 384, 390). Anal vein long (fig. 391). Female often with interfrontal bristles (fig. 392). **Anthomyiidae**
–	Ventral surface of scutellum in the middle bare, although in some cases bristles may be present laterally. Anal vein variable. Interfrontal bristles present or absent. **177**
177(176)	The part of the head behind and obliquely below the eyes with a few to many long, pale, thin hairs (fig. 393, 394); interfrontal bristles absent. Anal vein long, usually reaching the wing margin (fig. 394). **Scathophagidae**
–	If present, bristles behind and obliquely below the eyes not long, pale, thin or hair-like. Interfrontal bristles present or absent. Anal vein variable. **178**
178(177)	Anal vein reaching the wing margin (fig. 394), in some cases very thin or almost fold-like near the end. **179**

head Scathophagidae (Scathophaga)

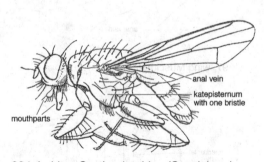

394 habitus Scathophagidae (Spaziphora)

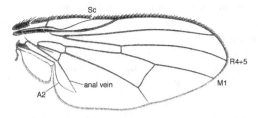

395 wing Fanniidae (Fannia)

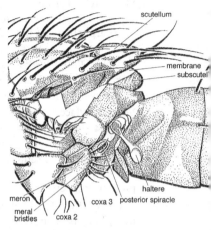

399 thorax Tachinidae (Exorista)

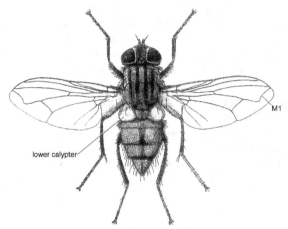

396 habitus Muscidae (Musca)

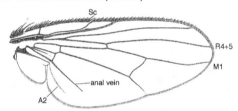

397 wing Muscidae (Muscina)

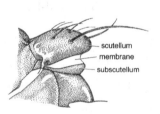

400 scutellum and subscutellum Rhinophoridae (Melanophora)

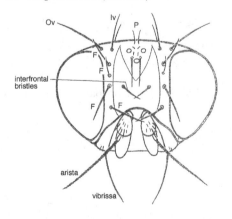

398 head Muscidae (Azelia) (female)

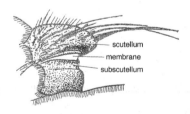

401 scutellum and subscutellum Calliphoridae (spec.)

–	Anal vein not reaching the wing margin (fig. 395-397).	**180**

179(178) Katepisternum with at least 2 large bristles, usually 3 (fig. 384). Wing clear. Female often with interfrontal bristles (fig. 392). **Anthomyiidae**

– Katepisternum with 1 large bristle (fig. 394) or wing darkened at the base, along the costa and at the tip. Interfrontal bristles absent.

Scathophagidae

180(178) Vein A2 curving toward the end of the anal vein which it would meet if it were extended; apical portion of vein Sc almost straight; vein M1 almost straight (fig. 395). Interfrontal bristles absent. **Fanniidae**

– Vein A2 and the anal vein would not meet if both were extended; apical portion of vein Sc distinctly curving (fig. 397); vein M1 distinctly curving forward (fig. 396) or not (fig. 397). Interfrontal bristles usually absent.

Muscidae

>>> [The genus *Azelia* (Muscidae) shows a venation similar to that of Fanniidae with respect to A2 and the anal vein; the female of *Azelia* possesses interfrontal bristles (fig. 398), which in the Fanniidae are always absent.]

181(175) Subscutellum strongly developed, inflated (convex), closely adpressed to the scutellum (membrane between scutellum and subscutellum distinctly narrower than the subscutellum itself; fig. 399, 374). Lower calypter usually rounded-triangular, rarely elliptical (*Cinochira, Catharosia*). Posterior spiracle without long hairs on anterior lappet (fig. 402).

Tachinidae

>>> [Vein M1 usually curving or geniculate toward the costa (fig. 405) or ending in vein R4+5 (cell r4+5 petiolate, fig. 406); rarely faintly curving, straight or not reaching the wing margin.]

– Subscutellum absent or weakly developed (fig. 401); if somewhat inflated, then subscutellum about as high as the membrane between scutellum and subscutellum (fig. 400) and posterior spiracle with long hair-like bristles along the anterior margin (fig. 403, 404). **182**

 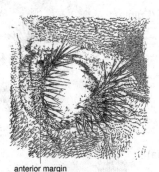 

anterior lappet

402 posterior spiracle Tachinidae (Peribaea)

anterior margin

403 posterior spiracle Rhinophoridae (Phyto)

anterior margin

404 posterior spiracle Rhinophoridae (Melanophora)

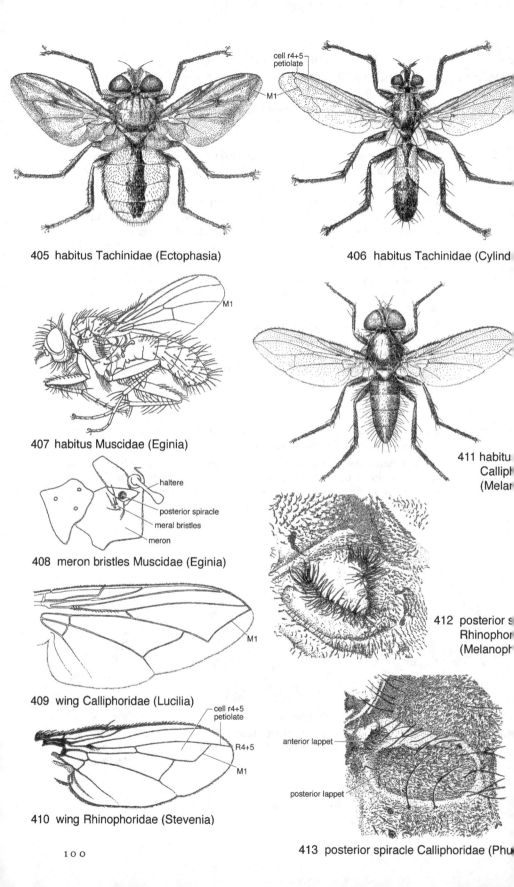

405 habitus Tachinidae (Ectophasia)

406 habitus Tachinidae (Cylind

407 habitus Muscidae (Eginia)

408 meron bristles Muscidae (Eginia)

409 wing Calliphoridae (Lucilia)

410 wing Rhinophoridae (Stevenia)

411 habitu
Callipł
(Melaɾ

412 posterior s
Rhinophoɾ
(Melanopł

413 posterior spiracle Calliphoridae (Phu

182(181) Vein M1 slightly curving backward (fig. 407). Meral bristles arranged in a small cluster (fig. 408). **Muscidae**: *Eginia*

>>> [Abdomen yellowish to orange. Lower calypter small, elliptical; anal vein continues up to just before the wing margin. Subscutellum absent; posterior spiracle without lappets, with bristles along the anterior margin.]

– Vein M1 curving toward the costa gently (fig. 385, 411) or geniculate (fig. 409) or ending in vein R4+5 (cell r4+5 petiolate, fig. 410). **183**

183(182) Meron with a row of delicate bristles along its hind margin, in size and thickness similar to the bristles along the upper margin of the meron. Vein M1 curving gently toward the costa. Habitus as in fig. 385.

Muscidae: *Graphomya*

>>> [Body pale with grey pruinosity, thorax with dark longitudinal stripes, abdomen checkered with round to oval spots in the middle (fig. 385).]

– Meron bristles strong, arranged in one (fig. 386, 399) or several (fig. 387) rows. Vein M1 usually curving sharply toward the costa or geniculate (fig. 409) or ending in vein R4+5 (cell r4+5 petiolate, fig. 410), only in few cases gently curving toward the costa (fig. 411). **184**

>>> [A key with good figures for the Tachinidae and the families and genera which key out in couplets 184-190 is given by Tschorsnig & Herting 1994 (in German).]

184(183) Posterior spiracle with a lining of long hair-like bristles, without lappets (fig. 403, 412). Lower calypter ovoid, spoon-shaped, erect, laterally extended (fig. 414). Subscutellum may be swollen or not. **185**

– Posterior spiracle at least partially covered by an anterior and/or posterior lappet, the latter in the form of a locking plate (operculum) (fig. 413). Lower calypter variable: usually large, triangular, in some cases small and spoon-shaped. Subscutellum narrow, flat, not swollen (fig. 417). **186**

185(184) Subscutellum distinctly swollen (convex) but not strongly developed, at most half as deep as high (fig. 415). **Rhinophoridae**

>>> [Cell r4+5 often petiolate (fig. 410). In some cases, bristles present on the gena, alongside the lower eye margin (fig. 416).]

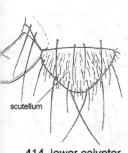

414 lower calypter and scutellum Rhinophoridae (spec.)

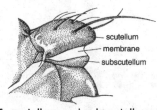

415 scutellum and subscutellum Rhinophoridae (Melanophora)

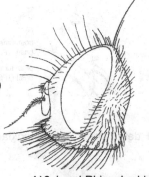

416 head Rhinophoridae (Rhinomorinia)

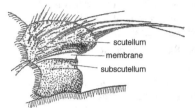

417 scutellum and subscutellum
Calliphoridae (spec.)

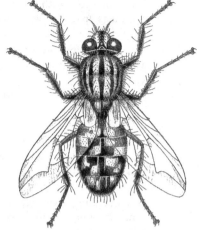

420a habitus Sarcophagidae (Sarcophag

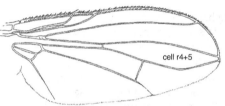

418 wing Calliphoridae (Angioneura)

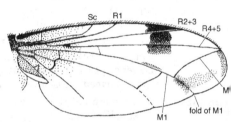

420b wing Sarcophagidae (Sphenometopa

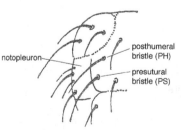

419 detail thorax Sarcophagidae (spec.)

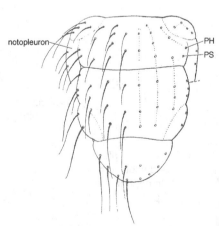

421 thorax Calliphoridae (spec.)

–   Subscutellum distinctly flat, not swollen (fig. 417).
                          Calliphoridae: *Morinia, Melanomya, Angioneura*
>>>   [Cell r4+5 open or closed at the wing margin, rarely with a short petiole
             (fig. 418). No bristles on the gena near the lower eye margin. Body black or
             black with grey pruinosity, no stripes on upper side of thorax and no
             checkered pattern on the abdomen.]
186(184)   Flies largely green or blue metallic lustrous.   **Calliphoridae**
–   Colour varying from yellow to black, but without metallic green or blue
             lustre.   **187**
187(186)   Presutural bristle (PS) inserted more outward than posthumeral bristle
             (PH) (fig. 419) or posthumeral bristle absent. Notopleuron with 2 bristles,
             usually interspersed with 1 or 2 smaller bristle(s) (fig. 419). Proepisternum
             usually bare.   **188**
–   Presutural bristle (PS) inserted more inward than posthumeral bristle
             (PH) (fig. 421, 422). In some genera, the presutural bristle (PS) and
             posthumeral bristle (PH) inserted at the same level (fig. 423) in which case
             the proepisternum is set with hairs. Notopleuron with 2 bristles (fig. 421-
             423).   **Calliphoridae**
188(187)   Flies not completely black and lustrous.   **Sarcophagidae**
>>>   [Often gena with bristles below the eye. Most species grey with black
             markings consisting of stripes on the thorax and a checkered pattern on
             the abdomen (fig. 420a) and with a fold or appendix at the corner of vein
             M1 (fig. 420b).]
–   Lustrous black flies.   **189**
189(188)   Flies over 7 mm body length. Vein M1 geniculate toward R4+5, carrying a
             fold or appendix at the corner; vein M1 ending in the costa near vein R4+5
             (fig. 420b) or in vein R4+5 (cell r4+5 with a short petiole).
                                                                       Sarcophagidae
–   Small flies (2-3 mm). Vein M1 regularly curving toward vein R4+5, with-
             out fold or appendix; cell r4+5 open or with a long petiole (as in fig. 410).
                                                                             190
190(189)   Cell r4+5 with a long petiole (as in fig. 410).
                          Tachinidae: *Litophasia hyalipennis* (Fallén)
–   Cell r4+5 open.   **Calliphoridae**: *Angioneura acerba* (Meigen)

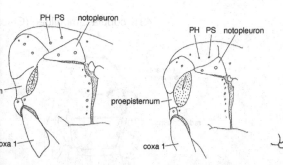
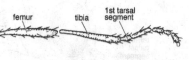

422 detail thorax   423 detail thorax   424 leg Cecidomyiidae (Frirenia)
    Calliphoridae   Calliphoridae
    (Protophormia   (Protophormia
    terraenovae)   atriceps)

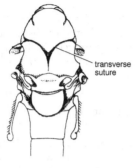

425 thorax Tipulidae (spec.)

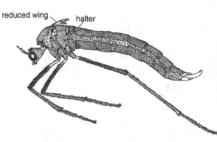

429 habitus Pediciidae (Tricypho

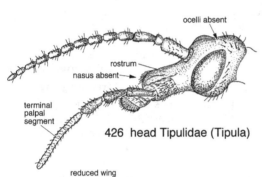

426 head Tipulidae (Tipula)

430 habitus Limoniidae (Dicranom

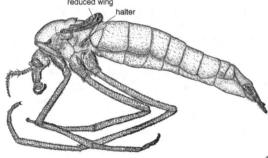

427 habitus Tipulidae (Tipula)

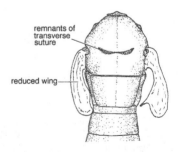

431 copula Limoniidae (Chion

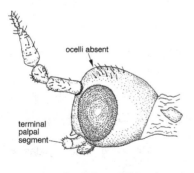

428 head Pediciidae (Tricyphona)

432 thorax Pediciidae (Tricypho

191(1)	Antenna usually long, filiform or resembling a string of beads, with 6 or (usually) more segments; usually all segments or those beyond the second more or less similar (fig. 426, 437); if less than 6 segments (4 segments in fig. 428), then ocelli absent and legs and abdomen lengthened (fig. 429). Palp usually with 3-5 segments. **Nematocera – 192**
–	Antenna not filiform or resembling a string of beads, with less than 6 segments; usually the first 2 segments short, the third segment large and the subsequent segments making up an arista or style. Palp usually with 1-2 segments. **Brachycera – 206**
192(191)	Ocelli absent (fig. 426, 428). **193**
–	Ocelli present. **199**
193(192)	Tarsi with less than 5 segments, or first tarsal segment far shorter than second segment (fig. 424). Small, delicate insects, generally with conspicuously long antennae. **Cecidomyiidae**
–	First tarsal segment not distinctly shorter than the second; tarsus with 5 segments. **194**
194(193)	Upper side of thorax with a V-shaped transverse suture (fig. 425). **195**
–	Upper side of thorax without V-shaped transverse suture. **197**
195(194)	Terminal (5th) segment of palp elongate, distinctly longer than the preceding segments; rostrum well developed, without (fig. 426) or with nasus (fig. 22). Body length generally over 8 mm (fig. 427). **Tipulidae**
–	Terminal (5th) palpal segment short; rostrum short, without nasus (fig. 428); body length usually less than 8 mm (fig. 429-431). **196**
196(195)	Eye pubescent, with ommatrichia in between the facets (fig. 428). **Pediciidae**
–	Eye bare. **Limoniidae**
197(194)	Upper side of thorax with remnants of a V-shaped transverse suture (fig. 432). **198**
–	Upper side of thorax without remnants of this suture (examples of habitus and wing venation cf. fig. 433, 434). **Chironomidae**
198(197)	Eye pubescent, with ommatrichia in between the facets (fig. 428). **Pediciidae**
–	Eye bare. **Limoniidae**
199(192)	Eyes connected by an eye bridge above the antennae, or touching each other (fig. 58, 73). **200**
–	Eyes separate. **202**

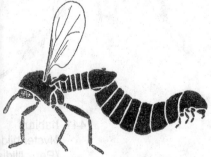

433 copula Chironomidae (Clunio)

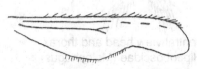

434 wing Chironomidae (Telmatogeton)

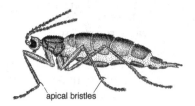

435 habitus Sciaridae (Epidapus)

apical bristles

436 antenna Scatopsidae
(Swammerdamella)

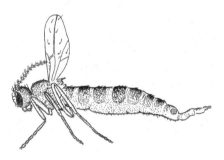

437 habitus Cecidomyiidae (Aprionus)

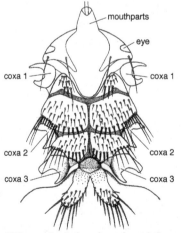

438 ventral view head and thorax
Hippoboscidae (Melophagus)

mouthparts

eye

coxa 1

coxa 1

coxa 2

coxa 2

coxa 3

coxa 3

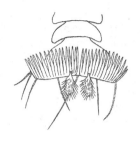

439 tarsal cl
Braulida
(Braula)

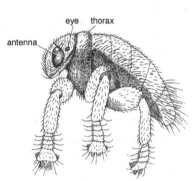

eye thorax

antenna

440 habitus Braulid
(Braula)

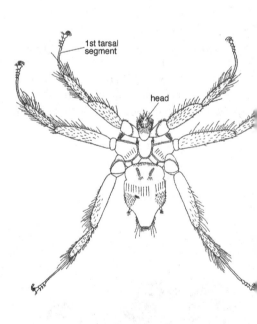

1st tarsal
segment

head

441 habitus
Nycteribiid
(Penicillidi
dorsal view

200(199)	Tibiae of mid and hind legs with apical bristles (fig. 435). **Sciaridae**
–	Tibiae of mid and hind legs without apical bristles. **201**
201(200)	Antennal segments more or less compact, with a strong terminal segment (fig. 436). Abdomen of even width throughout or even slightly wider to the rear. **Scatopsidae**
>>>	[The only scatopsid in Europe with rudimentary wings is *Thripomorpha paludicola* Enderlein.]
–	Antennal segments fragile, carrying a ring of long, thin verticils basally. Abdomen tapering to a point (fig. 437). **Cecidomyiidae**
202(199)	Tibiae of mid and hind legs without apical bristles. **Cecidomyiidae**
–	Tibiae of mid and hind legs with apical bristles (fig. 435). **203**
203(202)	Antenna more or less as long as head and thorax together; head and thorax black, abdomen blackish brown, size 4 – 5 mm, wing distinctly shortened (female of *Hesperinus imbecillus* Meigen). **Hesperinidae**
–	Combination of characters not as given above. **204**
204(203)	Antenna inserted halfway up the eye, or higher. **Sciaridae (& Mycetophilidae?)**
>>>	[The unknown female of the mycetophilid *Baeopterogyna mihalyii* Matile, which is believed to be short-winged, might key out here.]
–	Antenna inserted at lower part of the head, usually just above the lower margin of the face. **205**
205(204)	Vein R forked (fig. 67). **Pleciidae**
>>>	[Somewhat short-winged male of *Penthetria funebris* Meigen, fig. 68.]
–	Vein R not forked or wing absent. **Sciaridae**
206(191)	Coxae of mid legs, and usually of the other legs as well, far apart (fig. 438, 441). **207**
–	Coxae of mid legs not conspicuously wide apart, sometimes even touching each other. **210**
207(206)	Tarsal claws transformed into a comb-like structure, pointing inward (fig. 439, 440). Eye reduced to a spot behind the sunken antenna; ocelli, halteres and scutellum absent. Thorax small, more or less similar to the first abdominal segments (fig. 440). Flies living ectoparasitic on bees. **Braulidae**
	(In Europe 1 genus, *Braula*, with 3 species.)
–	Combination of characters not as given above. If thorax similar to the first abdominal segments, then the eyes well developed (e.g. fig. 445). **208**
208(207)	Head small, in resting position bent backward on to thorax. Eyes and ocelli small or absent. Legs long, with swollen femora and tibiae; first tarsal segment at least as long as all other tarsal segments together (fig. 441). Flies living ectoparasitic on bats. **Nycteribiidae**

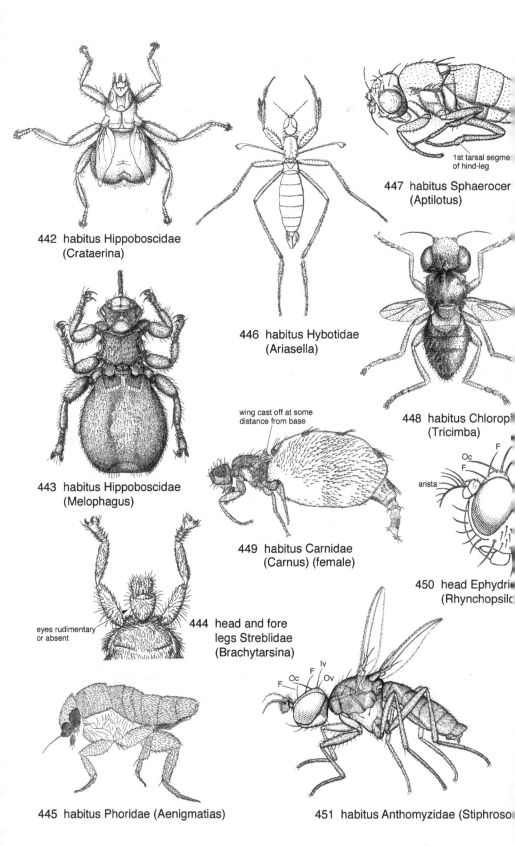

442 habitus Hippoboscidae (Crataerina)

447 habitus Sphaerocer (Aptilotus)

1st tarsal segme of hind-leg

446 habitus Hybotidae (Ariasella)

443 habitus Hippoboscidae (Melophagus)

448 habitus Chloropi (Tricimba)

wing cast off at some distance from base

449 habitus Carnidae (Carnus) (female)

arista

Oc
F
F

450 head Ephydri (Rhynchopsilo)

eyes rudimentary or absent

444 head and fore legs Streblidae (Brachytarsina)

445 habitus Phoridae (Aenigmatias)

F Oc F Iv Ov

451 habitus Anthomyzidae (Stiphroso

–	Head not bent backward on the thorax. Combination of other characters not as given above. Flies living ectoparasitic on birds and mammals (Hippoboscidae not on bats; Streblidae on bats only).	**209**
209(208)	Head adpressed to the thorax and eyes well-developed. Tarsal claws long, dentate at the base (fig. 442, 443). Flies living as ectoparasites on birds and mammals (except bats).	**Hippoboscidae**
–	Head not adpressed to the thorax. Eyes and ocelli rudimentary or absent. Last tarsal segment broadened (fig. 444). Flies ectoparasitic on bats.	**Streblidae**
210(206)	Antenna seems to consist of a single spherical segment, with a long arista. Wings reduced or wings and halteres absent (fig. 445).	**Phoridae**
–	Antenna distinctly with 3 segments.	**211**
211(210)	Third antennal segment ending more or less acute; ptilinal suture absent.	**212**
–	Third antennal segment ovoid or rounded; ptilinal suture present.	**213**
212(211)	Mouthparts long. Dorsal side of head not concave between the eyes (example of habitus: fig. 446).	**Hybotidae**
–	Mouthparts short; head with dorsal part, between eyes, concave.	**Dolichopodidae**
213(211)	First tarsal segment of hind leg swollen, usually shorter than the second tarsal segment (fig. 447).	**Sphaeroceridae**
–	First tarsal segment of hind leg not shorter than the second tarsal segment.	**214**
214(213)	Bristles on the head weakly developed or absent.	**Chloropidae**
–	Bristles on the head normally developed.	**215**
215(214)	Ocellar triangle strikingly large; 1 pair of interfrontal bristles immediately above the antennae. Wing cast off at some distance from the base. Habitus fig. 449.	**Carnidae**
–	Ocellar triangle not conspicuously large.	**216**
216(215)	Tibiae without dorsal preapical bristle.	**217**
–	Tibiae with dorsal preapical bristle.	**219**
217(216)	Face distinctly swollen, convex; if arista with long rays, then only on the upper side (fig. 450).	**Ephydridae**
–	Face not distinctly convex; if arista with long rays, then on both sides (plumose).	**218**
218(217)	At least 2 pairs of F-bristles present (fig. 451).	**Anthomyzidae**
–	Only 1 pair of F-bristles present (fig. 360).	**Opomyzidae**
219(216)	Subscutellum well developed, inflated (fig. 399).	**Tachinidae**
–	Subscutellum not inflated.	**220**
220(219)	Arista plumose (fig. 218).	**Drosophilidae**
–	Arista bare or pubescent only (fig. 10).	**Heleomyzidae**

FAMILY DESCRIPTIONS

CAT = Although they do not contain keys, the identification references include recent catalogues as valuable source on genera, species, distribution and references.
CMPD = Contributions to a Manual of Palaearctic Diptera.
Lindner = Chapter in Lindner, E., Die Fliegen der Paläarktischen Region.
( ) Family names between brackets refer to names as found in the literature, not recognised here as a separate family but, as indicated, considered part of another family.
et al. References with more than two authors are given as First author et al.

As far as not yet outdated, the number of genera and species in Europe is largely based on the Catalogue of Palaearctic Diptera, the CMPD and Fauna Europaea, the latter available online at: www.faunaeur.org (consulted was version 1.2, updated 7 March 2005).

As to size, the following categories are distinguished: minute: smaller than 2 mm; small: 2-5 mm; medium sized: 5-10 mm; large: 10-20 mm; very large: over 20 mm.

Acartophthalmidae (key couplet 113; fig. 243)

Systematics: Acalyptrate Brachycera; superfamily Opomyzoidea; in Europe 1 genus, *Acartophthalmus*, with 3 species.

Characters: Minute to small (1-2.5 mm), brownish grey flies. Arista pubescent, ocelli present; Oc-bristles present; P-bristles strong, far apart, diverging; 3 pairs of F-bristles, curving obliquely out-backward, increasing in size, the upper pair the largest; scattered interfrontal setulae present; vibrissae absent but with a series of strong bristles near the vibrissal angle. Wing unmarked or tinged along costa; costa with a humeral break only; vein Sc complete; crossvein BM-Cu present; cell cup closed. Tibiae without dorsal preapical bristle.

Biology: The larvae of this small family are presumed to feed on rotting organic matter. The adults appear to inhabit woodland and are found on rotting bracket fungi and mushrooms, decaying wood, droppings and carrion; deposition of eggs on a dead snail and carrion has been observed.

452 Acartophthalmus nigrinus
(Zetterstedt), female; Papp 197

Identification references: Ozerov 1986 (review); Stackelberg 1989al (former USSR); CMPD: Papp & Ozerov 1998; CAT: Papp 1984e.

Acroceridae (key couplet 45; fig. 98-99)

Systematics: Lower muscomorph Brachycera; superfamily Nemestrinoidea; in Europe 8 genera and about 35 species.

Characters: Despite variability in size (2.5-20 mm), colour and wing venation, these flies are easily recognised by their small head situated almost below the hunchbacked thorax, the large ear-shaped lower calypter and large abdomen. Body colour dull or lustrous, usually black with white, yellow, or orange markings on thorax and abdomen. The small head largely occupied by the relatively large eyes, holoptic in both sexes. Wing clear or tinged; venation varying from complete to strongly reduced. Legs simple, femora sometimes swollen; empodium pulvilliform.

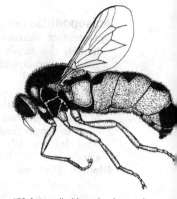

453 Astomella hispaniae Lamarck, male; Sack 1936

Biology: The larvae are endoparasitoids of spiders, usually one larva per spider, rarely two or more. The spider continues functioning normally until just prior to completion of the fourth instar. At the end of this stage the spider is largely eaten and the larva pupates outside the spider in the web the spider has woven at the very last. There are apparently no fixed host-parasitoid relationships at the species level. Adults may more or less abound locally but apart from species visiting flowers and feeding on nectar, they are rarely observed. They are mainly active on warm sunny days. Males may sometimes gather in large numbers at higher points in the landscape (hilltopping). Mating takes place in flight.

Identification references: Chvála 1980 (Central Europe); Van der Goot 1963, De Jong et al. 2000 (Netherlands), Nartshuk 1989f (former USSR); Oldroyd 1969; Stubbs & Drake 2001 (British Isles); Weinberg & Bächli 1997 (Switzerland); CMPD: Nartshuk 1997; CAT: Nartshuk 1988.

(Aenigmatiidae): part of the Phoridae.

Agromyzidae (key couplet 110, 123; fig. 241-242, 263-266)

Systematics: Acalyptrate Brachycera; superfamily Opomyzoidea; in Europe some 23 genera and about 910 species.

Characters: Minute to medium sized (1-6 mm) flies, varying in colour from all yellow to black to metallic green. Arista bare to pubescent; ocelli present; Oc-bristles present; P-bristles diverging; 2-8 pairs of F-bristles, the lower 1–3 pairs curving inward, the other pairs backward; interfrontal bristles absent; interfrontal setulae sometimes present; vibrissae present but in some cases weak. Wing unmarked; costa with subcostal break; vein Sc complete or incomplete, apically ending in vein R1 (Agromyzinae) or separate from vein R1 but reduced to a fold that may or may not reach the costa (Phytomyzinae); crossvein BM-Cu present; cell cup closed. Tibiae without dorsal preapical bristle. Female with oviscape, non retractable basal segment of the ovipositor.

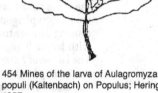

454 Mines of the larva of Aulagromyza populi (Kaltenbach) on Populus; Hering 1957

Biology: The larvae are mining, generally in the stalks or leaves of herbs, sometimes in roots, seeds or under bark. In the few species living on woody plants this may lead to gall formation. Most species are more or less host specific and several species cause damage to economically important plants. The shape of the mine is often characteristic of the species and therefore useful for identification. Adults occur in a variety of habitats, depending on the larval host plants.

Identification references: Dempewolf 2004 (World); Hering 1957 (mines); E.B. Rohdendorf 1989 (former USSR); Spencer 1972 (British Isles), 1976 (Northwestern Europe), 1987 (North America); CMPD: Darvas & Papp 2000 (genera); CAT: Papp 1984d (see also Spencer & Martinez 1987).

Anisopodidae (key couplet 38; fig. 38, 84)

Systematics: Nematocera; superfamily Anisopodoidea; traditionally this family includes the Mycetobiidae; in Europe 1 genus, *Sylvicola*, with about 10 species.

Characters: Small to large (4-12 mm), yellowish to brownish Nematocera with slender antennae and abdomen. Head small and rounded; mouthparts small; ocelli present; eyes dichoptic or holoptic; antenna with 16 segments varying from relatively short to longer than head and thorax together. Wing wide, with clear anal lobe and a pattern of smaller markings. Legs slender, relatively long; tibiae with apical spurs.

455 Sylvicola fenestralis (Scop female; Séguy 1940

Biology: The larvae live in, and feed on, all sorts of rotting and decomposing material like droppings, dung, vegetables, dying wood, cracks and fissures in trees filled with organic matter, bark, bracket fungi and mushrooms, leaves, roots, sap exuding from tree wounds and other liquids rich in organic matter. The adults are encountered mainly in summer in woodland, meadows, gardens, etc. They are found near the larval habitats, visiting flowers, near sap exuding from tree wounds, and are regularly seen on windows, both indoors and outside, or other vertical surfaces such as tree trunks, walls, etc. Males form swarms to attract females for mating.

Identification references: Haenni 1997c (Switzerland); Krivosheina & Menzel 1998 (review, key to species); Michelsen 1999 (systematics); Søli 1992 (Norway); Stackelberg 1989n (former USSR); CMPD: Krivosheina 1997c; CAT: Krivosheina 1986c.

Anthomyiidae (key couplet 176, 179; fig. 384, 390-392)

Systematics: Calyptrate Brachycera; superfamily Muscoidea; some 35 genera and about 480 species in Europe.

Characters: Small to large (4-12 mm), slender flies, usually coloured brownish, greyish or blackish, sometimes yellowish or with a distinct pattern on thorax and abdomen, never metallic lustrous. Arista bare to plumose; female often with interfrontal bristles. Wing generally clear, sometimes tinged, rarely marked; venation virtually always with straight vein M1 and long anal vein. Scutellum underneath in the middle usually with a patch of fine, pale hairs. Meron devoid of bristles near hind margin.

456 Anthomyia pluvialis (Linna male; Séguy 1951

Biology: The larvae of many genera are phytophagous, some of which are leaf miners. At the same time many species are saprophagous in diverse habitats (dung, nests of birds, etc.). Phytophagous species can be of economic importance, like various species of *Delia* causing damage to onions, beans and other vegetables. Some species are saprophagous, or with larvae living in mushrooms or as parasitoids of insects. The larvae of *Fucellia* live at the coast in rotting organic matter washed up by the sea. The adults visit flowers but can be found on exuding tree sap and on dung, carrion, rotting plants, etc.

Identification references: Elberg 1989 (former USSR); Hennig 1966-1976 (Lindner); CMPD: Suwa & Darvas 1998; CAT: Dely-Draskovits 1993.

Anthomyzidae (key couplet 105, 125, 218; fig. 232-235, 267-269, 451)

Systematics: Acalyptrate Brachycera; superfamily Opomyzoidea; 9 genera and about 30 species in Europe.

Characters: Minute to small (1.3-4.5 mm), slender flies, yellow to black, sometimes with dark spots or stripes; wing usually long and narrow. Arista pubescent or with longer rays; ocelli present; Oc-bristles present; P-bristles small, converging,

457 Typhamyza bifasciata (Wo male; Roháček 1992

rarely absent; 2-3 pairs of F-bristles, curving backward, usually preceded by 1 or more weaker bristles; interfrontal bristles absent or present; vibrissae present. Wing usually clear, sometimes marked; vein C with subcostal break; vein Sc incomplete; crossvein BM-Cu present but sometimes not complete; cell cup closed, rarely open. Some genera, including the more common *Anthomyza* and *Paranthomyza*, with femur of fore leg with a short spine on the ventral side, beyond the middle; tibiae without dorsal preapical bristle.

Biology: The larvae of most species develop in the stems and leaf sheaths of grasses and plants in marsh habitats such as species of *Typha, Juncus, Carex* and *Scirpus* without causing much damage; several species are known to mine in dicotyledons whereas others live in dead plants or as inquilines in existing galls. Two species of *Fungomyza* develop in the sporocarps of fungi. Adults of most species are usually found in moist habitats, both shaded and in the open such as damp meadows, marshes, bogs and damp deciduous or mixed forests with rich undergrowth; a preference for more dryer habitats such as dry grasslands is shown by some species of *Anthomyza* and the brachypterous *Stiphrosoma sabulosum* (Haliday).

Identification references: Andersson 1976, 1984a (Northwestern Europe); Soós 1981 (Central Europe); Roháček 1999, 2006 (revision); Stackelberg 1989ao (former USSR); CMPD: Roháček 1998b; CAT: Andersson 1984b.

Asilidae (key couplet 58; fig. 93, 123-124)

Systematics: Lower muscomorph Brachycera; superfamily Asiloidea; some 77 genera and about 540 species in Europe.

Characters: Small (3 mm) to very large (50 mm), but generally medium sized to large (8-20 mm) flies, varying in shape from long and slender to short and stout. The colour of the body and pubescence are variable from dark metallic lustrous to paler, with orange or yellow markings on body and wings, or resembling bees and wasps. Head dorsally concave in between the large, dichoptic eyes; ocelli implanted on an elevation between the eyes; antenna variable, third segment elongate, usually followed by 1 or 2 segment(s) which may, or may not, constitute an arista or style; face relatively long and nearly always with a distinct facial knob bearing long bristles, named the mystax or beard; mouthparts powerful and piercing. Wing usually clear, in some cases completely or partially tinged or darkened; cell m3 and cell cup open or closed. Legs long and powerful, often armed with strong bristles; tibiae with apical bristles or spurs; empodium bristle-like or absent.

458 Philonicus albiceps (Meigen), male; Verrall 1909

Biology: The larvae live in the soil and in decaying wood; as far as known they largely feed on the larvae of other insects. The adults are predators as well. Most species prefer open, sunny habitats and are mainly active during the warmest part of the day. They are able fliers that catch their prey - mainly other insects - on the wing. They are often seen on lookout posts like walls, branches and leaves, or on the ground, surveying their surroundings for possible prey flying by.

Identification references: Von der Dunk 1996 (Central Europe); Geller-Grimm 2003 (Germany); Hull 1962 (genera); Van der Goot 1985 (Northwestern Europe); Van Veen 1996 (Netherlands); Larsen & Meier 2004 (Denmark); Lehr 1996 (Asilinae); Richter 1989b (former USSR); Oldroyd 1969; Stubbs & Drake 2001 (British Isles); Weinberg & Bächli 1995 (Switzerland); CMPD: Majer 1997c; CAT: Lehr 1988.

Asteiidae (key couplet 104, 129; fig. 228-229, 258, 279-281)

Systematics: Acalyptrate Brachycera; superfamily Opomyzoidea; 4 genera and about 18 species in Europe.

Characters: Minute to small (1-3 mm), delicate, often weakly sclerotized flies with yellow-black or dark colours, male almost always with distinct interfrontal stripes. Arista sometimes bare or pubescent but usually with a characteristic zigzag

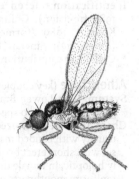

459 Asteia amoena Meigen, female; Papp 1973

of longer or shorter alternating rays, arista absent in the Mediterranean species *Asteia mahunkai* Papp; ocelli present; Oc-bristles present or absent; P-bristles weak, parallel, diverging or absent; 0-2 pairs of F-bristles, if present, curving forward or backward; interfrontal bristles sometimes present immediately above the antennae; scattered interfrontal setulae present; vibrissae absent to present. Wing unmarked; costa continuous; vein Sc incomplete; crossvein BM-Cu and cell cup absent. Tibiae without dorsal preapical bristle.

Biology: The larvae have been found under the bark of trees and have been reared from flower heads (inflorescences) and rotting or dry organic matter; larvae of *Leiomyza* have been associated with fungi. The adults are often found on flowers, on low vegetation, near sap exuding from tree wounds, and relatively often on windows indoors, or in case of one genus (*Leiomyza*) near or on mushrooms and bracket fungi in shaded places. The Mediterranean *A. mahunkai* is found near the sea, associated with *Inula crithmoides* (Compositae); other Mediterranean species (*Phlebosotera*, some *Asteia*) are often found under *Tamarix* trees. Adults of the species *Astiosoma rufifrons* Duda is attracted to wood ash after bonfires, especially during the evening.

Identification references: Chandler 1978 (British Isles); Merz 1996d (Northern and Central Europe); Stackelberg 1989as (former USSR); CMPD: Papp 1998e; CAT: Papp 1984h.

Atelestidae (key couplet 77, 79-80; fig. 165-166)

Systematics: Lower eremoneuran Brachycera; superfamily Empidoidea; traditionally part of the Empididae; in Europe 3 genera and 4 species. For a discussion about the phylogeny and classification see Sinclair & Cumming 2006.

Characters: Minute to small (1.5-3.5 mm), black flies. Head rather small; eyes big, holoptic in male; third antennal segment with an arista of 1-3 times its own length; mouthparts relatively short, directed forward or somewhat down. Wing clear or tinged, with a stigma spot between apex of veins SC and R1, stigma spot sometimes indistinct; cell dm present and vein M forked (*Meghyperus*), or cell dm absent and vein M not forked (*Atelestus*, *Nemedina*); cell cup normally larger than cells br and bm, if not then CuA2 convex and anal vein arched to wing margin; alula and large anal lobe usually present; cell bm and alula absent in *Nemedina*. Tibia of hind leg, and in some cases first tarsal segment as well, sometimes dilated and laterally compressed.

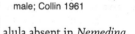

460 Atelestus pulicarius (Fallé male; Collin 1961

Biology: The biology of this small family is little known. The larvae are unknown. Adults of *Meghyperus* and *Atelestus* mainly occur in grassland or patches of herbs or on leaves in deciduous woodlands, often near water (in the Netherlands, *M. sudeticus* Loew was collected twice, in both cases near a river: on a gravel bank in the Geul and on a sandbar in the Slinge; Van Aartsen, pers. comm. 2005); adults of *Nemedina* have been collected along brooks. Both sexes fly in swarms. Because of the anatomy of their mouthparts, the adults are presumed to be flower visitors (Sinclair & Cumming 2006).

Identification references: Chandler 1981 (*Nemedina*); Chvála 1983 (revision); Collin 1961 (British Isles); Gorodkov & Kovalev 1989 (former USSR); Sinclair & Papp 2004 (*Nemedina*); Sinclair & Cumming 2006 (classification); CMPD: no contribution; CAT: Chvála 1989b.

Athericidae (key couplet 48; fig. 95, 105-106)

Systematics: Lower Brachycera; superfamily Tabanoidea; 4 genera and about 10 species in Europe.

Characters: Medium sized (7.5-10 mm), brownish to black flies, usually with banded markings on the abdomen. Body densely set with short erect bristles. Ocelli present; eyes in male nearly holoptic, in female dichoptic; antenna short, third segment reniform; mouthparts powerful, at most as long as the height of the head. Wing long and wide, often with dark markings;

461 Eggcluster of Atherix ibis (Tournier in Séguy 1950

veins R1 and R2+3 ending in the costa at the same point; cell m3 open; cell cup closed; legs relatively long; tibiae of mid and hind legs with apical spurs; empodium pulvilliform.

Biology: The larvae are aquatic and prey on larvae of other water insects. Prior to pupation, they move to nearby land. Adults are always encountered near running water, usually on the vegetation close by. Eggs are deposited in suitable spots over the water, such as bridges and overhanging branches. In some species, all females deposit their eggs together before they die. These clusters of eggs and dead flies may include up to several tens of thousands of females. The adults largely feed on nectar but some species feed on blood of humans and cattle.

Identification references: Nagatomi 1984 (genera); Nartshuk 1989b (former USSR); Stubbs & Drake 2001 (British Isles); Stuckenberg 1973 (revision); Thomas 1997 (species); CMPD: Rozkošný & Nagatomi 1997b; CAT: Majer 1988a.

Aulacigastridae (key couplet 115; fig. 249-250)

Systematics: Acalyptrate Brachycera; superfamily Opomyzoidea; in Europe 1 genus, *Aulacigaster*, with 4 species.

Characters: Small (2-5 mm), mainly dark coloured flies with an extensive grey pruinosity, pale tarsi and red eyes. Arista pubescent or with longer rays; ocelli present; Oc-bristles reduced or absent; P-bristles absent; 2 pairs of F-bristles, the lower pair curving forward and slightly inward, the upper pair backward and further from the eye margin than the lower pair; interfrontal bristles absent; vibrissae present. Wing clear; costa with humeral and subcostal breaks; vein Sc thin but complete, veins Sc and R1 reaching the costa separately, but just before they merge for a short distance or touch each other, apical part of vein Sc more pallid; crossvein BM-Cu absent; cell cup closed. Tibiae without dorsal preapical bristle.

462 Aulacigaster leucopeza (Meigen), male; Papp 1973

Biology: The larvae of *Aulacigaster* live in sap exuding from wounds in deciduous and coniferous trees, where they feed on micro-organisms. The adults are usually found near the larval sites but are also attracted to other fermenting fluids or substrates; they are known to hibernate in sheltered places in forests.

Identification references: Kassebeer 2001 (species); Papp 1998a (revision); Stackelberg 1989ar (former USSR); CMPD: Papp 1998d; CAT: Papp 1984f.

Axymyiidae (key couplet 21; fig. 56)

Systematics: Nematocera; superfamily Axymyioidea; in Eastern Europe 1 species, *Mesaxymyia kerteszi* (Duda).

Characters: Medium sized to large (9-12 mm), robust, dark, *Bibio*-like species. Ocelli conspicuous, eyes divided longitudinally in two by a distinct suture; male holoptic; antenna compact, with 14-17 segments. Wing darkened, with anal lobe and characteristic venation. Legs relatively short; tibiae without apical bristles or spurs.

Biology: The larvae of this small family live in tunnels in moist wood of tree trunks lying on the ground. Up to 150 tunnels per metre of trunk length have been recorded. They are bored from the outside inward, each mine being inhabited by a single larva. Pupation takes place within the trunk as well. Adults are found in woodland, in particular in shaded spots alongside rivers and in lowlands. On sunny days, they fly in swarms about 2 metres above the ground.

463 Axymyia furcata McAtee, male; Wood 1981 (North America)

Identification references: Mamaev 1989c (former USSR); CMPD: Krivosheina 2000; CAT: Mamaev & Krivosheina 1986c.

Bibionidae (key couplet 28; fig. 63-64)

Systematics: Nematocera; superfamily Bibionoidea; traditionally this family includes the Pleciidae and Hesperinidae as well; in Europe 2 genera, *Bibio* and *Dilophus*, and about 47 species.

Characters: Small to large (2-15 mm), robustly built Nematocera. Usually densely set with fine hairs; body colour pale brown to black, in some cases (partially) red or yellow. Ocelli conspicuous; in male eyes holoptic and consisting of two separate parts, in female eyes dichoptic and simple; antenna inserted rather low, with 6-12 segments. Wing wide, with distinct anal lobe; wing clear or darkened. Tibia of fore leg usually swollen, in *Bibio* with a long, stout spur and an apical projection; in *Dilophus* with several rows of short, stout spurs.

464 Bibio johannis (Linnaeus), male; Freeman & Lane 1985

Biology: The larvae are largely found in the topsoil of moist grasslands and woodland and in decaying wood. They appear to feed mainly on rotting plant matter; in some cases also on living roots or stems, thereby causing damage to valuable crops that can be substantial if they are present in large numbers. Densities up to 37,000 larvae per square metre have been recorded. The adults are mainly found in spring, usually present for only a few weeks every year, sometimes in very large numbers. The males often fly or hover together in swarms. The adults, those of *Dilophus* in particular, often visit flowers as well.

Identification references: Duda 1930 (Lindner); Freeman & Lane 1985 (British Isles); Haenni 1982 (*Dilophus*); Krivosheina 1989c (former USSR); Mikolajczyk 1977 (Poland); Tomasovic 2000 (Belgium); Zeegers 1997, 1998 (Netherlands); CMPD: Skartveit 1997; CAT: Krivosheina 1986b.

Blephariceridae (key couplet 19; fig. 54-55)

Systematics: Nematocera; superfamily Blephariceroidea; 5 genera and about 38 species in Europe.

Characters: Small to large (3-15 mm), slender, delicate, dull brown to grey coloured, long-legged Nematocera. Ocelli present; eyes often comprising two differently faceted parts; antenna compact, with 13-15 segments, sometimes less. Wing with a grid of secondary veins and a conspicuous anal lobe.

Biology: The larvae live in torrential streams or their splash zones and in waterfalls in hill and mountain areas. The larvae feed on algae, bacteria and organic matter they scrape from rocky substrates. The adults are often found near the larval habitats. When resting, they often hang by their fore legs, using leaves, projecting rocks, bridges etc. as their point of attachment. The females prey on small insects like Ephemeroptera, Plecoptera, Empididae (Diptera) or even Blephariceridae. In some cases the male is provided with long mouthparts used to feed on nectar, but most males have reduced mouthparts.

465 Liponeura brevirostris Loew, male; Séguy 1940

Identification references: Frutiger & Jolidon 2000 (larvae, pupae, Central Europe); Nartshuk 1989a (former USSR); CMPD: Courtney 2000; CAT: Zwick 1992.

Bolitophilidae (key couplet 36; fig. 81)

Systematics: Nematocera; superfamily Sciaroidea; traditionally part of the Mycetophilidae; in Europe 1 genus, *Bolitophila*, with about 36 species.

Characters: Small to medium sized (4-7 mm), delicate, greyish brown Nematocera with long, slender legs and abdomen and a characteristic wing venation (e.g. basal position of crossvein BM-Cu). Head and eyes small; ocelli present; antenna conspicuously long and slender, sometimes with long verticils.

466 Bolitophila cinerea Meigen, female; Hutson et al. 1980

Thorax much vaulted. Wing clear or with banded markings along the crossveins; anal lobe weak or absent. Coxae elongate; tibiae with apical bristles or spurs.

Biology. The biology of this small family is little known. The larvae all develop in terrestrial mushrooms. The adults are mainly found in moist surroundings and alongside water in well wooded areas.

Identification references: Chandler & Ribeiro 1995 (Atlantic islands); Hutson et al. 1980 (British Isles); Kurina & Schacht 2003 (synopsis identification references); Stackelberg 1989g (former USSR); A.I. Zaitzev (Eastern Europe); CMPD: Søli et al. 2000; CAT: Plassmann 1988.

Bombyliidae (key couplet 61; fig. 128-130)

Systematics: Lower muscomorph Brachycera; superfamily Asiloidea; traditionally this family includes the Mythicomyiidae; some 52 genera and about 340 species in Europe.

Characters: Minute to large (1-20 mm) flies, often thorax and abdomen covered with fine hairs or in part scale-like hairs. Body usually broad, bee-like but sometimes slender; colour brown, grey, black or with a conspicuous pattern of contrasting colours; in some cases the body is devoid of hair covering and shining. Eyes large, male often holoptic; antenna with 3-4 segments, the third segment largest and the fourth making up the flagellum, there may also be a very small transparent style. Mouthparts without or (usually) with a long, sucking proboscis. Wing occasionally clear but most often with a conspicuous pattern; veins R2+3 and R4 often strongly curving forward; cell below discal cell (cell cua1) open; cell cup usually open; wings held horizontally when at rest. Legs long and slender; tibiae with apical bristles or (rarely on mid leg) spurs; empodium bristle-like or absent.

467 Anthrax anthrax (Schrank), male; Tóth 1977

Biology: The larvae are parasitoids or predators of eggs, larvae and pupae of other insects and egg masses of spiders. Many species parasitise bees in which case the female deposits her eggs in or near the nest of the bees. The first instar enters the nest and feeds on the contents of the cells, including the bee larvae. Other species parasitise solitary wasps, antlions, pupae of moths and beetles in the soil or are predatory on eggs of grasshoppers. Adults usually feed on nectar and the females on pollen. In a few species, the short-lived adults do not feed at all. Bombyliidae are generally good fliers and many are able to hover at the same spot in front of a flower in order to feed while hovering. Species not possessing a long proboscis often settle on the flowers themselves. The adults are mainly active on warm, sunny days and prefer open, dry habitats.

Identification references: Van der Goot & Van Veen 1996 (Northwestern Europe); Greathead & Evenhuis 2001 (key genera); Oldroyd 1969; Stubbs & Drake 2001 (British Isles); Tóth 1977 (Hungary); Trojan 1967 (Poland); V.F. Zaitzev 1989b (former USSR); CMPD: Greathead & Evenhuis 1997; CAT: V.F. Zaitzev 1989d; Evenhuis & Greathead 1999.

(Borboridae): name used in the past for Sphaeroceridae.
(Borboropsidae): part of the Heleomyzidae.

Brachystomatidae (key couplet 61; fig. 131)

Systematics: Lower eremoneuran Brachycera; superfamily Empidoidea; separated from the Empididae by Sinclair & Cumming 2006; except for *Brachystoma*, the family keys out here under the Empididae by lack of suitable easy to use characters for inclusion in the key; in Europe 13 species in 5 genera: *Brachystoma* (2), *Gloma* (1), *Heleodromia* (7), *Pseudoheleodromia* (1) and *Trichopeza* (2).

Characters: Small to medium (2-7 mm), slender, yellowish to dark flies. Head small with relatively large eyes, in some cases

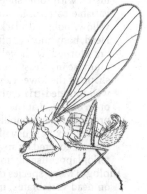

468 Heleodromia immaculata Haliday, male; Collin, 1961

male holoptic; antenna usually with 3 segments of which the third is the largest and bears a long usually apical arista or style; mouthparts stout, projecting downwards toward the fore legs. Wing clear or infuscate; vein R4+5 forked except in *Heleodromia*; cell dm always present. Legs strong, long and slender; fore legs somewhat raptorial.

Biology: The larvae are assumed to prey on other insects, although none are known; assumed to live in damp soil and likely not aquatic. The adults are predators, mainly on flying Diptera. Flies are associated with mature forests, some swept from wet depressions. Aerial mating swarms are not known.

Identification references: Collin 1961 (British Isles); Gorodkov & Kovalev 1989 (former USSR); Sinclair & Cumming 2006 (classification); Wagner 1985 (*Heleodromia*); CMPD: no contribution; CAT: Chvála & Wagner 1989.

Braulidae (key couplet 207; fig. 439-440)

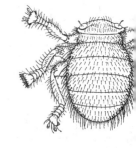

Systematics: Acalyptrate Brachycera; superfamily Carnoidea; in Europe 1 genus, *Braula*, with 3 species.

Characters: Minute to small (1.2-2.5 mm), dorsoventrally flattened, reddish to dark brown flies, with the body evenly covered by rather long bristles. Wings, halteres and scutellum absent. Eyes reduced to a pale spot just behind the sunken antenna; ocelli absent. Thorax small, closely adpressed to the abdomen. Legs relatively long and strong; tarsal claws transformed into a comb of over 20 teeth, curving inward.

Biology: The larvae and adults live as commensals in the nests of bees. The larvae feed on honey, pollen and organic debris. The adults attach themselves to the body of the queen (rarely of a worker), absorbing food and liquids from the mouth opening of the bee. The only species of importance to bee-keeping is *Braula coeca* Nitzsch. It was not a real nuisance but nevertheless has disappeared from many countries on account of measures taken against *Varroa* mites which cause severe damage in beehives.

469 Braula coeca Nitzsch, female; Séguy 1950

Identification references: Dobson 1999 (British Isles); Örösi Pál 1966 (revision); Stackelberg 1989ax (former USSR); CMPD: Papp 1998g; CAT: Papp 1984n.

Calliphoridae (key couplet 185-187, 190; fig. 12, 14, 179, 401, 409, 411, 413, 417-418, 421-423)

Systematics: Calyptrate Brachycera; superfamily Oestroidea; in Europe some 22 genera and about 115 species.

Characters: The family most probably does not represent a natural systematic grouping and hence is not easy to characterise unequivocally. It includes small to large (4-16 mm), usually robust flies. Many species have a greenish or bluish metallic lustre, with some silvery, yellowish, or golden pruinosity in a variable pattern; in some species the body colour is greenish yellow, polished black, occasionally without metallic sheen. Head, body and legs often with strong bristles, in *Pollenia* thorax clothed with yellow curly hairs. Arista usually long plumose in at least the basal two-thirds, rarely pubescent. The shape of the lower calypter is strongly variable, ranging from well rounded-triangular to elliptic; wing always with vein M1 curving forward; cell r4+5 open or closed at the wing margin, rarely with a very short petiole. Meral bristles always present and subscutellum distinctly absent, that is, the part below the scutellum is flat, not convex; it is usually separated from the scutellum by a membranous part; posterior spiracle covered by an anterior and posterior lappet, usually of unequal size with the posterior lappet usually forming an operculum.

470 Calliphora uralensis Villen male; Stackelberg 1956

Biology: Many species develop in carrion of various origin. Eggs are frequently deposited on dead vertebrates, making some calliphorids important in forensic entomology. The larvae can be: true obligate agents of myiasis in vertebrate animals (e.g. *Lucilia bufonivora* Moniez on amphibia) and humans (*Dermatobia*, in the Tropics); necrophagous to facultative agents of myiasis (*Calliphora, Cynomya, Lucilia*, e.g. *L. sericata* (Meigen) in sheep, *Phormia, Protophormia*); parasites of earthworms and other worms (*Bellardia, Onesia,* Polleniinae), snails and slugs (*Angioneura, Eurychaeta, Melinda, Melanomya*), or egg clusters of grasshoppers (*Stomorhina*); or parasites sucking blood from bird nestlings (*Protocalliphora*). The adults are often found on flowers, plant detritus, carrion and dung. Two species, *Pollenia rudis* (Fabricius) and *Protophormia terraenovae* (Robineau-Desvoidy) sometimes occur indoors in large numbers, looking for a shelter to survive the winter.

Identification references: Crosskey & Lane 1993 (medically important genera); Erzinclioglu 1996 (British Isles); Grunin 1989c (former USSR); Huijbregts 2002 (survey Netherlands); Rognes 1991 (Northwestern Europe), 2002 (Israel); CMPD: Rognes 1998; CAT: Schumann 1986.

(Calobatidae): part of the Micropezidae.

Camillidae (key couplet 95; fig. 191, 207, 209-210)

Systematics: Acalyptrate Brachycera; superfamily Ephydroidea; in Europe 1 genus, *Camilla*, with 8 species.

Characters: Small (2-3.5 mm), slender, dark and lustrous flies. Arista with long rays above and shorter rays below; ocelli present; Oc-bristles present; P-bristles converging; 2 pairs of F-bristles, the lower pair curving forward, the upper pair backward; scattered interfrontal setulae present; vibrissae present. Wing unmarked; costa with both humeral and subcostal breaks; vein Sc incomplete; crossvein BM-Cu absent; cell cup open. Femur of fore leg with a spine on its ventral side, beyond the middle; only tibia of mid leg with dorsal preapical bristle.

Biology: The biology of this small family is little known. Presumably the larvae feed on organic matter (dung and detritus); they have been collected in soil at the entrances of rabbit burrows and in the nest of a field vole (*Microtus agrestis* (Linnaeus)). The adults are frequently found on flowers, on low vegetation or on bare soil, near the entrance of burrows, and are often found on windows.

471 Camilla glabrata Collin, female; Papp 1973

Identification references: Beuk & De Jong 1994 (Netherlands); Papp 1985 (review); Stackelberg 1989ay (former USSR); CMPD: Papp 1998k; CAT: Papp 19840.

Campichoetidae (key couplet 100; fig. 208, 223-225)

Systematics: Acalyptrate Brachycera; superfamily Ephydroidea; traditionally part of the Diastatidae; in Europe 7 species in 2 genera, *Campichoeta* (6) and *Euthychaeta* (1).

Characters: Small (2.5-4 mm) flies, body colour dark with head and legs partially yellow. Arista short pubescent to moderately plumose; ocelli present; Oc-bristles present; P-bristles converging; 2-3 pairs of F-bristles, 1 pair curving forward, the other pair(s) backward; interfrontal bristles absent; vibrissae present. Wing partially tinged, in particular along the costa and with a central pale spot, or wing completely tinged brownish to greyish; costa with a subcostal break; vein Sc incomplete; crossvein BM-Cu present; cell cup closed. Tibiae with dorsal preapical bristle.

Biology: The larvae belonging to this family are unknown; they have been suggested to be saprophagous in rotten wood but this has not been confirmed. The adults are found on low vegetation in woodland areas; *Campichoeta* is found in moist woodlands in particular.

472 Campichoeta punctum Meigen, male; Papp 1973

Identification references: Chandler 1986 (British Isles), 1987 (revision); CMPD: Chandler 1998b; CAT: Papp 1984p (as subfamily of Diastatidae).

Canacidae (key couplet 100; fig. 238a-b)

Systematics: Acalyptrate Brachycera; superfamily Carnoidea; in Europe 4 species in 2 genera, *Canace* (3) and *Xanthocanace* (1).

Characters: Minute to small (1.6-5 mm), robust, yellow to greyish black, pruinose flies, generally with conspicuous whitish to greyish markings. Head large with small antenna; arista bare to pubescent; ocelli present; Oc-bristles present; P-bristles absent but often replaced by diverging pseudopostocellar bristles; 2-5 pairs of F-bristles, curving outward; interfrontal bristles present; gena with 1 or more upcurved bristles; vibrissae present. Wing unmarked; costa with a subcostal break; vein Sc parallel to vein R1 and merging with that vein just before the costa; crossvein BM-Cu present; cell cup closed. Tibiae without dorsal preapical bristle.

473 Canace salonitana Strobl, mal
Canzoneri & Meneghini 1983

Biology: All the European species live in the tidal zone where they feed on algae and organic matter.

Identification references: Canzoneri & Meneghini 1983 (Italy); Mathis 1982 (*Canace*), Mathis & Freidberg 1982 (*Xanthocanace*); Stackelberg 1989aw (former USSR); CMPD: Mathis 1998; CAT: Cogan 1984a.

Canthyloscelidae (key couplet 24; fig. 61-62)

Systematics: Nematocera; superfamily Scatopsoidea; the families Canthyloscelidae and Synneuridae are often considered one family, given various names; in Northern, Eastern and Central Europe 2 species: *Hyperoscelis eximia* (Boheman) and *H. veternosa* Mamaev & Krivosheina.

Characters: Small to medium sized (2.5-9 mm), stout Nematocera. Body colour brown to black with less dark spots. Ocelli present but medial ocellus small or absent; eyes large, holoptic and narrowly separated below the antennae; antenna with 12-16 segments, the terminal segment elongate. Wing slightly tinged, in particular along the costa and at the wing tip. Legs relatively short and stout; hind leg with apical half of femur swollen and tibia curved.

Biology: Almost nothing is known about the biology of the two European species of *Hyperoscelis*. The larvae have been found in moist, rotting wood of stumps and fallen tree trunks. The adults are very rare (at least in collections) and it is assumed that they occur especially in old undisturbed woodlands, in close vicinity of the breeding sites of the larvae.

474 Hyperoscelis veternosa Mama
Krivosheina; Haenni 1997

Identification references: Andersson 1982 (Sweden); Hutson 1977 (revision); Krivosheina 1989b (former USSR); CMPD: Haenni 1997b; CAT: Mamaev & Krivosheina 1986b.

Carnidae (key couplet 127, 131, 215; fig. 272a-b, 286-289, 449)

Systematics: Acalyptrate Brachycera; superfamily Carnoidea; 3 genera and about 40 species in Europe.

Characters: Minute to small (1-2.5 mm), more or less lustrous black flies. Arista pubescent; ocelli present; Oc-bristles present; P-bristles diverging, parallel, converging, P-bristles absent in *Carnus*; 4 pairs of F-bristles, lower 2 pairs

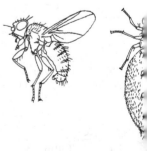

475 Carnus hemapterus Nitzsch, m
shortwinged female; Séguy 1951

curving inward, upper 2 pairs obliquely out-backward; interfrontal bristles present; vibrissae present. Wing not marked; costa with both humeral and subcostal breaks; vein Sc incomplete; crossvein BM-Cu absent, crossvein DM-Cu absent in *Carnus*; cell cup open or absent, in *Carnus* the wing might be shed off near the base. Tibia without dorsal preapical bristle.

Biology: The larvae of most species develop in all sorts of organic matter of animal origin, in particular in droppings and dung, carrion, and nests of birds; only a few species can develop in other decaying media, including rotten fungi or plant matter. Adults usually occur on or near the breeding substrates, but visit flowers as well, with a preference for umbellifers. The species of *Carnus* live exclusively in the nests of birds, especially those breeding in bushes, trees and in (tree)holes, where the larvae feed on nest debris. The adults emerge from the pupa with normal wings, fly about until they have found another suitable nest and then shed their wings. They attach themselves to the nestlings and feed on the flakes and moisture of their skin, and they are suspected to feed on blood as well.

Identification references: Grimaldi 1977 (*Carnus*); Hennig 1972 (review); CMPD: Papp 1998b; CAT: Papp 1984k.

Cecidomyiidae (key couplet 16, 22, 193, 201-202; fig. 49, 57, 424, 437)

Systematics: Nematocera; superfamily Sciaroidea; some 280 genera an about 1640 species in Europe.

Characters: Minute to small (0.5-3 mm), rarely larger (up to 8 mm, wing length up to 15 mm), delicate Nematocera. Eyes holoptic with the exception of a few genera with reduced wings; mouthparts reduced; antenna in general conspicuously long, usually with 12-14 segments, sometimes less, sometimes more, up to 40; antennal segments bearing differently-shaped sensoria, in some Porricondylinae and in all Cecidomyiinae these are thread-like, in the supertribe Cecidomyiidi the sensoria (some or most) in the shape of long loops; ocelli present (in the Lestremiinae) or absent. Wing usually clear, in a few species with a pattern; number of longitudinal veins reduced; costa usually with a break just beyond vein R5. Legs long, tibiae lacking apical bristles.

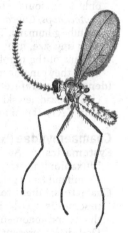

476 Contarinia tritici (Kirby), male; Balachowsky & Mesnil 1935

Biology: The larvae and adults can be found in a large array of habitats. Three main groups can be distinguished with respect to the feeding mode of the larvae: 1) larvae living in, and feeding on, mushrooms, bracket fungi, dying and decaying wood, or other organic substrates (soil, litter, plant remains, etc.); 2) larvae that cause damage to higher plants, either living free, in galls, or as leaf-miner; 3) larvae preying on invertebrates, in particular on (larvae of) small insects. Of the three subfamilies, Lestremiinae, Porricondylinae and Cecidomyiinae, only representatives of the last cause the formation of galls. Several of these species can cause severe damage in agriculture and horticulture.

Identification references: Berest & Mamaev 1989 (genera Lestremiinae); Jaschhof 1998 (Lestremiinae); Mamaev 1989b (former USSR), 1990 (genera Porricondylinae), 1991 (Holoneurini); Möhn 1966-1971 (Lindner); Nijveldt 1969 (species of economic importance); CMPD: Skuhravá 1997; CAT: Skuhravá 1986; Gagné 2004; Skuhravá et al. 2006 (Denmark).

Ceratopogonidae (key couplet 18; fig. 53)

Systematics: Nematocera; superfamily Chironomoidea; some 30 genera and about 590 species in Europe.

Characters: Minute to small (1-5 mm), slender to robust, often dark coloured Nematocera, most capable of piercing (e.g. *Dasyhelea* not capable), allied to the Chironomidae, but less slender and with broader wings. Ocelli absent; antenna with 13-15 segments which in the male are more or less pubescent.

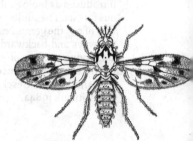

477 Culicoides pulicaris (Linnaeus); female; Séguy 1951

Wing usually clear, in species of some genera with a pattern; usually the R-veins shortened, enclosing 1 or 2 small cell(s), and more strongly developed than the other veins. Legs simple, sometimes with ventral spines on femora, especially of the fore leg, spines usually more developed in females.

Biology: The larvae of most genera are aquatic, living in wells, brooks, rivers, lakes (recorded as deep as 43 metres below the surface), ponds, pools, water in rot holes in trees, in leaf axils and other small bodies of water with a high content of organic matter; in the foam at the surface of aquariums, etc.; larvae of some species are known from saline waters. Larvae of other genera inhabit more terrestrial habitats such as wet meadows, peat bogs, fenland, moist soil, leaf litter, moss, e.g. *Sphagnum*, under tree bark, in rotting roots and bracket fungi, in exuding tree sap, in drying horse manure and cow dung, in ants nests. The larvae feed on a diversity of food sources; some feed on minute organic particles or micro-organisms, others are predators of insect larvae, worms and small invertebrates. The adults of most species are largely found at a distance from the larval habitats. Adults can occur in large numbers, in particular when evening approaches, but many species, e.g. those of *Leptoconops*, are mainly active by day. The males often fly in swarms. Adults of both sexes feed on flower nectar and honeydew. In some cases, e.g. *Dasyhelea*, this is the only food source, but usually the females feed on body fluids of insects; those of *Leptoconops, Culicoides* and *Forcipomyia* (*Lasiohelea*) feeding on the blood of mammals including humans. This blood serves as a source of protein for egg production. Despite their tiny size, man and animals alike can suffer substantially from ceratopogonids, on account of the number of ceratopogonids being so vast and because while feeding on blood they can transmit diseases.

Identification references: Boorman 1993 (genera feeding on blood); Remm 1989 (former USSR); Szadziewski et al. 1997 (genera); Wirth & Grogan 1988 (World); CMPD: Boorman 1997; CAT: Remm 1988; Borkent & Wirth 1999.

Chamaemyiidae (key couplet 119, 160; fig. 257a-b, 344-346)

Systematics: Acalyptrate Brachycera; superfamily Lauxanioidea; traditionally this family includes the Cremifaniidae; 9 genera and about 110 species in Europe.

Characters: Minute to small (1-5 mm), usually silvery grey flies, in some cases with brown, grey or black markings on the head, thorax and abdomen. Arista bare to pubescent; ocelli present; Oc-bristles present, in some cases curving backward, or absent; P-bristles converging or reduced to absent; 0-3 pairs of F-bristles, if present, curving backward; especially in the large genus *Leucopis* dorsal bristles of the head largely absent; scattered interfrontal setulae present; vibrissae present or absent, often a series of vibrissa-like bristles present. Wing unmarked, in some cases costal area darkened; costa continuous; vein Sc complete; crossvein BM-Cu present; cell cup closed. Tibiae without dorsal preapical bristle.

478 Chamaemyia polystigma (male; Papp 1979

Biology: The larvae prey on aphids, scale insects and their allies, e.g. gall-building Psyllidae. Some species have been artificially introduced as biological control agents against pest woolly aphids, especially in coniferous forests. The adults are found especially in grassland habitats and near the larvae where flies of e.g. the genus *Leucopis* feed on honeydew. The adults of this genus are able to walk sideways and backwards with ease, probably an adaptation enabling them to avoid the ants trying to chase off or kill everything that comes near 'their' aphids.

Identification references: Tanasijtshuk 1986, 1989d (former USSR), 1992 (genera); Tanasijtshuk & Beschovski 1991 (Eastern Europe); CMPD: McLean 1998a; CAT: Tanasijtshuk 1984a.

Chaoboridae (key couplet 14; fig. 39, 46)

Systematics: Nematocera; superfamily Culicoidea; 3 genera and 9 species in Europe.

Characters: Small to medium sized (2-10 mm), delicate, pale yellow, brown or grey Nematocera, resembling mosquitoes (Culicidae). Antennae, wings, legs and abdomen long and slender. Ocelli absent; antenna always with 15 segments, plumose in the male; mouthparts protruding but much less than in mosquitoes. Upper part of thorax with few to many bristles but with 2 pairs of bare stripes. Wing narrow; venation as in Culicidae.

Biology: Larvae with a remarkably transparent body, especially in *Chaoborus*, giving rise to the name of phantom larvae. They are aquatic and prey on small invertebrates. The larvae of *Cryophila*, *Mochlonyx* and *Chaoborus (Shadonophasma)* occur in temporary puddles, those of *Chaoborus (Chaoborus)*, *Chaoborus (Peusomyia)* and *Mochlonyx* in various permanent standing waters (lakes, ponds, swamps, peat bogs, etc.). The adults are largely found near the larval habitats and often fly in swarms.

Identification references: Saether 1997a (species), 2002 (species); Stackelberg 1989d (former USSR); CMPD: Saether 1997b; CAT: Wagner 1990b.

479 Chaoborus crystallinus (De Geer), male; Saether 1997b

Chironomidae (key couplet 18, 197; fig. 52, 433-434)

Systematics: Nematocera; superfamily Chironomoidea; in Europe some 180 genera and about 1190 species.

Characters: Minute to medium sized (1-10 mm) Nematocera with long, slender wings, abdomen and legs. Ocelli absent; antenna generally with 13 to 17 segments with conspicuously long verticils in the male. Wing usually clear, in some cases with dark markings; anterior veins more strongly developed than posterior ones.

Biology: Notwithstanding a considerable number of species of which the larvae are terrestrial, the larvae of the majority of the species are aquatic or semi-aquatic. They inhabit a large array of habitats ranging from the pristine to the most severely polluted and including brackish, marine and saline habitats. The larvae feed on algae, small or larger particles of organic matter and on plankton; some are carnivorous (Tanypodinae and some others). The composition of the larval fauna is a major factor in ascertaining the quality of the water. The adults, likewise, occur in a wide range of habitats, largely those near water. In some cases, the males fly in conspicuously large swarms.

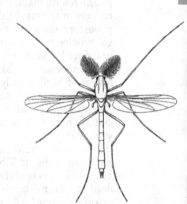

480 Microtendipes nervosus (Staeger), male; Séguy 1951

Identification references: Klink & Moller Pillot 2003 (larvae; Northwestern Europe); Langton & Visser 2003 (pupae; Northwestern Europe); Lindegaard 1997 (subfamilies, tribes); Pasini & Ferrarese 1998 (Italy); Pinder 1978a, 1978b (British Isles); Shilova 1989 (former USSR); Wiederholm 1983 (larvae), 1986 (pupae), 1989 (adults); Wilson & Ruse 2005 (pupae British Isles); CMPD: Saether et al. 2000; CAT: Ashe & Cranston 1990.

Chiropteromyzidae (key couplet 98; fig. 216a-b)

Systematics: Acalyptrate Brachycera; superfamily Sphaeroceroidea; traditionally part of the Heleomyzidae; in Europe 2 species: *Chiropteromyza wegelii* Frey and *Neossos nidicola* Frey.

481 Chiropteromyza wegelii Frey, male; original (drawing by Dick Langerak)

Characters: Minute (1.2-2 mm), black, metallic lustrous flies. Arista bare to pubescent; ocelli present; Oc-bristles present; P-bristles converging; 2 pairs of F-bristles, the lower pair curving outward; interfrontal setulae present; vibrissae present, in *Chiropteromyza* with a 5-6 stronger bristles on the gena behind the vibrissa. Wing unmarked, whitish; costa thin in front and after crossvein H and with a subcostal break; vein Sc incomplete, merging apically with vein R1; crossvein BM-Cu present; cell cup closed. Dorsal preapical bristle present, sometimes weaker on the tibiae of fore and hind legs. The original description of *C. wegelii* Frey, 1952, mentions that preapical bristles are absent. On our request, Pekka Vilkamaa, curator of the Finnish Museum of Natural History at Helsinki has examined the type material of *C. wegelii*. He has ascertained that the preapical bristles are present (see also Haenni 1988).

Biology: The biology of this small family is little known. The adults have been reared from nests of birds and from the excrement of bats.

Identification references: CMPD: Papp 1998i; CAT: Soós & Papp 1984.

Chloropidae (key couplet 130, 166, 214; fig. 240, 282-285, 361-362, 448)

Systematics: Acalyptrate Brachycera; superfamily Carnoidea; some 65 genera and about 395 species in Europe.

Characters: Minute to medium sized (1-5 mm. sometimes up to 8 mm), stout to slender flies, usually with a conspicuously large ocellar triangle and few bristles. Often yellowish or greenish with black, brown or red markings; sometimes body entirely or in part shining black or dusted grey. Arista bare to pubescent, the arista itself sometimes enlarged; ocelli present; Oc-bristles present or reduced; P-bristles parallel, converging or reduced; F-bristles usually reduced, sometimes several forward, outward, or backward curved pairs present; dorsal region of head usually closely packed with smaller setulae; vibrissae present in some cases, usually reduced or absent. Proepisternum with a sharp anterior carina. Wing generally

482 Camarota curvipennis (La male; Balachowsky & Mesnil 1

clear, in only a few species with a pattern; costa virtually always with a subcostal break; vein Sc incomplete; crossvein BM-Cu absent; vein CuA1 usually with a flexure halfway cell bm + dm; often vein R4+5 and sometimes also vein M1 curving toward the costa; cell cup open or absent. Tibiae without dorsal preapical bristle.

Biology: The larvae of the Chloropidae generally feed on living plant or animal tissues or on dead organic matter (insect frass, rotting plant material, mushrooms and bracket fungi, carrion, nests of birds, etc.). Some feed on root aphids or on eggs of spiders and grasshoppers. Some woodland species live in decaying wood or below the bark of trees. Larvae feeding on vegetable material prefer the stems and flowers of grasses, reeds and sedges where they may cause galls; a number of these gall-forming species can cause damage to agriculture. The adults are encountered chiefly in open habitats such as meadows and marshland, although some species are typical of woodland. Several species visit flowers; from late summer onward *Thaumatomyia notata* (Meigen) may occur indoors, where it seeks refuge for the winter.

Identification references: Andersson 1977 (genera); Collin 1945 (Oscinellinae British Isles); Ismay 1999 (Chloropinae British Isles); Nartshuk 1987 (general); Nartshuk et al. 1989 (former USSR); CMPD: Ismay & Nartshuk 2000; CAT: Nartshuk 1984c.

Chyromyidae (key couplet 119, 125; fig. 256, 270)

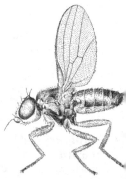

Systematics: Acalyptrate Brachycera; superfamily Sphaeroceroidea; 3 genera and about 60 species in Europe.

Characters: Minute to medium sized (0.5 -8 mm), usually yellow to pallid flies with yellow bristles and hairs, in some cases with brown, grey or black markings on head, thorax and

483 Aphaniosoma hungaricum female; Papp 1981

abdomen. Arista bare to pubescent; ocelli present; Oc-bristles present; P-bristles usually present, converging; 2-6 pairs of F-bristles, rarely absent, the lowermost pair sometimes curving inward; scattered interfrontal setulae present; vibrissae present but weak. Wing unmarked; costa with a subcostal break, vein Sc incomplete or complete, parallel to R1 up to the costa or apical part pallid; crossvein BM-Cu present, cell cup closed. Tibiae without dorsal preapical bristle.

Biology: The larvae are unknown but are believed to develop in a variety of organic detritus, such as nests and excrement of birds, mammal burrows, bat dung, rotting wood, below the bark of trees, etc. Adults are found on flowers, shrubs and grasses; in dry habitats, but also along river banks, in saltmarshes, in caves, and often indoors on windows. Species of *Aphaniosoma* and *Gymnochiromyia* preferentially occur in saltmarshes and semi-arid habitats and are frequent of flowering vegetation; *Aphaniosoma* species can also be found on *Tamarix* trees along the Mediterranean sea coast, often in very large numbers; *Chyromya* species seem to be associated with open deciduous woodland.

Identification references: Andersson 1971 (Sweden); Ebejer 1998a (*Aphaniosoma*), 1998b (*Gymnochiromyia*), 2005 (Italy); Merz 1998a (genera Europe, species Switzerland); Soós 1981 (Central Europe); Stackelberg 1989aq (former USSR); CMPD: Wheeler 1998; CAT: Soós 1984l.

Clusiidae (key couplet 88, 114; fig. 189-190, 245-246)

Systematics: Acalyptrate Brachycera; superfamily Opomyzoidea; 5 genera and about 15 species in Europe.

Characters: Minute to medium sized (1.5-8 mm), slender flies. Body dull to lustrous, yellow to dark brown or black, often with markings; near the eye often silvery white. Head small, arista pubescent or with longer rays; ocelli present; Oc-bristles present to absent; P-bristles diverging or absent; 2-5 pairs of F-bristles curving backward, the lowermost pair in some cases curving inward; interfrontal bristles present to absent; vibrissae present. Wing clear to somewhat infuscated, often the apical portion darker or crossveins with dark bands; costa with a subcostal break, vein Sc complete; crossvein BM-Cu present, cell cup closed. Legs all yellow or partially darker; tibiae with or without dorsal preapical bristle.

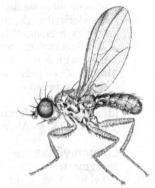

484 Clusia flava (Meigen), male; Soós 1981

Biology: The larvae of the few species of which something is known live in rotting wood of deciduous trees, largely in mines of other insects. Adults are mainly found in damp deciduous forests near dead or dying trees, with males displaying courtship behaviour on trunks and stumps. Adults are not frequently netted but at proper locations they are frequently collected in Malaise traps.

Identification references: Soós 1981 (Central Europe); Stackelberg 1989ak (former USSR); Soós 1987 (genera); Stubbs 1982 (British Isles); CMPD: Sasakawa 1998; CAT: Soós 1984i.

(Clythiidae): name used in the past for Platypezidae.

Cnemospathidae (key couplet 98; fig. 215a-c)

Systematics: Acalyptrate Brachycera; superfamily Sphaeroceroidea; traditionally part of the Heleomyzidae; in Europe 1 species, *Prosopantrum flavifrons* (Tonnoir & Malloch), of South American origin, known from the English and German coast.

Characters: Small (2.5-4 mm) flies; body dark with a brown-grey pruinosity. Antenna and part of the head yellow to orange; arista very short pubescent; ocelli present; Oc-bristles present; P-bristles converging; 2 pairs of F-bristles, the lower pair curving outward; strong vibrissae present. Wing unmarked, with a yellowish hue; costa weakened just past

485 Prosopantrum flavifrons (Tonnoir & Malloch), female; Ismay & Smith 1994

crossvein H, weak and with a subcostal break; vein Sc incomplete, apically merging with vein R1; crossvein BM-Cu present; cell cup closed. Tibiae with strong dorsal preapical bristle.

Biology: The biology of this small family is little known. *P. flavifrons*, of South American origin, is established on the south coast of England (Essex and Devon; Ismay & Smith 1994; Cole 1996) and in Germany (West Friesian Islands, where the species was collected in dry dune areas near colonies of gulls; Stuke & Merz 2005).

Identification references: Ismay & Smith 1994; CMPD: Papp 1998i.

Coelopidae (key couplet 92; fig. 202-203)

486 Coelopa pilipes Haliday, n
Séguy 1934

Systematics: Acalyptrate Brachycera; superfamily Sciomyzoidea; in Europe 3 species in 2 genera, *Coelopa* (2) and *Malacomyia* (1).

Characters: Small to medium sized (2.5-9 mm, usually 4-7 mm); robust, especially in *Coelopa* sometimes densely bristly or hairy, flattened and dark coloured. Eyes small, arista bare to pubescent; ocelli present; Oc-bristles present; P-bristles parallel or converging; 2 pairs of F-bristles, curving outward; scattered interfrontal setulae present; vibrissae absent, but strong bristles present near the vibrissal angle; in some cases many long, hair-like bristles present. Wing unmarked; costa continuous, vein Sc complete; crossvein BM-Cu present; cell cup closed. Legs usually densely hairy; tibiae with dorsal preapical bristle.

Biology: Coelopidae generally inhabit coastal habitats where the larvae largely develop in rotting seaweed washed up by the sea. Larvae and adults can be found throughout the year.

Identification references: Burnet 1960 (*Coelopa*); Stackelberg 1989z (former USSR); CMPD: D.K. McAlpine 1998b; CAT: Gorodkov 1984b.

Coenomyiidae (key couplet 43, 55; fig. 89, 89a, 117, 117a)

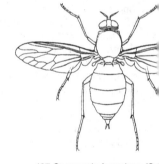

487 Coenomyia ferruginea (Sc
female; Séguy1926

Systematics: Lower Brachycera; superfamily Xylophagoidea; in Europe 1 widespread species, *Coenomyia ferruginea* (Scopoli).

Characters: The only European species, *Coenomyia ferruginea* is a rather large (14-20 mm) and robust fly, of light-to dark brown coloration, set with long golden yellow hairs on head and thorax, the abdomen being shorter pubescent and more lustrous. Head small; eyes holoptic in the male; antenna with 10 segments and tapering to a point. Wing with a yellow tinge; cell m3 open; cell cup much narrowing to the wing margin but nearly always open. Scutellum with a pair of short spine-like projections. Legs relatively short; tibiae with apical bristles; empodium pulvilliform.

Biology: The larvae prey on insect larvae in the upper layer of soil containing much organic matter, for example particles of dead wood, and have been found in dead wood itself as well. The adults mainly occur in woodland areas where they feed on nectar and honeydew. In some cases they emit a penetrating odour.

Identification references: Van der Goot 1985 (Northwestern Europe); Nartshuk 1989d (former USSR); CMPD: Rozkošný & Nagatomi 1997a; CAT: Majer 1988e.

Conopidae (key couplet 67, 134, 152; fig. 138, 143-145, 299-300, 333)

Systematics: Acalyptrate Brachycera; superfamily Conopoidea; some 14 genera and about 85 species in Europe.

Characters: Small to large (3-20 mm, usually 5-15 mm), slender to stout flies, often lustrous, with a black and yellow, wasp-like colour pattern or with reddish brown markings; mouthparts usually conspicuously long and geniculate. Arista bare; ocelli present or

absent (Conopinae); Oc-bristles small or absent; P-, F-, interfrontal bristles and vibrissae absent. Wing usually clear, in
some cases with dark markings, e.g. along the costa; costa
continuous; vein Sc complete; crossvein BM-Cu present, cell
cup closed. Tibiae with (Myopinae) or without dorsal preapical bristle.

Biology: Conopidae are parasitoids, in particular of bumblebees, bees and wasps in which the female deposits a single
egg in the abdomen. This happens in flight while the conopid
female holds the abdomen of her victim in a fixed position
using special structures on her own abdomen. The larva
develops inside, to the detriment of the host, pupating when
the host dies. The adults feed on nectar and frequently visit
flowers (for mating purposes as well). Species with very long
mouthparts visit flowers with deep-lying nectar, such as
Lamiaceae (Labiatae). The adults prefer warm and dry areas
and can be found flying near vegetation rich in flowers, near
blooming heather and on trees and shrubs in flower. The
females often fly slowly through the vegetation searching for

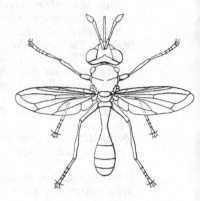

488 Physocephala chrysorrhoea (Meigen),
male; Séguy 1950

a suitable host. Many species show a striking resemblance to bees, wasps and their allies.
Identification references: Bankowska 1979 (Poland); Kormann 2002 (Central Europe);
Rivosecchi 1996 (Italy); Smith 1969b (British Isles); Van Veen 1984 (Belgium,
Netherlands); Zimina 1989 (former USSR); CMPD: no contribution; CAT: Chvála &
Smith 1988.

(Cordiluridae): name used in the past for Scathophagidae.
(Corynoscelidae): name used in the past for Canthyloscelidae.

Cremifaniidae (key couplet 160; fig. 347-348)
Systematics: Acalyptrate Brachycera; superfamily
Lauxanioidea; traditionally part of the Chamaemyiidae; 2
species in Europe: *Cremifania nigrocellulata* Czerny and *C.
lanceolata* Papp.
Characters: Minute to small (1.5-2.6 mm), grey or greyish
brown flies. Arista pubescent, in some cases arista of aberrant
shape; ocelli present; Oc-bristles present; P-bristles absent,
but a pair of diverging bristles present on the hind part of the
ocellar triangle; 2 pairs of F-bristles, curving obliquely out-
backward; scattered interfrontal setulae present; vibrissae
absent. Wing pale grey, wing base, subcostal cell or subcostal
and costal cells partially darkened; costa continuous; vein Sc
complete; crossvein BM-Cu present; cell cup closed. Tibiae
without dorsal preapical bristle.

489 Cremifania nigrocellulata Czerny,
male; Papp 1998h

Biology: In one of the two European species, viz., *C. nigrocellulata*, the biology is well known; its larvae live on balsam woolly aphids of the family Adelgidae, in coniferous woods. This
species, like some species of the Chamaemyiidae, was introduced in North America for the biological control of adelgid
pest species.
Identification references: CMPD: Papp 1998h; CAT:
Tanasijtshuk 1984b.

Cryptochetidae (key couplet 147; fig. 322)
Systematics: Acalyptrate Brachycera; superfamily Carnoidea;
in Europe 1 genus, *Cryptochetum*, with 3 species.
Characters: Small (2-4 mm), stout, black or metallic blue lustrous flies with a strikingly large third antennal segment lacking the arista. Ocelli present; Oc-, P- and F-bristles absent;

490 Cryptochetum grandicorne Rondani,
male; Hennig 1937b

scattered interfrontal setulae present; vibrissae absent. Wing unmarked, costa with both humeral and subcostal breaks; vein Sc complete up to the costa, but pallid, hardly visible; crossvein BM-Cu absent; cell cup closed. Tibiae without dorsal preapical bristle.

Biology: The larvae belonging to this family are endoparasitoids in coccoids (scale insects) living on Scots fir and allies (*Pinus*) and herbs. The adults in some cases crowding together on tree trunks, leaf stalks or leaves. Those of some Asian species can be a nuisance since they continually seek out the eyes of humans (eye gnats). One Australian species was introduced to North America and elsewhere in order to keep down the Australian coccoid, *Icherya purchasi* Williston; this is one of the best examples of successful biological pest control.

Identification references: Stackelberg 1989at (former USSR); CMPD: Nartshuk 2000; CAT: Nartshuk 1984b.

Culicidae (key couplet 14; fig. 6a, 17-18, 45)

Systematics: Nematocera; superfamily Culicoidea; 8 genera and about 105 species in Europe.

Characters: Small to medium sized (3-9 mm, rarely up to 15 mm), pale to dark coloured Nematocera with slender wings, delicate legs and abdomen; with long mouthparts, in the female piercing; wings, legs and often abdomen (partially) covered by scales. Head with elongate mouthparts that are much longer than the head; eyes reniform, dichoptic; ocelli absent; antenna with 15 segments, with long verticils in the male, verticils shorter in the female.

491 Aedes aegypti (Linnaeus) female; Séguy 1950

Biology: The larvae largely develop in shallow stagnant waters, predominantly in small or temporary bodies of water. They feed on particles of detritus, bacteria, algae, etc.; a few species are predatory, mainly on other culicid larvae. The females feed on the blood in order to obtain protein for the development of their eggs. Feeding preferences vary from one species to another and hosts include cold-blooded vertebrates but especially birds and mammals, including humans. The Culicidae occupy virtually all habitats and can occur locally, or temporarily, in vast numbers. Due to their blood feeding habits they can be such a nuisance that they interfere with human activities and have a negative impact on the yield of cattle breeding. Apart from this, they are vectors of several serious diseases of which malaria is the best known and claims the greatest number of casualties.

Identification references: Becker 2003 (genera feeding on blood); Cranston et al. 1987 (British Isles); Dahl 1997 (genera and subgenera); Gutsevich 1989 (former USSR); Haren & Verdonschot 1995 (Netherlands); Huang 2002 (genera); Mohrig 1969 (Germany); Schaffner et al. 2001 (Europe); Service 1993 (genera feeding on blood); Skierska 1977 (Poland); Snow 1990 (British Isles); Stojanovich & Scott 1995a (illustrated key Fennoscandia), 1995b (European Russia); CMPD: Minář 2000a; CAT: Minář 1990; Tóth 2004 (Hungary).

Curtonotidae (key couplet 93; fig. 204)

Systematics: Acalyptrate Brachycera; superfamily Ephydroidea; in Europe 1 widespread species, *Curtonotum anus* (Meigen).

Characters: Medium sized (5-7 mm), yellow to pale grey flies with a somewhat hunchbacked thorax. Arista long plumose; ocelli present; Oc-bristles present; P-bristles converging; 2 pairs of F-bristles, the lower pair curving forward, the upper pair backward; interfrontal bristles or setulae absent; vibrissae or vibrissa-like bristles present. Wing unmarked; costa with humeral and subcostal breaks; vein Sc complete; crossvein BM-Cu absent; cell cup closed. Tibiae with dorsal preapical bristle.

Biology: Little is known about the biology of this small family of 3 genera. The adults are found especially in moist areas (e.g. in Israel along shores of streams, Freidberg pers. comm.). They

492 Curtonotum anus Meigen, female; Papp 1973

are mainly active at dusk, during the day they hide in shadowy spots such as animal burrows. The larvae of non-European species have been found in egg-pouches of the migratory locust.

Identification references: Stackelberg 1989bb (former USSR); CMPD: Papp 1998j; CAT: Papp 1984q.

Cylindrotomidae (key couplet 7; fig. 26, 34-37)

Systematics: Nematocera; superfamily Tipuloidea; traditionally part of the Tipulidae; 4 genera and 6 species in Europe.

Characters: Large (11-16 mm), yellowish to pale brownish Nematocera with long, slender antennae, wings, legs and abdomen. Head with a short rostrum; nasus absent; antenna with 16 segments; ocelli absent. Upper side of the thorax with 3 wide dark stripes, or thorax mainly grey (*Triogma*); transverse suture distinct in median third of thorax, fading out laterally. Wing with 2 anal veins. Abdomen in male with bipartite (*Diogma*) or tripartite distiphallus, in female with broad and short cercus.

Biology: Larvae phytophagous and, *Cylindrotoma* excepted, living on terrestrial, semi-aquatic and aquatic mosses. The larvae of *Cylindrotoma* living on various herbs (*Caltha, Viola, Stellaria, Anemone*, etc.) in moist woodland habitats with a life mode similar to that of caterpillars. The adults predominantly in marshy areas and moist woodland, resting on the vegetation or slowly flying about near the sites where the larvae are found.

493 Cylindrotoma distintissima (Meigen), male; Papp & Schumann 2000

Identification references: Brinkmann 1997 (genera); Peus 1952 (Lindner); Savchenko 1989a (former USSR); Stubbs 1994 (British Isles); CMPD: no contribution; CAT: Soós & Oosterbroek 1992; Oosterbroek 2006.

(Cypselidae): name used in the past for Sphaeroceridae.
(Cyrtidae): name used in the past for Acroceridae.
(Dacidae): part of the Tephritidae.

Diadocidiidae (key couplet 37; fig. 82)

Systematics: Nematocera; superfamily Sciaroidea; traditionally part of the Mycetophilidae; in Europe 1 genus, *Diadocidia*, with 5 species.

Characters: Small (3-4.5 mm), yellowish brown, delicate Nematocera with pale brown to grey wings with a characteristic venation. Head and eyes small; eyes dichoptic; ocelli present; antenna with 17 segments. Thorax vaulted. Legs slender, coxae elongate; tibiae with apical bristles or spurs.

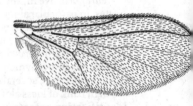

494 Wing Diadocidia ferruginosa (Meigen); Søli et al. 2000

Biology: The larvae feed on mycelium, encapsulated in mucous tubes in and underneath rotting wood and in mushrooms and bracket fungi. In some cases the adults are found in and around hollow tree stumps in great numbers.

Identification references: Hutson et al. 1980 (British Isles); Kurina & Schacht 2003 (synopsis identification references); Laštovka & Matile 1972 (review); Stackelberg 1989j (former USSR); A.I. Zaitzev (Eastern Europe); CMPD: Søli et al. 2000; CAT: Krivosheina 1988a.

Diastatidae (key couplet 100; fig. 221-222)

Systematics: Acalyptrate Brachycera; superfamily Ephydroidea; traditionally this family includes the Campichoetidae; in Europe 1 genus, *Diastata*, with 9 species.

Characters: Small (2.5-4 mm), stout, dark flies with grey pruinosity, head partially dull yellow and lustrous silvery grey; antennae and legs largely yellow. Arista short to medium plumose; ocelli present; Oc-bristles present, P-bristles converging; 2 pairs of F-bristles, the outer pair curving forward, the inner pair backward; interfrontal bristles absent; vibrissae present. Wing with various patterns of dark markings, in some cases clear, but cell c always darkened; costa with both humeral and subcostal breaks and with several small spines among the customary setulae; vein Sc incomplete; crossvein BM-Cu present; cell cup closed. Tibiae with dorsal preapical bristle.

495 Diastata fuscula Fallén, male; Papp 1973

Biology: The larvae of this small family are unknown; they are thought to be saprophagous because the adults are frequently encountered near litter and soil, in all types of woodland, marshes, bog areas and alongside stagnant waters, sometimes in great numbers.

Identification references: Chandler 1986 (British Isles), 1987 (revision); Stackelberg 1989az (former USSR); CMPD: Chandler 1998c; CAT: Papp 1984p.

Diopsidae (key couplet 155; fig. 337)

Systematics: Acalyptrate Brachycera; superfamily Diopsoidea; 1 species, *Sphyracephala europaea* Papp & Földvári, in Central Europe.

Characters: The single species of European Brachycera with eyestalks, *S. europaea*, is small (3-4 mm). Body largely lustrous black and covered with grey, medium length hairs; head partially yellow. Arista bare; ocelli present; Oc- and P-bristles absent; 1 pair of F-bristles, curving backward; interfrontal bristles absent, interfrontal setulae present; vibrissae absent. Wing with faint markings in the central part and at the tips; costa continuous; vein Sc complete; crossvein BM-Cu absent; cell cup closed. Tibiae without dorsal preapical bristle.

496 Sphyracephala europaea Földvári), male; Simova-Tosic Stojanovic 1999

Biology: Most larvae of Diopsidae are root borers or miners, especially in grasses, while some species, probably including *S. europaea*, live in, and feed on, rotting organic matter. The adults prefer moist, shadowy sites, in particular alongside waters and may occur in great numbers. It is known that they seek refuge in great numbers in caves and other sheltered places during winter. In October 1996 *S. europaea* was found in Hungary (and Europe) for the first time. Hundreds of individuals were observed at the wall of a high river bank, mainly at the entrances of small holes made by hymenopterans. The species has also been found in Serbia where in November 1999 several thousands of individuals were observed sitting on the ground on a high bank of loess along a river.

Identification references: Papp et al. 1997 (Hungary); Simova-Tosic & Stojanovic 1999 (Serbia); CMPD: Hilger 2000.

Ditomyiidae (key couplet 33; fig. 76)

Systematics: Nematocera; superfamily Sciaroidea; traditionally part of the Mycetophilidae; in Europe 2 genera, *Ditomyia* and *Symmerus*, with 2 species each.

Characters: Medium sized (6-8 mm), yellowish brown Nematocera with long and slender legs and abdomen, and a characteristic venation (including a long vein R2+3). Head and eyes small; eyes almost touching each other or distinctly dichoptic; ocelli present; antenna with 15 segments, in some

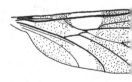

497 Wing Ditomyia fasciata (M Hutson et al. 1980

cases laterally compressed and/or long pubescent. Thorax vaulted. Wing clear or slightly tinged (*Symmerus*) or with transverse bands (*Ditomyia*); anal lobe weak or absent. Coxae elongate; tibiae with long apical bristles or spurs.

Biology: The larvae live in bracket fungi and in dead parts of deciduous trees. The adults are found in woodland areas where most of their activity probably takes place at dusk and in the evening.

Identification references: Hutson et al. 1980 (British Isles); Kurina & Schacht 2003 (synopsis identification references); Stackelberg 1989h (former USSR); A.I. Zaitzev (Eastern Europe); CMPD: Søli et al. 2000; CAT: Mamaev & Krivosheina 1988a.

Dixidae (key couplet 13; fig. 43-44)

Systematics: Nematocera; superfamily Culicoidea; in Europe 2 genera, *Dixa* and *Dixella*, and about 30 species.

Characters: Small to medium sized (3-5.5 mm), yellowish to brownish grey Nematocera. Antennae, wings, legs and abdomen long and slender. Ocelli absent; antenna with 16 segments. Wing usually clear, with an anal lobe and in some cases characteristic pigment spots; vein R2+3 strongly curving. Tibiae of fore and mid leg with short spurs.

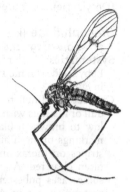

498 Dixa spec., female; Wagner 1997

Biology: The larvae are found at or near the water surface where they take on a characteristic body posture, a reversed U shape, near the edge of the water or just above it, holding themselves up by objects projecting from the water surface. Pupation also takes place there. They are filter feeders, living on microorganisms and tiny particles of organic matter. *Dixa* larvae prefer running water, those of *Dixella* are usually found in stagnant water, while both genera occur in slow-moving water. The adults mainly occur near the larval habitats where the males may fly in swarms.

Identification references: Disney 1999 (British Isles); Stackelberg 1989c (former USSR); Wagner 1997e (genera); CMPD: Wagner 1997b; CAT: Rozkošný 1990.

Dolichopodidae (key couplet 76, 212; fig. 97, 155, 162-164)

Systematics: Lower eremoneuran Brachycera; superfamily Empidoidea; some 60 genera and about 775 species in Europe. For a discussion about the phylogeny and classification see Sinclair & Cumming 2006.

Characters: Minute to medium sized (1-9 mm) flies with long and slender legs. Body slender, the smaller species stouter. Body colour usually greenish metallic lustrous, in some cases dull yellow, brown or black. Eyes virtually always dichoptic; third antennal segment largest and with a usually long, apical or dorsal arista; mouthparts usually short and with a wide aperture adapted for sucking small prey. Wing clear or tinged, in some cases partially strongly coloured, marked or with distinct spots; cells br, bm and cup small, cell cup and/or the anal vein may be absent; crossvein BM-Cu absent; 2 veins branching from crossvein DM-Cu go in the direction of the wing margin; the upper one in some cases strongly curving or forked into M1 and M2. Legs, the tibiae in particular, usually with long bristles; in some genera legs raptorial.

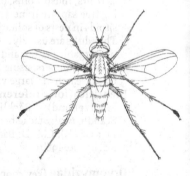

499 Dolichopus popularis Wiedemann, male; Verrall 1909

Biology: The larvae prey on small invertebrates and are found in particular in moist soil, in rotting organic matter, below tree bark, etc.; larvae of *Thrypticus* mine the stalks of Monocotyledons. Truly aquatic species are lacking but many are semi-aquatic (living in the water margin zone); a restricted number of species develop at the shores of saline inland bodies of water or the intertidal zone along seashores. The adults occur in a wide range of habitats, for example near water or more in the open, e.g. in meadows and gardens. They are often sitting high on their legs on the ground or on vegetation, on tree trunks, rocks, walls etc.; some species walk about on the water surface. The adults, like the larvae, are predators, largely feeding on small invertebrates. The species of *Dolichopus*, for example, prefer to prey on the larvae of Culicidae.

Identification references: D'Assis Fonseca 1978 (British Isles); Grichanov 2004 (Sweden); Meuffels 2001 (genera Netherlands); Negrobov & Stackelberg 1989 (former USSR); Parent 1938 (France); Sinclair & Cumming 2006 (classification); Stackelberg & Negrobov 1930-1979 (Lindner, incomplete); Weber 1989 (Central Europe); CMPD: no contribution; CAT: Negrobov 1991.

(Dorylaidae): name used in the past for Pipunculidae.

Drosophilidae (key couplet 99, 126, 133, 220; fig. 217-220, 271, 297-298)

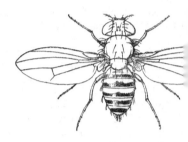

Systematics: Acalyptrate Brachycera; superfamily Ephydroidea; some 17 genera and about 120 species in Europe.

Characters: Minute to medium sized (1.5-7 mm) flies, dull or lustrous with body colour varying from yellow to brownish black, often with some whitish markings on head and thorax and/or a pattern of stripes on thorax and abdomen; when alive eyes often clear red. Arista usually with long rays, in some cases pubescent; ocelli present, Oc-bristles present, except in *Acletoxenus*; P-bristles usually present, converging; 3 pairs of F-bristles, 1 pair curving forward or obliquely inward, 2 pairs backward; scattered interfrontal bristles present; vibrissae present. Wing usually unmarked, in some cases with dark markings or extensively infuscated; costa with both humeral and subcostal breaks; vein Sc incomplete; crossvein BM-Cu usually absent, in some cases present; cell cup open or closed. Tibiae virtually always with dorsal preapical bristle, sometimes small or absent (e.g. *Cacoxenus*).

500 Drosophila funebris (Fabricius); Séguy 1951

Biology: The larvae of most species feed on micro-organisms in rotting organic matter such as fruits, mushrooms, bracket fungi, cacti, etc. Other larvae live in slime fluxes or exuding tree sap, are miners, or prey on other insects (e.g. larvae of *Cacoxenus* living as parasitoids in nests of solitary bees; *Acletoxenus* larvae are predators of larvae of Aleurodidae). The adults are usually attracted by the scent of mushrooms, bracket fungi, ripening or rotting organic matter, and are very frequently found indoors. Some species are easy to rear in large numbers, using artificial substrates. Hence Drosophilidae are much used in scientific research in a large variety of fields.

Identification references: Bächli & Burla 1985 (Switzerland); Bächli et al. 2004 (Denmark, Scandinavia); Stackelberg 1989ba; Gornostaev 2001 (former USSR); CMPD: Bächli 1998; CAT: Bächli & Teresa Rocha Pité 1984.

Dryomyzidae (key couplet 145; fig. 314-316)

Systematics: Acalyptrate Brachycera; superfamily Sciomyzoidea; traditionally this family includes the Helcomyzidae and Heterocheilidae; in Europe 4 species in 2 genera, *Dryomyza* (3) and *Neuroctena* (1).

Characters: Medium sized to large (5-18 mm), yellow, orange, or brown coloured, bristled or quite 'hairy' flies. Arista bare to pubescent; ocelli present; Oc-bristles present, P-bristles parallel to diverging; 1-2 pairs of F-bristles, curving obliquely out-backward; scattered interfrontal setulae present; vibrissae absent. Wing clear or tinged and often with dark bands along the crossveins;

501 Neuroctena anilis (Fallén), male; Stackelberg 1956

costa continuous; vein Sc complete; crossvein BM-Cu present; cell cup closed. Tibiae with dorsal preapical bristle.

Biology: The larvae develop in rotting organic matter such as mushrooms and bracket fungi, droppings, carrion, etc. The adults are mainly found in moist, shaded places in woodland areas, usually on putrefying matter, including carrion, decaying mushrooms, excrement and tree sap.

Identification references: Falk 2005 (British Isles); Stackelberg 1989aa (former USSR); CMPD: Ozerov 1998; CAT: Soós 1984g.

(Eginiidae): part of the Muscidae.

Empididae (key couplet 63, 78; fig. 133-134, 156, 170, 172)

Systematics: Lower eremoneuran Brachycera; superfamily Empidoidea; traditionally this family includes the Atelestidae, Brachystomatidae, Hybotidae and Microphoridae; some 23 genera and about 810 species in Europe. For a discussion about the phylogeny and classification see Sinclair & Cumming 2006.

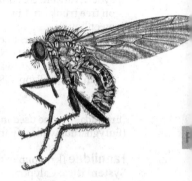

Characters: Minute to large (1-12 mm, sometimes larger), slender to robust, pallid to dark flies, in several genera with powerful, piercing mouthparts. Head small with relatively large eyes, in some cases holoptic, often with an incision at level of antennae; antenna usually with 3 segments of which the third is the largest and bears a short or long, usually apical arista or style; mouthparts short to strongly elongate, in the latter case projecting forward or downward toward the fore legs. Wing clear or (partially) tinged, in some cases with a stigma spot, or with a distinct wing pattern; vein R4+5 often forked; cell dm almost always present. Legs powerful, usually long and slender; in some cases fore legs raptorial, adapted to catching and holding on to prey.

502 Empis tesellata Fabricius, male; Colyer & Hammond 1968

Biology: The larvae prey on other insects, mainly larvae of Diptera living in the ground, on organic matter, or in decaying wood; some are aquatic. The adults are originally predators as well, mainly on flying Diptera, but some species forage by walking about on tree trunks or prey on aquatic insects, including those that have just pupated near the shoreline. Some adults visit flowers and partially or entirely feed on pollen and nectar. Many species form swarms, the males offering a prey or a silken balloon, which may or may not contain a prey, to the female as part of the mating ritual. In *Hormopeza* large swarms of both sexes have been observed in smoke of wood fires.

Identification references: Barták 1982 (*Rhamphomyia* Central Europe); Chvála 1994 (*Empis*), 2005 (*Hilara*); Collin 1961 (British Isles); Van der Goot 1990 (*Empis* Netherlands, Belgium, Luxembourg); Van der Goot et al. 2000 (*Hilara* Netherlands); Gorodkov & Kovalev 1989 (former USSR); Niesiolowski 1992 (Poland); Sinclair & Cumming 2006 (classification); Wagner 1997g (genera); CMPD: no contribution; CAT: Chvála & Wagner 1989.

Ephydridae (key couplet 95, 132, 167, 217; fig. 211-213, 290-293, 365-366, 450)

Systematics: Acalyptrate Brachycera; superfamily Ephydroidea; some 60 genera and about 335 species in Europe.

Characters: Minute to medium sized (1-11 mm), usually dull, dark coloured flies. Face often swollen and convex; arista sometimes bare to pubescent but usually with long rays dorsally only; ocelli present; Oc-bristles present, P-bristles absent, pseudopostocellar-bristles, if present, parallel to diverging; F-bristles pairs variable, curving forward, outward, or backward, in some cases absent; scattered interfrontal setu-

503 Parydra aquila (Fallén), male; Zatwarnicki 1997

lae present or absent; vibrissae present. Wing usually unmarked, in some cases darkened, marked or spotted; costa with humeral and subcostal breaks; vein Sc incomplete; crossvein BM-Cu absent; cell cup open or absent. Only tibia of mid leg in some cases with dorsal preapical bristle; in some cases fore legs raptorial.

Biology: The Ephydridae usually inhabit aquatic, semi-aquatic and coastal or non-coastal saline habitats where they may be found in large numbers. The larvae feed on micro-organisms such as bacteria, algae and yeasts; some species prey on insects and other small aquatic life forms while others prefer rotting animal tissue, excrement, eggs of spiders or frogs; species feeding on plant juices or mining in plants are known as well. Adults are usually found, sometimes in very large numbers, on or near the soil in humid or wet habitats where they mate, oviposit, or feed on a wide variety of organic matter like algae, honeydew, carrion, etc.; otherwise, adults may be found at some distance from humidity, e.g. on tree trunks and in vegetation. Some species have been recorded as pests in crops and greenhouses.

Identification references: Canzoneri & Meneghini 1983 (Italy); Hollmann-Schirrmacher 1988 (Ilytheinae); Nartshuk 1989i (former USSR); Olaffson 1991 (*Scatella, Lamproscatella*); Zatwarnicki 1997 (genera); CMPD: Mathis & Zatwarnicki 1998; CAT: Cogan 1984b; Papp 1984m (Risinae); Mathis & Zatwarnicki 1995.

(Erinnidae): name used in the past for Xylophagidae.
(Eurygnathomyiidae): part of the Pallopteridae.

Fanniidae (key couplet 180; fig. 395)
Systematics: Calyptrate Brachycera; superfamily Muscoidea; 3 genera and about 82 species in Europe.
Characters: Small to medium sized (2-5 mm, some up to 9 mm) flies, usually with grey to black body colour. Occiput without white or pallid, thin hair-like bristles; arista bare to pubescent; interfrontal bristles absent. Meral bristles absent; lower side of scutellum bare in its middle region; subscutellum absent. Lower calypter small. Apical half of vein Sc straight; vein M1 not strongly curving; anal vein not reaching the wing margin; vein A2 and the anal vein would intersect if A2 would continue further. Tibia of hind leg with a submedian dorsal bristle.

504 Fannia canicularis (Linnae male; Stackelberg 1956

Biology: The larvae are saprophagous and feed mostly among decaying organic materials. Many are surface scrapers, feeding on micro-organisms on the surface of the substrate, which include a wide range of vegetable and animal detritus, mushrooms, nests or burrows of birds, mammals and insects, dead insects, molluscs and vertebrates, animal droppings and dung, human faeces, etc. The adults occur mainly in or near woodland areas and less in more open habitats. The males of several species fly in swarms. Some species have successfully adapted to the human environment and are frequently met with indoors. Good examples are *Fannia canicularis* (Linnaeus) (the lesser housefly) and *F. scalaris* (Fabricius) (the latrine fly).

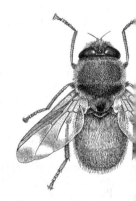

Identification references: Crosskey & Lane 1993 (synanthropic species of *Fannia*); D'Assis Fonseca 1968 (British Isles); Gregor & Rozkošný 1995 (Central Europe); Hennig 1955-1964 (Lindner); Rozkošný et al. 1997 (revision); CMPD: Pont 2000; CAT: Pont 1986a; Rozkošný et al. 1997.

(Fungivoridae): name used in the past for Mycetophilidae.

Gasterophilidae (key couplet 173; fig. 376)
Systematics: Calyptrate Brachycera; superfamily Oestroidea; traditionally part of the Oestridae; in Europe 1 genus, *Gasterophilus*, with 6 species.
Characters: Medium sized to large flies (9-16 mm). Body and

505 Gasterophilus pecorum (F Grunin 1969

legs covered with soft, hair-like bristles, giving them the aspect of bumblebees or bees. Mouth opening small; mouthparts rudimentary; antenna small. Wing with vein M1 rather straight, slightly curving backward, reaching the wing margin beyond the wing tip; lower calypter small. Subscutellum absent; meron lacking strong bristles along its hind margin near the posterior spiracle. Female with a long ovipositor which can be folded below the abdomen in rest.

Biology: The larvae live as parasites in the stomach and intestine of horses and donkeys. The eggs are deposited directly on the host's skin and the first larval instar penetrates the skin and works its way inside the animal, they first migrate to the mouth and later to the stomach and intestine. The larva leaves the host along with the excrement and pupates in the soil. The adults possess reduced mouthparts, they do not feed and are found near horses and stables.

Identification references: Colwell et al. 2004 (general); Grunin 1969 (Lindner), 1989a (former USSR); CMPD: Minář 2000b; CAT: Soós & Minář 1986a.

Helcomyzidae (key couplet 146; fig. 317-320)

Systematics: Acalyptrate Brachycera; superfamily Sciomyzoidea; traditionally part of the Dryomyzidae; 2 species in Europe: *Helcomyza ustulata* Curtis and *H. mediterranea* (Loew).

Characters: Medium sized to large (6-11 mm), stout, predominantly grey pruinose flies, densely covered with fine setae. Head large with relatively small eyes; arista bare to pubescent; ocelli present; Oc-bristles reduced or absent; P-bristles parallel to somewhat diverging; 1-2 pairs of F-bristles, curving outward; scattered interfrontal setulae present; vibrissae absent. Wing unmarked, crossvein often accompanied by dark bands; costa continuous; vein Sc complete; crossvein BM-Cu present; cell cup closed. Tibiae with dorsal preapical bristle.

Biology: The species of this small family occur along the coast where the adults are often found on the beach, making a short flight when disturbed. The larvae live in seaweed washed up on the shore and which has dried to a greater extent than the seaweed in which Coelopidae, Heterocheilidae or *Orygma* species are living.

506 Helcomyza ustulata Curtis, male; McAlpine 1998c

Identification references: Stackelberg 1989y (former USSR); CMPD: D.K. McAlpine 1998c; CAT: Gorodkov 1984a.

(Heleidae): name used in the past for Ceratopogonidae.

Heleomyzidae (key couplet 92, 117, 220; fig. 10, 201, 253-254)

Systematics: Acalyptrate Brachycera; superfamily Sphaeroceroidea; traditionally this family includes the Chiropteromyzidae, Cnemospathidae and Trixoscelididae; some 23 genera and about 150 species in Europe.

Characters: Robust, minute to large (1.2-12 mm), pallid yellow to dark grey or brown, in some cases reddish flies. Arista bare or plumose; ocelli present; Oc-bristles present; P-bristles converging; 1 or 2 pair(s) of F-bristles, curving outward or backward; interfrontal bristles absent, interfrontal setulae present or absent; vibrissae present. Wing clear, tinged or with a pattern, especially along the crossveins; costa with a subcostal break; vein Sc complete; crossvein BM-Cu present; cell cup closed. Tibiae usually with dorsal preapical bristle which may be small or lacking in the Borboropsinae and in *Oldenbergiella*.

507 Eccoptomera microps (Meigen), male; Séguy 1950

Biology: The larval habitats of the various subfamilies can be summarised as follows: In the Heleomyzinae and Heteromyzinae the larvae mainly feed on rotting organic matter and can be found in carrion, excrement, mushrooms, bracket fungi, animal burrows, nests of birds, etc. The larvae of the Suilliinae have chiefly been found in mushrooms and bracket fungi while *Suillia lurida* (Meigen) is a well-known pest in onions and garlic. The adults are often found near the larval habitats, e.g. in woodland areas, but also in caves and in or near nests of birds and mammal burrows.

Identification references: Gorodkov 1989b (former USSR); Papp 1998l (*Nidomyia*); Soós 1981 (Central Europe); Withers 1978 (*Suillia*); CMPD: Papp 1998i; CAT: Gorodkov 1984c.

(Helomyzidae): name used in the past for Heleomyzidae.

Hesperinidae (key couplet 26, 203; fig. 65-66)

Systematics: Nematocera; superfamily Bibionoidea; traditionally part of the Bibionidae; in Europe (Balkan) 1 species, *Hesperinus imbecillus* (Loew).

Characters: A slender, dark coloured, small to medium sized (4-6 mm) species with long antennae, legs and abdomen. Ocelli conspicuous; antennal insertion near lower margin of the face; antenna with typically elongate third segment. Wing darkened, long in the male, short in the female where it reaches to halfway down the second abdominal segment. Tibiae with apical bristles or spurs.

Biology: Little is known about the biology of this small family. The larvae live in rotting wood of deciduous trees.

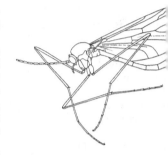

508 Hesperinus imbecillus (Loew), male; Séguy 1951

Identification references: Stackelberg 1989m (former USSR); CMPD: Krivosheina 1997b; CAT: Krivosheina & Mamaev 1986.

Heterocheilidae (key couplet 146; fig. 321)

Systematics: Acalyptrate Brachycera; superfamily Sciomyzoidea; traditionally part of the Dryomyzidae; in Europe, along North Sea and Baltic Sea, 1 species, *Heterocheila buccata* (Fallén).

Characters: Medium sized (6-9 mm), stout, largely greyish brown species, densely covered with fine setae. Head large with small eyes; arista bare to pubescent; ocelli present; Oc-bristles present; P-bristles more or less parallel; 3 pairs of F-bristles, curving outward; scattered interfrontal setulae present; vibrissae absent. Wing unmarked; costa continuous; vein Sc complete; crossvein BM-Cu present; cell cup closed. Tibiae with dorsal preapical bristle.

509 Heterocheila buccata (Fallén), male; McAlpine 1998d

Biology: The sole representative of this family in Europe, *Heterocheila buccata*, is a species of coastal habitats. The larvae develop in seaweed washed up on the shore; the adults are usually found near the larval breeding sites.

Identification references: Stackelberg 1989y (former USSR); CMPD: D.K. McAlpine 1998d; CAT: Gorodkov 1984b (as part of the Helcomyzidae).

Hilarimorphidae (key couplet 57; fig. 122)

Systematics: Lower muscomorph Brachycera; superfamily Asiloidea; traditionally part of the Rhagionidae; 2 species in Europe: *Hilarimorpha singularis* Schiner and *H. tristis* Egger.

Characters: Small to medium sized (2-7 mm), stout, brown or black coloured flies resembling species of

510 Hilarimorpha spec., male; Webb 1974

Hilara (Empididae), with pubescence and a characteristic venation. Male holoptic; antenna short, with 4 distinct segments of which the third is the largest, and with a small apical style; mouthparts small. Wing clear or tinged, in some cases with dark markings along vein R1; discal cell absent; veins R4+5 and M1+2 forked in more or less similar ways; cell cup closed. Legs without bristles or spurs and without empodium.

Biology: The larvae of this small family are unknown. The adults were chiefly collected near bodies of water, for example near puddles in cart tracks, at open sandy spots near fenland, or on willows alongside creeks.

Identification references: Lindner 1925 (species); Nartshuk 1989c (former USSR); Webb 1974 (revision); CMPD: Nagatomi 1997c; CAT: Majer 1988d.

Hippoboscidae (key couplet 169, 209; fig. 174a-b, 370-371, 438, 442-443)

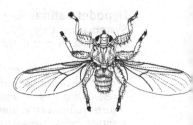

511 Lipoptena cervi (Linnaeus), female; Hutson 1984

Systematics: Calyptrate Brachycera; superfamily Hippoboscoidea; 11 genera and about 30 species in Europe.

Characters: Small to medium sized (2.5-10 mm) flies. Most species are dorsoventrally flattened as an adaptation to their ectoparasitic mode of life. Head closely adpressed to thorax and eyes well-developed. Most species possess well-developed wings and are able fliers; some have reduced wings or shed their wings after finding a suitable host. Wing venation reduced and concentrated at the wing base and anterior region. Legs short and powerful; coxae of mid legs and usually also hind legs far apart; tarsi with hook-like claws, enabling these insects to cling to the hairs or feathers of their hosts.

Biology: All species are ectoparasites of birds or mammals and feed on blood. The larvae develop inside the abdomen of the female fly and are deposited one by one, immediately followed by pupation, generally on the ground or in the soil. Most species visit a limited number of host species but some species are found on species of birds sharing the same habitat. Flightless species can find new hosts by contact between hosts or through infection at nest sites.

Identification references: Borowiec 1984 (Poland); Büttiker 1994 (Switzerland); Grunin 1989b (former USSR); Hutson 1984 (British Isles); Maa 1963 (genera); Schuurmans Stekhoven jr. & Van den Broek 1969 (Netherlands); Theodor & Oldroyd 1964 (Lindner); CMPD: no contribution; CAT: Soós & Hůrka 1986.

Hybotidae (key couplet 77, 80, 212; fig. 96, 154, 167-169, 171, 173, 446)

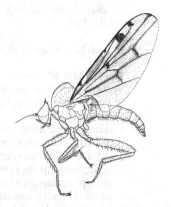

512 Synechus muscarius (Fabricius), male; Chvála 1983

Systematics: Lower eremoneuran Brachycera; superfamily Empidoidea; traditionally part of the Empididae; some 30 genera and about 440 species in Europe. For a discussion about the phylogeny and classification see Sinclair & Cumming 2006.

Characters: Minute to medium sized (1-6 mm, rarely larger) flies, usually dark but colour variable, yellow to black. Eyes relatively large, dichoptic or holoptic, with an incision near the antennae; antenna short or with a large third segment and long arista; mouthparts varying between short (Ocydromiinae) to long and directed straight or obliquely downward (Hybotinae) or directed downward (Tachydromiinae). Wing clear or tinged, in some cases with markings, along the costa and around or in between veins R1 and R2+3 in particular; cell dm absent (Tachydromiinae) or present. Legs long, slender and often partially adapted to catching and holding prey; femora of fore legs and/or mid legs swollen in the Tachydromiinae while in the other subfamilies those of the hind legs are swollen.

Biology: The larvae of only a few species are known. They live in soil, rotting wood, organic matter, dung, etc., and prey on small invertebrates. The adults prey on small insects and other arthropods, catching their prey in flight (Hybotinae, Ocydromiinae, some

Tachydromiinae) or while walking or running about on the ground, on tree trunks, leaves, etc. (most Tachydromiinae). Only a few species visit flowers to feed on nectar and pollen. Swarming has been observed in the Ocydromiinae.

Identification references: Chvála 1975 (Tachydromiinae), 1983 (Hybotinae), 1989a (*Platypalpus*); Collin 1961 (British Isles); Van der Goot 1990 (*Hybos* Netherlands, Belgium, Luxembourg); Gorodkov & Kovalev 1989 (former USSR); Papp & Földvári 2001 (Central Europe); Sinclair & Cumming 2006 (classification); CMPD: no contribution; CAT: Chvála & Kovalev 1989.

(Hyperoscelididae): name used in the past for Canthyloscelidae.

Hypodermatidae (key couplet 174; fig. 377-378)

Systematics: Calyptrate Brachycera; superfamily Oestroidea; traditionally part of the Oestridae; 4 genera and 8 species in Europe.

Characters: Large to very large (10-22 mm) flies. Body with scant short bristles or body and legs with soft pile of hairs, thereby resembling bees or bumblebees. Mouth opening small, mouthparts rudimentary; antenna small; the ends of the ptilinal suture, below the antennae, not strongly curved toward each other. Vein M1 with part beyond crossvein DM-Cu gradually curving or more or less straight toward the costa; cell r4+5 open or closed at the wing margin, without a petiole; lower calypter large. Subscutellum narrow, swollen; meron usually with bristles along its hind margin, near the posterior spiracle.

513 Hypoderma diana Brauer; 1965

Biology: The larvae live as subcutaneous endoparasites in mammals, especially deer and cattle but also occasionally mammals (e.g. horses, guinea pigs). Most species are specialised in one or only a few host species. The eggs are deposited on the skin of the host and the first instar larva penetrates the skin. Second and third instar larvae accumulate on the host's back and form large warbles. Just prior to pupation the fully grown larva leaves the host through the skin, pupating in the soil. The adults have reduced mouthparts and do not feed. They are often found sitting on walls, plants, etc., or near their host. The species that are of economic importance to cattle breeding have been virtually exterminated during the second half of the 20th century.

Identification references: Colwell et al. 2004 (general); Grunin 1964-1969 (Lindner), 1989e (former USSR); Zeegers 1992 (Netherlands); CMPD: Minář 2000d; CAT: Soós & Minář 1986c.

(Itonididae): name used in the past for Cecidomyiidae.

Keroplatidae (key couplet 35; fig. 77-80)

Systematics: Nematocera; superfamily Sciaroidea, traditionally part of the Mycetophilidae; some 16 genera and about 110 species in Europe.

Characters: Medium to large (4-15 mm), slender and delicate to stout, in some cases conspicuously coloured Nematocera with characteristic venation (e.g. crossvein R-M absent). Head small; mouthparts sometimes elongated; ocelli present; eyes dichoptic; antenna with 16-17 segments; antenna ranging from relatively short, consisting of laterally compressed segments to very slender and longer than the body. Thorax vaulted. Wing relatively wide, clear or with markings or crossbands, in *Macrocera* with a striking anal angle. Coxae elongate; tibiae with apical bristles or spurs.

514 Cerotelion lineatus (Fabric male; Matile 1990

Biology: The larvae are usually found in moist and dark spots in woodland. Keroplatinae larvae often occur underneath bracket fungi where they construct a web to collect spores (mycophagous species) or catch small invertebrates (predatory species). The latter group includes species provisioning the strands of their web with acid fluids in order to kill their prey. *Macrocera* larvae are found in rotting wood, underneath tree bark and below tree trunks. They are covered in a thin layer of mucus and feed on animal detritus or prey upon small invertebrates. The adults are found near the larval habitats; some species visit flowers where they feed on nectar; species of *Macrocera* are often found on the upper side of leaves, standing high on their legs and holding their wings slightly spread out.

Identification references: Chandler & Ribeiro 1995 (Atlantic islands); Hutson et al. 1980 (British Isles); Kurina & Schacht 2003 (synopsis identification references); Matile 1990 (revision, excluding Orfeliini); Stackelberg 1989i, 1989k (former USSR); A.I. Zaitzev (Eastern Europe); CMPD: Søli et al. 2000; CAT: Krivosheina & Mamaev 1988b.

(Larvaevoridae): name used in the past for Tachinidae.

Lauxaniidae (key couplet 91, 138; fig. 186-187, 193, 200, 303-304)
Systematics: Acalyptrate Brachycera; superfamily Lauxanioidea; some 18 genera and about 160 species in Europe.

Characters: Small to medium sized (2-7 mm), often rather plump, dull or partially lustrous flies; body colour varying from yellow to brown or black, or with a combination of colours. Arista bare, pubescent, or with long rays; ocelli present; Oc-bristles present; P-bristles converging, in rare cases parallel (e.g. among a few specimens of *Cnemacantha muscaria* (Fallén)); 1-2 pairs of F-bristles, curving backward, the lower pair in some cases curving inward; interfrontal bristles absent; vibrissae absent but some genera, e.g. *Trigonometopus*, with strong bristles near the vibrissal angle. Wing usually unmarked, in a number of species with spots along the veins, in some cases entirely marked; costa continuous; vein Sc complete; crossvein BM-Cu present; cell cup closed. Usually all tibiae with dorsal preapical bristle but in a few cases tibia of hind legs without such bristle.

515 Eusapromyza multipunctata (Fallén); male; Papp 1981

Biology: As far as known the larvae live on micro-organisms in bracket fungi and mushrooms, in rotting vegetable matter, in some cases as miners or in nests of birds. The majority of species are found in shaded spots in moist woodlands, marshes, banks, etc. Some species of *Homoneura*, *Minettia* and *Calliopum* are adapted to drier, more open habitats like dunes and grasslands.

Identification references: Merz 2002 (genera Central Europe); Shatalkin 2000 (species); Stackelberg 1989ad (former USSR); Stuckenberg 1971 (genera); CMPD: Papp & Shatalkin 1998; CAT: Papp 1984a.

(Leptidae): name used in the past for Rhagionidae.
(Leptoconopidae): part of the Ceratopogonidae.
(Leptogastridae): part of the Asilidae.
(Lestremiidae): part of the Cecidomyiidae.
(Limnobiidae): name used in the past for Limoniidae.

Limoniidae (key couplet 7, 196, 198; fig. 31-32, 430-431)
Systematics: Nematocera; superfamily Tipuloidea; traditionally part of the Tipulidae; some 70 genera and about 560 species in Europe.

Characters: Pale yellow to black, small to very large (2-11 mm, a single species up to 30 mm) Nematocera with slender antennae, legs and abdomen. Rostrum usually short and without

516 Limonia phragmitidis (Schrank), male; Papp & Schumann 2000

nasus; ocelli absent; antenna usually with 14-16 segments, in some cases (*Hexatoma*) less or (*Ludicia*) more. Thorax with a V-shaped transverse suture. Wing with 2 anal veins; in some cases wing with characteristic markings or wing reduced in females or in both sexes (*Chionea*).

Biology: The larvae are generally found in various aquatic and semi-aquatic habitats. Some species are terrestrial, living in mosses, mushrooms, bracket fungi, or in dying or decaying wood. A few species are found in brackish or saline habitats. Most larvae feed on rotting vegetable matter, algae, mosses, mushrooms, etc. The subfamily Limnophilinae has many species of which the larvae prey on insect larvae, on worms and other small invertebrates. In the subfamily Limoniinae most larvae live in a tube made of strands they have spun and in which they also pupate. Adults are found in several, usually moist habitats, often near those inhabited by their larvae. Males often fly in swarms.

Identification references: Dienske 1987 (genera and subgenera); Geiger 1986 (Limoniinae Switzerland); Reusch & Oosterbroek 1997 (genera and subgenera); Savchenko 1989c, 1989d (former USSR); Stubbs 1994-1999 (British Isles); CMPD: no contribution; CAT: Savchenko et al. 1992; Oosterbroek 2006.

(Liriopeidae): name used in the past for Ptychopteridae.

Lonchaeidae (key couplet 83; fig. 182-183)

Systematics: Acalyptrate Brachycera; superfamily Tephritoidea; 8 genera and about 100 species in Europe.

Characters: Usually stout, metallic green or bluish black, small to medium sized (3-6 mm) flies. Head large, hemispherical, showing considerable sexual dimorphism with frons narrower in the male; lunula large; arista bare or with shorter to longer pubescence; ocelli present; Oc-bristles present; P-bristles weak, diverging; 1 pair of F-bristles, situated at the level of the ocellar triangle and curving backward; scattered interfrontal setulae present; vibrissae absent but in

517 Lonchaea chorea (Fabricius) female; Colyer & Hammond 19

some genera with a series of vibrissa-like bristles near the vibrissal angle. Wing unmarked; cell c wide; vein Sc complete; costa with a subcostal break; crossvein BM-Cu present; cell cup closed. Haltere dark brown to black. Tibiae without dorsal preapical bristle or present only on tibia of mid legs. Abdomen short, wide, dorsoventrally flattened; the female with a lanceolate, incompletely retractable ovipositor.

Biology: The larvae of most species are found in mouldering wood and underneath the bark of dying or dead trees, with several species associated with certain tree species. In other species the larval development takes place in rotting vegetable matter such as onions, conifer seeds, fruits, vegetables, galls on grasses or other substances. The larvae of most species feed on vegetable matter, although some are carrion eaters or prey on the larvae of other insects, e.g. those of beetles. Males congregate in swarms in openings in forests. Lonchaeidae are usually less frequently encountered and in many cases have been obtained by rearing; however, they are by no means rare as is clear from the larger series obtained by beer/wine traps and Malaise traps.

Identification references: MacGowan & Rotheray 2000 (*Lonchaea*); J.F. McAlpine 1964 (revision), 1975 (*Lonchaea*); J.F. McAlpine & Steyskal 1982 (genera); Morge 1959-1974 (review); Stackelberg 1989ai (former USSR); CMPD: no contribution; CAT: Kovalev & Morge 1984.

Lonchopteridae (key couplet 68; fig. 146-147)

Systematics: Lower Cyclorrhapha; superfamily Lonchopteroidea; in Europe 2 genera, *Lonchoptera* and *Neolonchoptera*, with about 13 species.

Characters: Small (2-5 mm), yellow to brownish black flies with strong bristles and long, subacute wings. Eyes broadly dichoptic; third antennal segment rounded, with a long arista; 1 pair of strong, diverging bristles just above the antennae. Wing clear to tinged, rarely with markings; venation characteristic and sexual dimorphic: in male anal vein reaching the hind margin, in the female ending in vein CuA1.

Biology: The larvae feed on rotting vegetable matter, perhaps including mushrooms and bracket fungi. They can be found among dead leaves, in plant debris and other decaying material in various habitats ranging from gardens to woodland and meadows. The adults feed on nectar and are in some cases found in large numbers, often in moist places alongside running or stagnant waters, in woodlands, marshes, some species in the tidal zone; other species prefer drier habitats, some are ubiquitous and occur in a variety of habitats from lowlands to high mountain areas. Most species are bisexual, but at least *L. bifurcata* (Fallén) is parthenogenetic in most areas of its nearly cosmopolitan range.

518 Lonchoptera lutea Panzer, male; Séguy 1950

Identification references: Bährmann & Bellstedt 1988 (Germany); Barták 1986 (Europe); De Meijere 1906 (species); Smith 1969a (British Isles); Stackelberg 1989o (former USSR); Vaillant 1989 (species), 2002 (species); CMPD: Barták 1998; CAT: Andersson 1991.

(Lycoriidae): name used in the past for Sciaridae.
(Macroceridae): part of the Keroplatidae.
(Manotidae): part of the Mycetophilidae.

Megamerinidae (key couplet 154; fig. 335-336)

Systematics: Acalyptrate Brachycera; superfamily Diopsoidea; in Europe 1 widespread species, *Megamerina dolium* (Fabricius).

Characters: The only European species, *Megamerina dolium*, is medium sized (6-9 mm), elongate, black lustrous with pale to reddish femora. Head small and spherical; arista pubescent; ocelli present; Oc-, P-, F-, interfrontal bristles and vibrissae absent. Wing pale brown, unmarked; costa continuous; vein Sc complete; crossvein BM-Cu present; cell cup closed. Femora of hind legs swollen, with 2 rows of spines on the lower side; tibiae without dorsal preapical bristle.

Biology: The larvae live under the bark of dying or dead deciduous trees and are predatory or necrophagous on the larvae of other insects. The adults are largely found near the larval habitats, chiefly on tree trunks and leaves.

519 Megamerina dolium (Fabricius), female; Séguy 1934

Identification references: Van der Goot & Van Veen 1996 (Northwestern Europe); Stackelberg 1989t; CMPD: no contribution; CAT: Nartshuk 1984a.

(Melusinidae): name used in the past for Simuliidae.

Micropezidae (key couplet 158; fig. 339-341)

Systematics: Acalyptrate Brachycera; superfamily Nerioidea; 5 genera and about 22 species in Europe.

Characters: Strikingly slender flies of small to large size (3-16 mm) with narrow wings and long legs. Head small, rounded to elongate; antenna small with arista bare or pubescent; ocelli present; Oc-bristles absent; P-bristles diverging or absent; 0-3 pairs of F-bristles, if present, curving forward or backward; interfrontal bristles absent; vibrissae absent. Wing clear to fumosely patterned; costa continuous; vein Sc complete, its ending in the costa may be close to that of vein R1; crossvein BM-Cu present or absent (in the Micropezinae); cell cup closed. Tibiae without dorsal preapical bristle.

Biology: With respect to their biology, the European

520 Raineria calceata (Fallén), female; Mihályi 1972

141

Micropezidae can be divided in three groups. Species of *Micropeza* have phytophagous larvae feeding in the root nodules of leguminous plants and are associated with open habitats; species of *Rainieria* develop in the rotting wood of deciduous trees and are restricted to old forests; the majority of the other species have saprophagous larvae living in decayed vegetation, old manure or fungi and their adults usually prefer humid forests along river and creeks. Adult Micropezidae prey on small insects and other invertebrates and can be found on flowers, leaves, rotting fruit, excrement, etc.

Identification references: Andersson 1989 (Scandinavia); Van der Goot & Van Veen 1996 (Northwestern Europe); Greve & Nielsen 1991 (Norway); Merz 1997 (Switzerland); Ozerov 1991 (Calobatinae); Roháček & Barták 1990 (Central Europe); Stackelberg 1989r, 1989s (former USSR); CMPD: no contribution; CAT: Soós 1984a.

Microphoridae (key couplet 75; fig. 161)

Systematics: Lower eremoneuran Brachycera; superfamily Empidoidea; traditionally part of the Empididae, considered part of the Dolichopodidae by Sinclair & Cumming 2006; 4 genera and about 16 species in Europe.

Characters: Minute to small (1.5-3 mm), greyish to black flies. Head and eyes relatively large; male holoptic; third antennal segment elongate, with terminal arista; mouthparts somewhat elongate, directed straight forward or obliquely downward. Wing clear or tinged, in some cases with a stigma spot; costa surrounding the wing entirely; cell cup small; anal vein short or absent; cell dm present with 3 veins from cell dm to the wing margin. Legs short, hind leg in some cases compressed and with conspicuous widened tibia and first tarsal segment.

521 Microphor holosericeus (N male; Collin 1961

Biology: The larvae of this small family are little known; they are assumed to prey on organisms living in rotting organic material in sandy soil in particular. The adults are predators but can also frequently be found on flowers where they feed on pollen and nectar. The females of the genus *Microphor* are known to feed on insects caught in webs of spiders. Both males and females fly in swarms. Species of *Schistosoma* and *Parathalassius* prefer sandy habitats, coastal areas and deserts.

Identification references: Chvála 1983-1988a (revision); Collin 1961 (British Isles); Gatt 2003 (Mediterranean *Microphorella*); Gorodkov & Kovalev 1989 (former USSR); Sinclair & Cumming 2006 (classification); CMPD: no contribution; CAT: Chvála 1989c.

Milichiidae (key couplet 127-128; fig. 273-278)

Systematics: Acalyptrate Brachycera; superfamily Carnoidea; 9 genera and about 45 species in Europe.

Characters: Minute to medium sized (1-6 mm), usually dark coloured, sometimes lustrous black flies and/or with extensive silvery dusting. Mouthparts sometimes elongate; arista bare to pubescent; ocelli present; Oc-bristles present; P-bristles parallel or converging; 2-7 pairs of F-bristles, nearly always the lower 2 pairs curving inward, the upper pairs curving forward, outward, or backward; interfrontal bristles present, in some cases in distinct rows, in other cases more scattered; vibrissae or vibrissa-like bristles present, sometimes very thin. Wing unmarked; costa with humeral and subcostal breaks; vein Sc incomplete; crossvein BM-Cu usually present, in some cases vague or absent; cell cup closed. Tibiae without dorsal preapical bristle.

522 Madiza glabra Fallén, mal Stackelberg 1956

Biology: The species in this family show diverse modes of life; most species feed on organic matter or animal waste, several species feed on dung, sap exuding from tree wounds, seeds, in nests of birds, in nests of ants and other social insects, etc. Adults can be expected in all sorts of habitats, including indoors; they also regularly visit flowers, carrion and dung. Adults of some species are known to attach themselves to spiders and predatory insects (e.g. Asilidae) in order to suck juices from the prey caught by these animals.

Identification references: Brake 2000 (key World genera); Sabrosky 1983 (*Desmometopa*); Stackelberg 1989av (former USSR); CMPD: Papp & Wheeler 1998; CAT: Papp 1984j.

Muscidae (key couplet 180, 182-183; fig. 104, 178, 385, 396-398, 407-408)

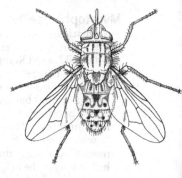

523 Stomoxys calcitrans (Linnaeus), female; Séguy 1951

Systematics: Calyptrate Brachycera; superfamily Muscoidea; some 45 genera and about 575 species in Europe.

Characters: Small to large (2-18 mm) flies, usually coloured grey to black, in some cases yellowish brown or with an extensive green to blue metallic lustre. Mouthparts well developed; occiput lacking white or pallid hairs; arista bare to long plumose; interfrontal bristles usually absent, in some cases present in female (e.g. in *Azelia*). Scutellum bare underneath; subscutellum absent; meral bristles virtually always absent although *Eginia* has a patch of rather strong bristles and *Graphomya* has a row of rather weak bristles. Wing with apical half of vein Sc distinctly curving; vein M1 straight or with a faint to distinct forward curve; cell r4+5 open; anal vein usually well developed, but not continuing to wing margin; vein A2 and the anal vein would not intersect if virtually continued further (with the exception of some species of *Azelia*). Tibia of hind leg lacking a submedian dorsal bristle.

Biology: The following life modes of the larvae are known: carnivorous (Phaoniinae, Mydaeinae, *Coenosia, Limnophora, Lispe, Graphomya*), saprophagous in decaying organic matter (vegetable, animal, or excrement) where the third instar may or may not be carnivorous (Muscini, Stomoxyni, *Hydrotaea*) or, rarely, phytophagous (*Atherigona*). The larvae of *Phaonia exoleta* (Meigen) are known to live in rot holes in trees where they feed on the larvae of Culicidae; those of *Limnophora riparia* (Fallén) live in running water and the splash zones along canals, larvae of other genera are semi-aquatic (e.g. *Limnophora, Lispe, Graphomya*). Many species have adapted to human environments. The adults of most species feed on nectar but some species are predators (Coenosiinae), feed on blood (Stomoxyni) or on exuding (animal) wound liquor (*Hydrotaea*); the latter two groups are vectors of diseases. *Musca autumnalis* De Geer is a species that can occur indoors in large numbers from late summer onward, seeking shelter for the winter.

Identification references: Couri & Pont 1999 (genera Coenosiini); Crosskey 1993b (species feeding on blood); Crosskey & Lane 1993 (synanthropic species of *Musca*); D'Assis Fonseca 1968 (British Isles); Gregor et al. 2002 (Central Europe); Hennig 1955-1964 (Lindner); Kabos 1964 (Netherlands); Rozkošný & Gregor 1997, 2004 (species); Stackelberg 1989bc; Oosterbroek & De Jong 2005 (*Eginia*); Zimin & Elberg 1989 (former USSR); CMPD: no contribution; CAT: Pont 1986b.

(Musidoridae): name used in the past for Lonchopteridae.

Mycetobiidae (key couplet 30; fig. 24, 72)

Systematics: Nematocera; superfamily Anisopodoidea; traditionally part of the Anisopodidae; in Europe 4 species in 2 genera, *Mycetobia* (3) and *Trichomycetobia* (1).

Characters: Small to medium sized (4-7 mm), dark coloured Nematocera. Head small and almost spherical; ocelli present; eyes almost holoptic; antenna with 16 simple, uniform segments. Wing clear, with anal lobe. Legs rather sturdy but long, often less dark; coxae elongated; tibiae with apical bristles or spurs.

524 Mycetobia pallipes Meigen, male; Krivosheina 1997d

Biology: The larvae usually occur in small groups and are found in moist habitats in rotting or fermenting organic matter, such as underneath the bark or in wood of dead deciduous and coniferous trees, in exuding tree sap, in rot holes filled with organic matter, etc. The adults mainly occur near the larval habitats.

Identification references: Mamaev 1989a (former USSR); Pedersen 1971 (Scandinavia); CMPD: Krivosheina 1997d; CAT: Mamaev & Krivosheina 1988b.

Mycetophilidae (key couplet 32, 204; fig. 75)

Systematics: Nematocera; superfamily Sciaroidea; traditionally this family includes the Bolitophilidae, Diadocidiidae, Ditomyiidae and Keroplatidae; some 70 genera and about 945 species in Europe.

Characters: Slender to relatively stout Nematocera, minute to large (2-15 mm) but more usually small to medium sized (4-8 mm). The colour varies from pale, dull yellow to brown or black. Head small; eyes dichoptic; 2 or 3 ocelli present; antenna usually with 16 segments but sometimes segments close together resembling a single third segment; mouthparts short, usually half as long as the height of the head, but sometimes longer or even extremely elongate as in *Gnoriste*. Thorax vaulted. Wing usually clear, in some cases tinged or with markings; venation

525 Mycetophila fungorum (De female; Mihályi 1972

characteristic (vein CuA1 and stem of veins M1 and M2 not connected or connected as far up as crossvein H); anal lobe absent or weak. Legs with elongate coxae; tibiae with long apical spurs.

Biology: The larvae of most species feed on mushrooms, bracket fungi or mycelium in moist, dead wood or in other moist organic substrates. They rarely are predators or feed on organic detritus in nests of birds, on mosses or on liverworts (Hepaticae). Some larvae live in a web they have spun themselves underneath tree trunks, bark, mushrooms and bracket fungi, etc. The adults ("fungus gnats") occur chiefly in moist habitats, in woodlands in particular although some species prefer rather different habitats like grassland, marshes, moist heather, etc. They often are active in the evening while during the day they hide in dark and shaded spots; some species are mainly found in caves and in burrows of rodent. Some species visit flowers and are important as pollinators of certain plants.

Identification references: Chandler & Ribeiro 1995 (Atlantic islands); Hutson et al. 1980 (British Isles, excluding Mycetophilinae); Kurina & Schacht 2003 (synopsis identification references); Ostroverkhova & Stackelberg 1989; Stackelberg 1989l (former USSR); A.I. Zaitzev (Eastern Europe, 5 subfamilies), 2003 (Eastern Europe, Manotinae and Mycetophilinae); CMPD: Søli et al. 2000; CAT: Hackman et al. 1988; Krivosheina & Mamaev 1988c (Manotinae).

Mydidae (key couplet 56; fig. 120-121)

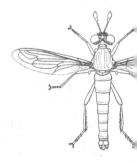

Systematics: Lower muscomorph Brachycera; superfamily Asiloidea; in Europe 4 genera and 7 species.

Characters: Large to very large (10-35 mm), dark coloured flies with characteristic head and wing venation, and with bristles on the legs only. Eyes dichoptic; head dorsally concave in between the eyes, as in the Asilidae; antenna with 4 segments, the terminal segment swollen. Wing clear or tinged, veins R4, R5 and M curving forward, hence at least vein R4 (in European species also vein R5) ending in vein R1; cell m3 and cell cup closed. Legs long and powerful, tibiae with apical spurs; empodium bristle-like or absent.

Biology: The larvae prey on the larvae of Coleoptera living in wood and the soil. The males in particular feed on nectar but in some species the mouthparts are so much reduced that it is likely that they are unable to feed at all. Mydidae mainly occur in open, hot, dry, sandy habitats.

526 Syllegomydas cinctus (Ma male; Verrall 1909

Identification references: Carles-Tolrá 2006; CMPD: Richter 1997b; CAT: Richter & Zaitzev 1988.

Mythicomyiidae (key couplet 66; fig. 101, 141-142)

Systematics: Lower muscomorph Brachycera; superfamily Asiloidea; traditionally part of the Bombyliidae; 7 genera and about 30 species in Europe.

Characters: Minute to small (0.5-5 mm) flies, body bare or with short sparse hairs, mostly black often with a pattern of yellow markings; very variable in shape from broad and compact to slender and elongate. Eyes large, male sometimes holoptic; antenna with four segments, the third segment largest and with the fourth forming the flagellum, the bristle-like style minute. Mouthparts elongate forming a sucking proboscis, palps absent or minute. Wing usually clear and without markings; wing venation reduced, costa incomplete, R2+3 unbranched, discal cell open or closed. Wings held vertically above the body when at rest. Legs slender; empodium bristle-like or absent.

527 Glabellula arctica (Zetterstedt), female; Engel 1932

Biology: Little is known about the biology, since only six species have been reared worldwide. Larvae of the European *Glabellula arctica* (Zetterstedt) have been found in the nests of wood ants (*Formica*) and are presumed to be predatory. Elsewhere species have been reared from grasshopper egg-pods and as a parasite of solitary bees. Adults are found resting on flowers on which they feed and flying close by. Most species are found in warm arid environments.

Identification references: Van der Goot & Van Veen 1996 (Northwestern Europe); Greathead & Evenhuis 2001 (key genera); V.F. Zaitzev 1989b (former USSR); CMPD: Greathead & Evenhuis 1997 (as Mythicomyiinae); CAT: V.F. Zaitzev 1989d; Evenhuis 2002.

Nannodastiidae (key couplet 167; fig. 363-364)

Systematics: Acalyptrate Brachycera, not yet assigned to a superfamily; in Europe 1 genus, *Azorastia*, with 3 species in the Mediterranean region and the Azores.

Characters: Minute (0.7-1.3 mm), largely dull brown flies. Arista pubescent or with short rays; ocelli present; Oc-bristles present; small, weak pseudopostocellar bristles present, curving forward or converging; 3 pairs of F-bristles, curving backward; vibrissae absent but weak bristles present along the lower margin of the genae. Wing clear; costa with humeral and subcostal breaks; vein Sc incomplete; crossvein BM-Cu absent; crossvein DM-Cu present and vein M1 reaching the wing margin; cell cup absent. Tibiae without dorsal preapical bristle.

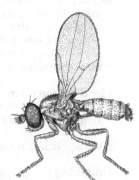

528 Azorastia mediterranea Papp, male; Papp 1980

Biology: The biology of this small family is very little known; all species are found in coastal regions where they often occur in caves and below overhanging cliffs. The larvae are presumed to feed on organic detritus.

Identification references: Papp & Mathis 2001 (review); CMPD: Papp & Mathis 1998.

Nemestrinidae (key couplet 49; fig. 107-109)

Systematics: Lower muscomorph Brachycera; superfamily Nemestrinoidea; 6 genera and about 13 species in Europe.

Characters: Medium sized to very large (9-24 mm) flies devoid of bristles, body often covered with relatively long pubescence; body colour variable, thorax and abdomen sometimes striped or banded. Ocelli present; eyes dichoptic; mouthparts often elongate. Wing clear or tinged, large and long, with complete venation; R-veins curving parallel to the hind margin, in some cases with a large number of small cells; cell m3 closed; cell cup open

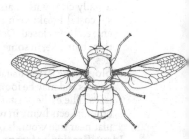

529 Nemestrinus reticulatus Latreille, female; Verrall 1909

or closed. Legs long and slender, lacking conspicuous bristles; empodium pulvilliform.

Biology: The larvae are endoparasitoids of grasshoppers or of beetles of the family Scarabaeidae and play an important part in regulating grasshopper populations. They feed on the fats and body juices of their host. Larvae that are parasitoids on grasshoppers are connected with the air outside through a long respiratory tube. The host remains alive until the larva reaches the fourth instar and emerges from the host to spend the winter in the soil. Pupation takes place in the next season. Adults often hover in the air over flowers or in sunlit, open spots; those with elongate mouthparts frequently visit flowers.

Identification references: Bernardi 1973 (genera); Paramonov 1945-1956 (species); Richter 1989a (former USSR); CMPD: Richter 1997a; CAT: Richter 1988.

(Nemopalpidae): part of the Psychodidae.
(Neottiophilidae): part of the Piophilidae.

Nycteribiidae (key couplet 208; fig. 441)

530 Phthiridium biarticulatum female; Theodor 1975

Systematics: Calyptrate Brachycera; superfamily Hippoboscoidea; 4 genera and about 15 species in Europe.

Characters: Minute to medium sized (1.5-5 mm), yellowish to brown, spider-like, flattened flies without wings. Halteres present. Head small, not adpressed to the thorax but, in resting position, bent backward on to thorax; eyes and ocelli small or absent. Legs long with swollen femora and tibiae; first tarsal segment at least as long as all other tarsal segments combined.

Biology: The adults are ectoparasites of bats, feeding on their blood. The larvae develop inside the abdomen of the female fly. Just prior to deposition the female leaves her bat host and glues the mature, almost pupated larva (prepupa) to a solid substrate near the resting place of the bats, after which the prepupa fairly rapidly forms a puparium. Nycteribiidae show various degrees of host specificity, ranging from species-species associations, through associations with closely related hosts to apparent absence of any host preference.

Identification references: Hutson 1984 (British Isles); Schuurmans Stekhoven Jr. & Van den Broek 1969 (Netherlands); Stackelberg 1989be (former USSR); Theodor 1967 (revision); CMPD: Hůrka 1998a; CAT: Hůrka & Soós 1986a.

(Ochthiphilidae): name used in the past for Chamaemyiidae.

Odiniidae (key couplet 96, 121; fig. 214, 261-262)

531 Odinia maculata (Meigen) Papp 1978

Systematics: Acalyptrate Brachycera; superfamily Opomyzoidea; 3 genera and about 14 species in Europe.

Characters: Small (2-5 mm), stout flies, body usually grey with dark markings, legs sometimes banded. Arista pubescent; ocelli present; Oc-bristles present; P-bristles diverging; 3 pairs of F-bristles, the lower pair curving inward, the other pairs backward; interfrontal bristles absent; vibrissae present. Wing usually clear with markings along the crossveins and near the subcostal break; vein Sc incomplete; crossvein BM-Cu present; cell cup closed. Dorsal preapical bristle on the tibiae often thin, hair-like, in some cases similar to the other hair-like bristles.

Biology: The larvae usually live in association with woodboring insects and may be found in tunnels and underneath the bark of deciduous and coniferous trees. The earlier instars feed on exuding juices and insect frass, the later instars prey on other insects living in wood. The adults live in the vicinity of the larval habitats, in particular near tree wounds or on wood-associated mushrooms and bracket fungi.

Identification references: Stackelberg 1989aj (former USSR); CMPD: Papp 1998c; CAT: Krivosheina 1984b.

Oestridae (key couplet 174; fig. 102, 153, 379-383)

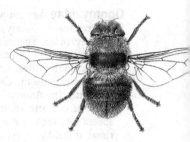

Systematics: Calyptrate Brachycera; superfamily Oestroidea; traditionally this family includes the Gasterophilidae and Hypodermatidae; 4 genera and 8 species in Europe.

Characters: Medium sized to large flies (9-18 mm). Body largely without bristles, nearly bare, or body and legs covered in a soft pilosity, making these insects resemble bees or bumblebees. Mouth opening small, mouthparts rudimentary; antenna small; ptilinal suture usually with the ends converging below the antennal insertions. Lower calypter large. Vein M1 strongly curving or with an angle in the direction of the wing margin or ending in vein R4+5; in the latter case cell r4+5 petiolate. Subscutellum flat or strongly swollen; meron in some cases with bristles along its hind margin, near the posterior spiracle (often difficult to see because of the dense pilosity).

532 Cephenemyia trompe (Modeer); Grunin 1966

Biology: The larvae are parasites in the nostrils or nasal cavities of mammals, especially sheep and goats but also occasionally antelopes, deer, camels and horses. The host relation is rather specific. Oestridae are larviparous which means that larvae (instead of eggs) are deposited into the nostrils of the host, usually from a distance. The female hovers just in front of the host and ejects the larvae into the host's nostrils. The larvae migrate to the nasal cavities and, in most cases, later into the pharynx. Full-grown larvae leave their host through the nostrils or mouth opening and pupate in the soil. The adults have reduced mouthparts and do not feed. They seem to be rather inactive insects and are relatively rarely observed. A conspicuous phenomenon in this family is hilltopping, i.e., the habit of the adults seeking out high places or landmarks, in order to raise the chances of males and females meeting each other.

Identification references: Colwell et al. 2004 (general); Grunin 1966-1969 (Lindner), 1989d (former USSR); Zeegers 1992 (Netherlands); CMPD: Minář 2000c; CAT: Soós & Minář 1986b.

(Olbiogastridae): part of the Anisopodidae.
(Omphralidae): name used in the past for Scenopinidae.
(Oncodidae): name used in the past for Acroceridae.

Opetiidae (key couplet 73; fig. 157-158)

Systematics: Lower Cyclorrhapha; superfamily Platypezoidea; traditionally part of the Platypezidae; in Europe 1 widespread species, *Opetia nigra* Meigen.

Characters: A small (2-5 mm), slender, all dark species with a characteristic wing venation. Male with holoptic eyes; antenna with elongate second and third segments, arista long, terminal. Wing darkened along the costa, in particular in the male; vein Sc reaching the wing margin; crossvein DM-Cu absent; vein M with a long fork; small cell cup narrowing to a point; anal lobe large in the male, virtually absent in the female. Legs slender, simple.

Biology: The only European species, *Opetia nigra*, has been reared from a piece of birch wood in an advanced state of decay. The ovipositor is peculiarly shaped, suggesting a specialised mode of ovipositing and a distinctive larval habitat. The adults often occur in wooded areas but in open habitats as well. Adult males are often found on leaves of trees and bushes, the much less frequently found females in lower vegetation. It appears that nectar feeding may occur. The males sometimes congregate in swarms.

533 Opetia nigra Meigen, male; Chandler 1998

Identification references: Chandler 2001 (revision); CMPD: Chandler 1998a; CAT: Chandler 1991a.

Opomyzidae (key couplet 105, 164, 218; fig. 236-237, 358-360)
Systematics: Acalyptrate Brachycera; superfamily Opomyzoidea; 3 genera and about 33 species in Europe.
Characters: Small (2-5 mm), slender, yellow, brown, reddish or black coloured flies. Arista pubescent or with relatively long rays; ocelli present; Oc-bristles present; P-bristles absent or diverging (in *Anomalochaeta*); 1 pair of F-bristles, curving backward; scattered interfrontal setulae present; vibrissae absent although *Geomyza* with a strong bristle near the vibrissal angle. Wing nearly always with an apical spot and/or darkly marked crossveins or marked all over; costa with a subcostal break; vein Sc usually incomplete but apical part sometimes visible as a thin line reaching the costa; crossvein BM-Cu present but usually incomplete; cell cup closed. Tibiae without dorsal preapical bristle.

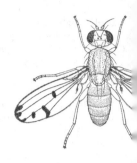

534 Opomyza florum (Linnaeu Balachowsky & Mesnil 1935

Biology: In view of the larval food plants, i.c., grasses, the adults are mainly found in open habitats. The larvae live in the stems of grasses, a few species being a pest in agriculture.
Identification references: Czerny 1928 (Lindner); Drake 1992, 1993 (British Isles); Soós 1981 (Central Europe); Stackelberg 1989ap (former USSR); Carles-Tolrá 1993, 1994 (Spain); Van Zuijlen 1999 (Fallén collection); CMPD: Brunel 1998; CAT: Soós 1984k.

(Ortalidae): name used in the past for Ulidiidae: Otitinae.
(Otitidae): part of the Ulidiidae.

Pachyneuridae (key couplet 29; fig. 71)
Systematics: Nematocera; superfamily Pachyneuroidea; in Northern Europe 1 species, *Pachyneura fasciata* Zetterstedt.
Characters: *Pachyneura fasciata* is a large species (10-13 mm), resembling Tipulidae on account of its long and slender antennae, wings, legs and abdomen. Ocelli prominent; antenna nearly as long as the thorax, with 17 segments. Wing partly tinged and with a distinct stigma spot. Tibiae with apical bristles.
Biology: Larvae and pupae of the only European species, *Pachyneura fasciata*, are found in rotting wood lying on the ground, in particular in parts with black fungous tissue. Adults occur in early summer on tree trunks, predominantly on birch.

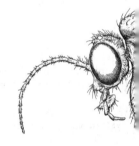

535 Head Pachyneura fasciata Zetterstedt; Krivosheina 1997a

Identification references: Stackelberg 1989f (former USSR); CMPD: Krivosheina 1997a; CAT: Krivosheina & Mamaev 1988a.

Pallopteridae (key couplet 144, 162; fig. 311-313, 352-353)
Systematics: Acalyptrate Brachycera; superfamily Tephritoidea; 5 genera and about 23 species in Europe.
Characters: Small to medium sized (2.5-7 mm), slender flies varying in colour from pale yellow to reddish, grey or black; some parts of the head often silvery grey to white. Arista bare or with short rays, rays at most twice as long as basal diameter of arista; ocelli present; Oc-bristles present; P-bristles parallel to somewhat diverging; 1 pair of strong F-bristles, curving backward, sometimes with a weaker F-bristle in front; scattered interfrontal setulae present, sometimes almost invisible or indeed lacking; vibrissae absent but usually a series of fine setulae present along the lower margin of the genae. Wing often with a pattern, seldom clear; costa weak just beyond crossvein H and with a subcostal break; vein Sc complete;

536 Palloptera quinquemacula (Macquart), female; Balachows Mesnil 1935

crossvein BM-Cu present; cell cup closed. Tibiae without dorsal preapical bristle, except in *Eurygnathomyia*.

Biology: In some species the larvae live under tree bark where they feed on mycelium or prey on beetles (Carabidae, Cerambycidae and Scolytidae). Larvae of other species live in plants, such as grasses, Apiaceae or the flower heads (inflorescences) of Asteraceae (Compositae); it is unknown whether the species living in plants are really phytophagous, carnivorous, or both. The adults are generally found in shaded habitats, often near water. Depending on the larval habitat, the adults are found on flowers, often those of umbellifers, or on tree trunks and leaves.

Identification references: Andersson 1990 (Sweden); Greve 1993 (Norway); J.F. McAlpine 1981c (World genera); Morge 1963-1974 (review); Stackelberg 1989ah (former USSR); CMPD: Merz 1998b; CAT: Morge 1984.

Pediciidae (key couplet 5, 196, 198; fig. 30, 428-429, 432)

Systematics: Nematocera; superfamily Tipuloidea; until recently part of the Limoniidae which is traditionally included in the Tipulidae; 5 genera and about 60 species in Europe.

Characters: Nematocera varying from medium sized (5 mm in *Dicranota*, subgenus *Paradicranota*) to very large (35 mm in *Pedicia*). Antennae, wings, legs and abdomen long and slender. Ocelli absent; eyes pubescent, with ommatrichia in between the facets; antenna with 12-17 segments. Thorax with a V-shaped transverse suture. Wing with 2 anal veins, wing in some cases with a characteristic pattern.

Biology: The larvae of the genus *Ula* feed on mushrooms and bracket fungi; all other larvae, as far as known, are aquatic to semi-aquatic and prey on worms, insect larvae, mites and other small invertebrates. Predatory larvae are mobile and rapidly moving; because of their massive head capsule they are not able to swallow large prey as they are but have to tear them to pieces. Of the three more common genera, the larvae of

537 Tricyphona immaculata (Meigen), male; Papp & Schumann 2000

Tricyphona are found in diverse wet and semi-aquatic habitats (from the shores of springs, streams and lakes to bogs and forest litter in wet places), those of *Dicranota* tend to occur along and in running waters but develop also at the shores of standing water bodies; larvae of *Pedicia* are restricted to running water of wells, brooks and small rivers. Pupating takes place inside a silken tube. Adults are found in an array of moist habitats.

Identification references: Dienske 1987; Reusch & Oosterbroek 1997 (genera and subgenera); Savchenko 1989c, 1989d (former USSR); Stubbs 1994-1999 (British Isles); CMPD: no contribution; CAT: Savchenko et al. 1992; Oosterbroek 2006.

Periscelididae (key couplet 121; fig. 259-260)

Systematics: Acalyptrate Brachycera; superfamily Opomyzoidea; in Europe 1 genus, *Periscelis*, with 4 species.

Characters: Minute to small (1-5 mm), grey or dark coloured flies, often with banded legs. Antenna with second segment overhanging third segment; arista with long rays; ocelli present; Oc-bristles present; P-bristles diverging; 1 pair of F-bristles, curving backward; interfrontal bristles absent; a row of vibrissa-like bristles present. Wing clear or with infuscated spots; costa continuous; vein Sc incomplete; crossvein BM-Cu present; cell cup open or closed. Tibiae without dorsal preapical bristle.

Biology: The larvae of this small family live in sap exuding from deciduous trees. The adults are often found near the larval sites and feed on tree sap. Males may be seen hovering near sap runs.

538 Periscelis annulata (Fallén), female; Papp 1973

Identification references: Bächli 1997 (Switzerland); Stackelberg 1989ae (former USSR); CMPD: Mathis & Papp 1998; CAT: Papp 1984b.

(Petauristidae): name used in the past for Trichoceridae.

Phaeomyiidae (key couplet 143; fig. 310)

Systematics: Acalyptrate Brachycera; superfamily Sciomyzoidea, traditionally part of the Sciomyzidae; in Europe 1 genus, *Pelidnoptera*, with 3 species.

Characters: Small to large (3-11 mm), yellowish brown to brown coloured, slender flies with brownish tinged wings. Arista pubescent; ocelli present; Oc-bristles present; P-bristles diverging; 2 pairs of F-bristles, curving backward; interfrontal bristles absent; vibrissae absent. Wing tinged brown, without markings; costa continuous; vein Sc complete, crossvein BM-Cu present; cell cup closed. Tibiae with dorsal preapical bristle.

539 Pelidnoptera nigripennis (♈ male; Rivosecchi 1992

Biology: One species in this small family is known to be a parasitoid of Diplopoda (millipedes) of the genus *Ommatoiulus*. The female deposits her eggs on the millipede after which the young larva enters its host. Usually only one larva per millipede will emerge. The host remains alive until the larva is fully grown and pupates in the host which, by that time, is no more than an empty shell.

Identification references: Revier & Van der Goot 1989 (Northwestern Europe); Rivosecchi 1992 (Italy); Stackelberg 1989ac (former USSR); CMPD: Rozkošný 1998b; CAT: Rozkošný & Elberg 1984 (as subfamily of Sciomyzidae).

(Phlebotomidae): part of the Psychodidae.

Phoridae (key couplet 69, 210; fig. 148-149, 445)

Systematics: Lower Cyclorrhapha; superfamily Platypezoidea; some 35 genera and about 605 species in Europe.

Characters: Minute to medium sized (0.5-6 mm) flies, often somewhat arched and more or less stout, with a characteristic wing venation. Head, palps and legs usually with strong, dentate or feathered setae. The body colour varies from black through brown, orange and yellow to pale grey or pale white. Eyes dichoptic; antenna with a large, rounded or elongate third segment bearing a long apical or somewhat dorsal arista directed sideways. Wing clear or tinged, rarely with markings; venation reduced, the strong R-veins ending in the costa about halfway up the wing; the other veins are weaker and usually follow a diagonal course, often parallel to each other. Legs typically with stout femora; hind femora often somewhat laterally compressed.

540 Megaselia scalaris (Loew)♈ Disney 1983

Biology: Rather variable, showing the greatest diversity of all Dipterous families. Larvae are found in several terrestrial habitats, in the nests of social insects, and in some aquatic habitats, feeding on organic detritus such as dung, carrion, insect frass, dead snails, etc.; some feed on bracket fungi and mushrooms, on mycelium or on living plants (among others as miners), they may be predators or parasites of a wide array of organisms such as earthworms, snails, spiders and their eggs, centipedes, millipedes, insect larvae and pupae. Most adults feed on nectar, honeydew and juice exuding from fresh carrion and dung; some feed on body juices of living beetle larvae and pupae, others prey on small insects. The adults are conspicuous on account of their fast and somewhat abrupt manner of locomotion. In some species, the males fly in swarms.

Identification references: Disney 1983-1989 (British Isles); Disney 1994 (genera); V.F. Zaitzev 1989c (former USSR); CMPD: Disney 1998; CAT: Disney 1991.

(Phryneidae): name used in the past for Anisopodidae.
(Phthiriidae): part of the Bombyliidae.

Piophilidae (key couplet 114; fig. 247-248)

Systematics: Acalyptrate flies; superfamily Tephritoidea; some 16 genera and about 30 species in Europe.

Characters: Minute to medium sized (1.5-7 mm), yellow or brown to black, in some cases lustrous flies, some species with a striking pale coloration of the face or with more or less extensive yellow or reddish coloured parts. Arista bare to pubescent; ocelli present; Oc-bristles present; P-bristles diverging, rarely parallel; 0-4 pairs of F-bristles, if present curving outward or backward, in *Actenoptera* the lower 2 pairs curving inward; scattered interfrontal setulae present; vibrissae present. Wing usually unmarked, in some genera with spots along the crossveins and/or at the ends of the longitudinal veins; costa with a subcostal break; vein Sc complete; crossvein BM-Cu virtually always present; cell cup closed. Tibiae without dorsal preapical bristle.

541 Piophila casei Linnaeus, male; Stackelberg 1956

Biology: The larvae of most species live in rotting wood, bracket fungi, or mushrooms, some species live in nests of birds where they feed on the blood of the nestlings (*Neottiophilum*), or in carrion, mainly that of larger vertebrates (*Thyreophora* and its allies), or in substrates rich in protein. A well-known representative of the latter group is the larva of *Piophila casei* (Linnaeus), a pest in a large variety of animal products such as cheese, dried fish, meat, etc. The adults are not very active fliers and are mainly found near the larval habitats for oviposition or on tree trunks and logs, possibly for mating.

Identification references: J.F. McAlpine 1977 (genera); Merz 1996b (Switzerland); Gorodkov 1989a; Stackelberg 1989af, 1989ag (former USSR); CMPD: Ozerov 2000; CAT: Zuska 1984; Papp 1984c (Thyreophorinae); Soós 1984h (Neottiophilinae).

Pipunculidae (key couplet 65; fig. 139-140)

Systematics: Lower Cyclorrhapha; superfamily Syrphoidea; some 14 genera and about 200 species in Europe.

Characters: Small to large (2-12 mm), dark coloured flies with long, narrow wings and very large eyes occupying nearly all of the semi-spherical to nearly entirely spherical head. Antenna short, third segment usually the largest, with a dorsal arista or style; mouthparts small. Wing clear or tinged brown; vein M1 and cell dm sometimes absent; cell cup elongate, ending acutely just before the wing margin. The female has a lance-like ovipositor.

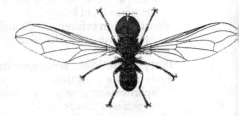

542 Pipunculus campestris Latreille, male; Verrall 1909

Biology: The larvae are predominantly parasitoids of leafhoppers and planthoppers (Hemiptera: Auchenorrhyncha) except *Nephrocerus* which has been reared from adult Tipulidae (Koenig & Young 2007). The large eyes of the adult pipunculids apparently play an important role in locating a suitable host, as does the ability to hover in one spot. Once located, the female uses her lance-like ovipositor to deposit an egg directly inside the host. More than one larva may be present in a single host. When nearing completion of development, the larvae eat the host until only an empty shell remains. The larvae then leave through the cuticle to pupariate in the soil.

Identification references: Albrecht 1990 (*Dorylomorpha*); Coe 1966 (British Isles); Von der Dunk 1998 (Central Europe); Földvári & De Meyer 2000 (*Tomosvaryella*); Jervis 1992 (*Chalarus*); Kehlmaier 2005 (tribe Eudorylini); De Meyer 1989 (*Cephalops, Beckerias*); Tanasijtshuk 1989c (former USSR); CMPD: Kozánek et al. 1998; CAT: Tanasijtshuk 1988; De Meyer 1996.

Platypezidae (key couplet 74; fig. 159-160)

Systematics: Lower Cyclorrhapha; superfamily Platypezoidea; traditionally this family includes the Opetiidae; some 12 genera and about 45 species in Europe.

Characters: Minute to medium sized (1.5-6 mm), slender to stout flies, male and female often differing in colour, the male in particular often all black; one or both sexes partially to all grey or yellow, in some cases with orange, grey or silvery markings, or abdomen bicolourous. Eyes holoptic in the male; antenna with 3 segments, the third of which is largest and bears a long, apical arista. Wing clear or tinged, in some genera the area along the wing margin, in between and around veins Sc and R1 dark-

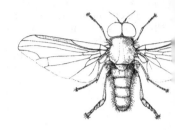

543 Protoclythia modesta (Zetterst⟨ male; Verrall 1909

ened; wing with conspicuous anal lobe; anal vein reaching the wing margin; cell cup ending in an acute angle, often elongate; in most genera vein M forked; crossvein DM-Cu virtually always present (absent in *Microsania* where crossvein R-M is also lacking). Legs relatively short, stout; first tarsal segment of hind leg in the subfamily Callomyiinae long and cylindrical, usually swollen in the male, slender in the female, the subsequent tarsal segments cylindrical; in the subfamily Platypezinae the first tarsal segment of hind leg short, laterally compressed, subsequent tarsal segments laterally compressed as well, in particular in the female.

Biology: As far as known the larvae develop on and in mushrooms and bracket fungi. The adults are mainly found in woodland areas, walking about on leaves of trees and shrubs where they feed on honeydew and other organic matter on the leaf surface. The males sometimes fly in swarms; those of *Microsania* are known as 'smoke flies' since they swarm in the smoke of forest fires (and other fires as well). The larval mode of life of *Microsania* is not known.

Identification references: Chandler 2001 (revision); Tanasijtshuk 1989b (former USSR); CMPD: Chandler & Shatalkin 1998; CAT: Chandler 1991b.

Platystomatidae (key couplet 112, 163; fig. 306, 354-355)

Systematics: Acalyptrate Brachycera; superfamily Tephritoidea; 2 genera, *Platystoma* and *Rivellia*, and about 20 species in Europe.

Characters: Small to large (3-11 mm), dark flies, in *Platystoma* with brown wings with numerous hyaline dots and stout, microtrichose body and usually bright yellow ventral part of abdomen, in *Rivellia* with dark banded or spotted wings and shiny, narrow body. Head round to distinctly flattened; arista bare to pubescent; ocelli present; Oc-bristles reduced or absent; P-bristles reduced or absent; 1-2 pairs of F-bristles; scattered interfrontal setulae present; vibrissae absent but sometimes vibrissa-like bris-

544 Platystoma seminationis (Fabr⟨ male; Colyer & Hammond 1968

tles present. Costa with a humeral break, otherwise continuous; vein Sc complete, crossvein BM-Cu present; cell cup closed. Tibiae without dorsal preapical bristle. Female with oviscape, non retractable basal segment of the ovipositor.

Biology: The larval habitats are rather diverse, most species feeding on rotting plant materials or on roots in the soil, some species living under the bark of deciduous trees; larvae of *Rivellia* feed on the root nodules in legumes. The adults are also found in various habitats such as woodland, moist meadows, dunes, etc., often resting on tree trunks, stones and leaves, in the latter case preferring the lower side. The adults feed on the juice exuding from dung and carrion, on nectar, honeydew, fruit and on sap exuding from tree wounds.

Identification references: Clements 1990 (British Isles); Korneyev 2001 (genera); Merz 1996c (Switzerland); Richter 1989c (former USSR); CMPD: D.K. McAlpine 1998a; CAT: Soós 1984e.

Pleciidae (key couplet 27, 205; fig. 67-68)
Systematics: Nematocera; superfamily Bibionoidea, traditionally part of the Bibionidae; in Europe 2 species, *Penthetria
funebris* Meigen (native, widespread) and *P. heteroptera* (Say)
(probably introduced, Germany).
Characters: The only native European species, *Penthetria funebris*, is medium sized (6-8 mm), *Bibio*-like, dark to black
coloured, with short antennae, an elongate abdomen and short
and narrow wings, especially so in the male. Ocelli prominent;
eyes in the male almost touching and divided into two parts;
in the female the eyes are simple and dichoptic; antenna
inserted below the middle of the face, short, with 11 segments.
Wing tinged; vein R forked into R2+3 and R4+5; vein M forked
into M1 and M2; anal lobe especially present in the female. Legs
simple; tibiae with distinct but relatively short apical spurs.

545 Penthretia funebris (Meigen),
male; Séguy 1940

Biology: The larvae feed on decaying organic matter in the top-
soil of deciduous woodlands, often near little streams or near tree trunks; the larvae of *P.
funebris* have been found developing in alder litter, the Nearctic *P. heteroptera* was found
at a composting facility in Germany where the larvae apparently developed in vegetable
refuse (Fitzgerald & Werner 2004). The adults are usually rare but can occur in large numbers locally; they do not fly but walk on the ground or crawl through the vegetation near
the larval habitats.
Identification references: Duda 1930 (Lindner); Fitzgerald & Werner 2004 (key, review);
Mikolajczyk 1977 (Poland); Zeegers 1997, 1998 (Netherlands); CMPD: Skartveit 1997;
CAT: Krivosheina 1986a.

Pseudopomyzidae (key couplet 133; fig. 294-296)
Systematics: Acalyptrate Brachycera; superfamily Nerioidea;
in Europe 1 widespread species, *Pseudopomyza atrimana*
(Meigen).
Characters: A minute to small (1.7-2.5 mm), dark to black
coloured species. Arista pubescent; ocelli present; Oc-bristles
present; P-bristles converging; 3 pairs of F-bristles, curving
backward; interfrontal bristles absent; vibrissae present.
Wing unmarked, costa with both subcostal and humeral
breaks or a subcostal break and a humeral weak spot; vein Sc
incomplete; crossvein BM-Cu absent; cell cup closed. Tibiae
without dorsal preapical bristle.

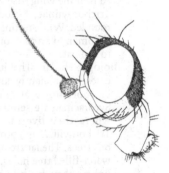

546 Head Pseudopomyza atrimana
(Meigen); Chandler 1983

Biology: The biology of this small family is little known. The
only European species, *Pseudopomyza atrimana*, occurs in
woodland habitats, and adults have been recorded to gather
over rotting logs, a habitat similar to that of an East Asian
species whose larvae were found living under bark of various
trees.
Identification references: Chandler 1983 (British Isles);
Stackelberg 1989an (former USSR); CMPD: D.K. McAlpine &
Shatalkin 1998; CAT: Krivosheina 1984c.

Psilidae (key couplet 150; fig. 325-327)
Systematics: Acalyptrate Brachycera; superfamily Diopsoidea;
6 genera and about 50 species in Europe.
Characters: Small to medium sized (2.5-10 mm), slender to
stout, yellow to reddish, brown or black coloured flies with
few bristles. Head with relatively small eyes, most often with
receding face; antenna small or with third antennal segment
conspicuously elongate; arista short or longer pubescent;
ocelli present; Oc-bristles present; P-bristles diverging or

547 Psila merdaria (Collin), male; van der
Goot & van Veen 1996

absent; 0-2 pairs of F-bristles; scattered interfrontal setulae present; vibrissae absent. Wing usually clear to slightly tinged, rarely marked, in some cases along costa, the wing tip or the crossveins with dark band(s); wing with a transverse weakening in basal half; costa with a subcostal break at some distance from where vein R1 meets the costa; vein Sc incomplete, its apical part transparent and more or less perpendicular to the basal part; crossvein BM-Cu present; cell cup closed. Tibiae without dorsal preapical bristle.

Biology: The larvae are nearly all phytophagous. They live in roots, tubers and stems of a range of non-woody herbs, a few species live under tree bark. Several species are pests in agriculture and horticulture (such as the root fly *Psila rosae* Fabricius, living in several crops nearly all over the world; and *Psila nigricornis* Meigen which inhabits the roots of chrysanthemums and other garden plants). The adults prefer a low, dense vegetation.

Identification references: Van der Goot & Van Veen 1996 (Northwestern Europe); Stackelberg 1989w (former USSR); CMPD: Iwasa 1998b; CAT: Soós 1984c.

Psychodidae (key couplet 12; fig. 42-42a)

Systematics: Nematocera; superfamily Psychodoidea; some 40 genera and about 500 species in Europe.

Characters: Small to medium sized (2-6 mm), often stout, usually densely pubescent Nematocera, non-piercing except for the subfamilies Sycoracinae and Phlebotominae. Ocelli absent; in the Psychodinae, the largest and most common subfamily, the eyes are reniform and with an eye bridge, eyes circular in the other subfamilies; antenna usually with 12 to 16 segments, the flagellomeres bottle- or barrel-shaped. Wing hairy, with 9-10 longitudinal veins and the R-fork and M-fork always present, though they may be incomplete basally; crossveins situated near the wing base or (partly) absent; wing usually acute, in Trichomyiinae, Sycoracinae and some Psychodinae wing rounded. While resting or walking the wings are hold above the abdomen like a roof, or besides the abdomen in a broad V-shape. Legs ranging from short to elongate.

548 Phlebotomus papatasi (S(male; Séguy 1950

Biology: The larval habitats of the five European subfamilies are rather diverse (and in some cases insufficiently known) but truly aquatic species are lacking. The larvae of the Phlebotominae are known from soil dug up from rodent burrows. Those of the Sycoracinae are semi-aquatic in mosses, spring brooks, or waterfalls and fast streams where they live on rocks near the surface or in the splash zone. Those of the Trichomyiinae and probably the Bruchomyiinae live in rotting and dead wood of deciduous trees. The larvae of the Psychodinae occur in a wide range of habitats: semi-aquatic, water-filled tree holes, leaf axils, kitchen sink deposit catchers, water purification plants (where they can be a pest), or more terrestrially in humus and leaf litter, as miners in leaves, in excrement, in public lavatories and toilets, in sand and mud alongside water, in moss, mushrooms and bracket fungi, in dead wood, etc. The adults have a short and erratic flight and are largely found in woodland areas near water, in marshland, or near (indoor) larval habitats. They are predominantly nocturnal and by day they rest in dark places. Certain species are regularly found on window panes at dusk or occasionally swarm round sink overflows. Mouthparts are non-functional except in females of the Sycoracinae, which feed on the blood of amphibians, and the Phlebotominae, the 'sand flies' which feed on human blood and, in the tropics, can be important disease vectors. Some Psychodinae do occasionally become a nuisance if they, for example, breed in numbers in domestic drains.

Identification references: Jung 1958 (Lindner: Bruchomyiinae); Lane 1993 (Phlebotominae); Perfilyev 1989 (former USSR); Szábo 1983 (Central Europe); Tanasijtshuk 1989a (former USSR); Theodor 1958 (Lindner: Phlebotominae); Vaillant 1971-1983 (Lindner: Psychodinae); Wagner 1982 (Trichomyiinae), 1997d (subfamilies, genera Psychodinae); Withers 1989 (British Isles); CMPD: Wagner 1997a; CAT: Wagner 1990a.

Ptychopteridae (key couplet 11; fig. 40-41)

Systematics: Nematocera; superfamily Ptychopteroidea; in Europe 1 genus, *Ptychoptera*, with about 15 species.

Characters: Medium sized to large (7-15 mm), slender, black lustrous Nematocera, often with lighter markings on thorax and/or abdomen. Antennae, wings, abdomen and legs long and slender. Ocelli absent; antenna with 15 to 21 segments. Thorax with deep, posteriorly directed transverse suture. Wing with markings, in particular along the crossveins and where veins bifurcate; spurious vein present on either side of crossvein R-M and wing membrane with a distinct fold between veins A1 and CuA2.

Biology: The larvae are aquatic to semi-aquatic, living in muddy water, shallow pools in marshes, *Sphagnum* pools, or the margins of streams. They filter-feed on small organic particles. They are connected with the air through a long ventilation tube at the end of the abdomen. The adults live in marshy and moist habitats, near suitable substrates for the larvae as well as quite some distance from the larval habitat.

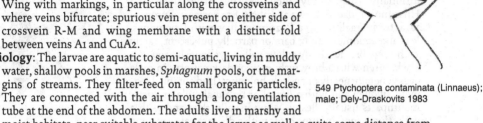

549 Ptychoptera contaminata (Linnaeus); male; Dely-Draskovits 1983

Identification references: Andersson 1997a (species); Krzeminski 1986 (Poland); Peus 1958 (Lindner); Stackelberg 1989b (former USSR); Dely-Draskovits 1983 (Central Europe); CMPD: Rozkošný 1997a; CAT: Rozkošný 1992.

Pyrgotidae (key couplet 148; fig. 323-324)

Systematics: Acalyptrate Brachycera; superfamily Tephritoidea; in Central Europe 1 species, *Adapsilia coarctata* Waga.

Characters: The only European species, *Adapsilia coarctata,* is medium sized (8-9 mm), brightly to reddish yellow, the female with a conspicuously long ovipositor. Arista pubescent, ocelli absent, Oc-bristles small or absent, P-bristles small and diverging or absent; 1-2 pairs of small F-bristles; scattered interfrontal setulae present; vibrissae absent. Wing venation as in fig. 324, the intensity and size of the markings variable; costa continuous, with a weak spot near where vein Sc ends in it; vein Sc nearly complete, the apical part near the costa being pallid; crossvein BM-Cu present; cell cup closed. Tibiae without dorsal preapical bristle.

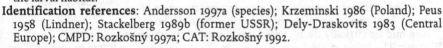

550 Adasilia coarctata Waga, male; Séguy 1934

Biology: As far as known, the larvae of Pyrgotidae are endoparasitoids of beetles of the superfamily Scarabaeoidea. Similar to what is found in the Conopidae, an egg is deposited inside the abdomen of the host during flight. The adults appear mainly to be nocturnal, but they are found by day as well, for example when they visit flowers.

Identification references: Enderlein 1942 (genera); Hendel 1933 (Lindner); Korneyev 2004 (genera); Merz 1996c (Switzerland); Stackelberg 1989x (former USSR); CMPD: no contribution; CAT: Soós 1984d.

Rachiceridae (key couplet 39; fig. 85)

Systematics: Lower Brachycera; superfamily Xylophagoidea; in Southwestern Europe (France, Spain) 1 species, *Rachicerus tristis* (Loew).

Characters: *Rachicerus tristis,* the only European species, is medium sized (6.5-8.5 mm), slender and dark coloured; legs yellow with at least tarsal segments darkened; pubescence yellow on thorax, black on abdomen. Eyes dichoptic; antenna

551 Rachicerus varius Nagatomi, male; Nagatomi 1970 (Japan)

extraordinary, with 34-35 segments. Wing with a grey tinge; both the cell m3 and cell cup closed. Legs slender; tibiae with short apical spurs; empodium pulvilliform.
Biology: The biology of this small family is virtually unknown. The larvae live in rotting wood, probably as predators.
Identification references: CMPD: Nagatomi 1997a; CAT: Majer 1988c.

Rhagionidae (key couplet 55; fig. 94, 118-119)

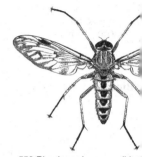

552 Rhagio scolopaceus (Linr male; Oldroyd 1969

Systematics: Lower Brachycera; superfamily Tabanoidea; traditionally this family includes the Athericidae and Hilarimorphidae; 6 genera and about 85 species in Europe.
Characters: Small to large (2-20 mm) flies with long wings, legs and abdomen. Body bare or partially pubescent, rarely with a few bristles. Body colour yellow, yellowish brown or black, often with a brownish, grey or dark pattern. In male eyes generally nearly or entirely holoptic; third antennal segment swollen and with a dorsal or terminal style or arista; mouthparts usually fleshy. Wing usually clear and with a dark stigma spot, in some cases with a distinct pattern; wing venation usually with 3 veins from the discal cell in the direction of the wing margin; cell m3 open; cell cup open or closed. Legs simple, apical spurs on the tibiae variable, however, mid leg virtually always with 2 apical spurs; empodium pulvilliform.
Biology: The larvae of most species breed in moist soil, some are found in mosses, liverworts, etc., with the larvae of many *Chrysopilus* species breeding inside decaying wood of deciduous trees. The larvae prey on insect larvae, worms and other invertebrates; some species are known to feed on organic detritus and carrion. The adults are generally less abundant, being found in shaded habitats near the verges of woodlands and near bodies of water. Some Rhagionidae, including the well known *Rhagio scolopaceus* (Linnaeus), are often seen on leaves, tree trunks and other vertically oriented objects. *Symphoromyia* species are known to feed on blood of mammals. Most of the other species feed on nectar, pollen and honeydew in the adult stage; many species are presumed to prey as well on small insects and other invertebrates.
Identification references: Van der Goot 1985 (Northwestern Europe); Nartshuk 1989b (former USSR); Oldroyd 1969; Stubbs & Drake 2001 (British Isles); Thomas 1997 (species); CMPD: Majer 1997a; CAT: Majer 1988b.

Rhinophoridae (key couplet 185; fig. 400, 403-404, 410, 412, 414-416)

553 Phyto melanocephala (Me male; Mihályi 1986

Systematics: Calyptrate Brachycera; superfamily Oestroidea; some 11 genera and about 45 species in Europe.
Characters: Small to large (2-11 mm) flies, often with many bristles; usually grey to black with some pruinosity or brownish yellow. Arista bare to plumose. Wing in some cases distinctly marked; vein M1 always curving forward; cell r4+5 sometimes petiolate. Lower calypter always small, ovoid, spoon-shaped, erect, laterally extended (i.e., not recumbent on the thorax). Subscutellum narrow and distinct, only weakly convex; meral bristles always present although fewer in number and less strong than in related families; posterior spiracle with a lining of long hair-like bristles, without lappets.
Biology: As far as known, the larvae live as endoparasitoids in terrestrial woodlice (Isopoda: Oniscidea). The eggs are deposited on trees, walls, underneath stones, etc., and after hatching the young larva actively searches for its host. Pupation takes place within the exoskeleton of the woodlouse which by that time is no more than an empty shell. The adults are often found at sunlit spots, on walls, flowers, windows, etc.
Identification references: Draber Monko 1989 (Poland); Herting 1961 (Lindner); Mihályi 1986 (Central Europe); Stackelberg 1989bf (former USSR); Zeegers & Van Veen 1993 (Netherlands); CMPD: Pape 1998b; CAT: Herting 1993.

(Rhyphidae): name used in the past for Anisopodidae.
(Risidae): part of the Ephydridae.
(Sapromyzidae): name used in the past for Lauxaniidae.

Sarcophagidae (key couplet 188-189; fig. 419-420)

Systematics: Calyptrate Brachycera; superfamily Oestroidea; some 30 genera and about 310 species in Europe.

Characters: Small to very large (3-22 mm), usually stout and bristly flies. Most species grey to greyish black with a characteristic, shifting, checkered pattern on their abdomen; some species yellow-grey, green-grey or almost entirely black lustrous, never metallic blue or green. Arista bare to plumose. Wing usually hyaline, sometimes infuscated anteriorly or variously ornamented with dark markings; vein M1 always curving forward, sometimes ending in vein R4+5 at the wing margin or before, making cell r4+5 petiolate. Lower calypter usually broadly rounded-triangular, in a very few cases smaller and more or less round to elliptical. Subscutellum narrow and flat, not swollen; meral bristles always present; posterior spiracle covered by an anterior and posterior lappet, usually the anterior lappet a narrow fringe and the posterior one forming an operculum.

554 Sarcophaga carnaria (Linnaeus), female; Oldroyd 1970

Biology: The life mode of the larvae is most variable: brood parasitoids of Hymenoptera such as Sphecidae, Pompilidae and, in the Miltogramminae, Apoidea; feeding on excrement (coprophagous larvae); inquilines preying on larvae and pupae of bumblebees and social wasps; feeding on carrion (necrophagous larvae) of fish, crustaceans, amphibians, reptiles, birds and mammals; necrophagous with all transitional stages into endoparasitism on grasshoppers, beetles and caterpillars and pupae of butterflies, on true bugs (Hemiptera), sawflies and wood wasps (Hymenoptera: Symphyta), on centipedes and millipedes, on earthworms and snails; parasitoids in pupae of Lepidoptera and Hymenoptera: Symphyta, on egg clusters of grasshoppers and spiders, on eggs of reptiles (including those of lizards); facultative and mandatory agents of myiasis in vertebrates and man. Adults can be found in a wide range of habitats; some species prefer coastal habitats. The males in particular often visit flowers.

Identification references: Pape 1987 (Northwestern Europe); Povolny & Verves 1997 (Central Europe); B.B. Rohdendorf 1930-1982 (Lindner, mainly Miltogramminae), 1937, 1989 (former USSR); Verves 1982-1993 (Lindner), 1994 (Miltogramminae); CMPD: Pape 1998a; CAT: Verves 1986; Pape 1996; Pape & Carlberg 2001.

Scathophagidae (key couplet 177, 179; fig. 393-394)

Systematics: Calyptrate Brachycera; superfamily Muscoidea; some 38 genera and about 160 species in Europe.

Characters: Small to large (3-12 mm), usually slender flies. Colour ranging from a dull yellowish brown to lustrous black or yellow, in some species bicolourous. Body and legs often with many bristles, sometimes densely covered by fine hairs. Occiput usually with some to many pale, long hairs; arista bare to plumose; interfrontal bristles absent. Wing usually clear, sometimes distinctly marked or darkened at the tip or along the crossveins; anal vein long, usually reaching the wing margin. Meron without bristles along the hind margin, near the posterior spiracle.

555 Scathophaga stercoraria (Linnaeus), male; Stackelberg 1956

Biology: The biology of the species in this family varies widely. In most species the larvae are phytophagous; some are leaf-miners or living in flower heads (inflorescences), seed capsules, stems, etc. In other species the larvae feed or prey on small organisms in rotting organic matter such as seaweed, nesting material and dung; or they prey on eggs of aquatic insects. The adults prey mainly on small soft bodied insects and other organisms. Some species are attracted to dung in considerable numbers. One of the best-known examples

of this family is *Scathophaga stercoraria* (Linnaeus). This species, of which the reddish brown, densely pilose males gather on cow dung, may be observed nearly all year round.
Identification references: Andersson 1997b (species); Dely-Draskovits 1981; Gorodkov 1989c (former USSR); Sifner 2003 (Central Europe); CMPD: De Jong 2000; CAT: Gorodkov 1986.

(Scatomyzidae): name used in the past for Scathophagidae.

Scatopsidae (key couplet 23, 201; fig. 58-59, 436)

Systematics: Nematocera; superfamily Scatopsoidea; some 22 genera and about 100 species in Europe.
Characters: Minute to small (0.5-4 mm), stout Nematocera. Body colour dull or lustrous, dark to black, occasionally with lighter spots or more reddish. Ocelli present; eyes reniform and with an eye bridge present in almost all species; antenna short, with 7-12 segments, the terminal segment often somewhat longer. Wing short and wide, with an anal lobe, clear or slightly tinged; the R-veins well developed and ending far short of the wing tip, the other veins weaker. Legs relatively short and stout; tibiae without apical bristles or spurs.

Biology: The larvae live in rotting organic matter, in all sorts of substrates such as rotting plants, vegetables and fruits, in or underneath decaying wood and bark, rot holes in trees, soil, leaf litter, mushrooms and bracket fungi, dung, carrion, ants nests, etc. The adults are found in a wide range of habitats but

556 Ectaetia clavipes (Loew), le; Haenni 2004

generally prefer marshy and more open areas where they often visit flowers. Some species form swarms consisting of thousands of individuals on low vegetation and shrubs.
Identification references: Andersson 1982 (Sweden); Cook 1969-1974 (revision); Freeman & Lane 1985 (British Isles); Krivosheina 1989a (former USSR); CMPD: Haenni 1997a; CAT: Krivosheina & Haenni 1986.

Scenopinidae (key couplet 59, 63; fig. 125-126, 132)

Systematics: Lower muscomorph Brachycera; superfamily Asiloidea; in Europe 2 genera, *Caenoneura* and *Scenopinus*, and about 16 species.
Characters: Small to medium sized (2-7 mm) flies, usually dark coloured, stout and somewhat dorsoventrally flattened. Head and thorax pubescent, devoid of bristles. Eyes nearly always holoptic in the male and consisting of two differently faceted parts; antenna usually oriented downward, with 3 segments; the third segment is the largest and in some cases split at the end and/or bearing a minute style. Wing clear or tinged; vein M1 curving well forward, ending in the costa before the wing tip, or in vein R5; cell m3 open; cell cup closed. Legs relatively short and powerful, empodium bristle-like or absent.

557 Scenopinus fenestralis (L female; Tóth 1977

Biology: The larvae live in a variety of dry substrates such as soil, dust, nests of birds and other animals, tunnels mined by beetles and other organisms, holes in wood, under bark, etc., where they prey on small invertebrates, in particular on mites and beetles. Two species with a world-wide distribution, *Scenopinus fenestralis* (Linnaeus) and *S. glabrifrons* Meigen, are well-known predators of carpet beetles. The adults feed on nectar and visit flowers. Adults are not often encountered in the wild but more regularly found indoors on windows or in rooms with lots of dust and dry organic matter (disused rooms, attics, church towers).
Identification references: Kelsey 1969 (revision); Nartshuk 1989g (former USSR); Oldroyd 1969; Stubbs & Drake 2001 (British Isles); CMPD: Krivosheina 1997e; CAT: Kelsey & Soós 1989.

Sciaridae (key couplet 28, 31, 200, 204-205; fig. 69-70, 73-74, 435)

Systematics: Nematocera; superfamily Sciaroidea; some 28 genera and about 620 species in Europe. The genera *Heterotricha* and *Sciarosoma*, mentioned under key couplet 37, are sometimes tentatively placed in the Sciaridae.

Characters: Delicate, often black or rarely black and yellow to orange coloured Nematocera, minute to medium sized (1-6 mm, rarely up to 10 mm). Ocelli present; eyes usually with an eye bridge; antenna with 16 simple segments. Wing clear, in some cases tinged; wing venation rather constant: costa ending in between R4+5 and M1, the only crossveins present are H and R-M; vein M forked at about half its length; CuA1 and CuA2 with a basal stem. Legs relatively long, tibiae with apical bristles or spurs.

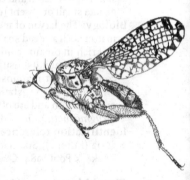

558 Lycoriella solani (Winnertz), male; Freeman 1983

Biology: The larvae are largely terrestrial but restricted mainly to wet habitats. They are an important faunal element in soil and litter layers, but can also be found in dead wood, excrement, mushrooms, bracket fungi, etc.; a number of species develop in semi-aquatic habitats. Larvae of Sciaridae mainly feed on mycelium. A limited number of species feed on living plant tissues such as roots or, as miners, leaves and stems, or live in mushrooms and bracket fungi. Some species are pests in forestry, horticulture and agriculture, mainly in mushroom culture and in glasshouses. The larvae of some species inhabit caves, others live in colonies of social insects. Larvae may occur in very high densities; massive processions of 2 to 10 metres in length, so-called 'army worms', are well-documented. The adults are found in a wide range of habitats as well, mainly near the larval habitat. They are often found indoors, in which case they probably hatched from larvae that developed in potting soil of indoor plants.

Identification references: Freeman 1983 (British Isles); Gerbachevskaya 1989 (former USSR); Menzel 2000 (Germany); Menzel & Mohrig 2000 (revision); CMPD: Menzel & Mohrig 1997; CAT: Gerbachevskaya 1986; Menzel et al. 2006 (British Isles).

Sciomyzidae (key couplet 137, 141, 143, 154; fig. 188, 302, 307-309, 334)

Systematics: Acalyptrate Brachycera; superfamily Sciomyzoidea; traditionally this family includes the Phaeomyiidae; some 24 genera and about 140 species in Europe.

Characters: Small to large (2-14 mm), generally slender flies with a body colour varying from black lustrous to dull grey, brown, reddish or yellow. Antenna sometimes elongate; arista pubescent or with shorter or longer rays; ocelli present; Oc-bristles usually present, rarely (*Sepedon*) absent; P-bristles diverging to parallel, sometimes slightly converging in *Pteromicra*; 1-2 pairs of F-bristles, curving backward, the lower pair sometimes curving inward; interfrontal bristles absent; interfrontal setulae sometimes present; vibrissae absent. Wing clear or with conspicuous markings; costa continuous; vein Sc complete; crossvein BM-Cu present; cell cup closed. Tibiae almost always with dorsal preapical bristle.

559 Euthycera chaerophylli (Fabricius), male; Rivosecchi 1992

Biology: The larvae of nearly all species feed on snails or slugs, some feeding on the eggs of snails. Most of the known larvae are semi-aquatic, some are aquatic; a number of species have terrestrial larvae. Larvae mainly prey on snails lacking an operculum; in general, aquatic larvae have to visit the surface frequently in order to breathe but some species preying on bivalves have adapted to breathing under water. The larvae of some species preying on terrestrial snails and slugs are parasitoids that develop inside a single snail or slug; in other terrestrial species the penultimate instar emerges from the snail or slug it developed in and are as last instar predatory on several snails. The adults are often found sitting on the vegetation, their heads down. Depending on the larval habitat, they are often found near water, in marshy vegetation, but also in woodland or even dry open habitats.

Identification references: Revier & Van der Goot 1989 (Northwestern Europe); Rozkošný 1984 (Fennoscandia and Denmark), 1987 (review); 1997d (genera), 2002 (species); Stackelberg 1989ac (former USSR); Vala 1989 (France, Southern Europe); CMPD: Rozkošný 1998a; CAT: Rozkošný & Elberg 1984.

(Scopeumatidae): name used in the past for Scathophagidae.

Sepsidae (key couplet 91, 113, 138, 156; fig. 194-199, 244, 305, 338)

560 Sepsis cynipsea (Linnaeus) male; Oldroyd 1970

Systematics: Acalyptrate Brachycera; superfamily Sciomyzoidea; 9 genera and about 50 species in Europe.

Characters: Small to medium sized (2-6 mm), slender, ant-like flies with relatively few bristles or hairs. Body colour generally black, in some cases lustrous and/or with a distinct pattern of silvery pruinosity on the side of the thorax. In general, walking about on vegetation while waving their wings. An exception is *Orygma luctuosum* Meigen, a robust, hairy, somewhat flattened fly. The Sepsidae, including *Orygma*, are distinct from all other Acalyptrate flies on account of the presence of 1 or more bristles at the posteroventral margin of the posterior spiracle. Head rounded, arista bare; ocelli present; Oc-bristles present; P-bristles diverging or absent; 0-3 pairs of F-bristles, if present curving outward or (obliquely) backward; interfrontal bristles absent, interfrontal setulae present in *Orygma*; vibrissae or vibrissa-like bristles usually present. Wing clear or tinged, often with a dark spot or clouding at the wing tip; costa continuous; vein Sc complete; crossvein BM-Cu present; cell cup closed. In the male femur and tibia of fore legs often with extrusions, spurs, teeth and the like. Tibiae usually with dorsal preapical bristle, but in several genera small or absent (e.g. *Meroplius*, *Nemopoda*, *Themira*).

Biology: The larvae of most Sepsidae largely develop in dung, excrement, rotting organic matter such as dead snails and insects, mushrooms and bracket fungi, in nests of birds, in soil rich in organic compounds, in compost, etc. The adults are generally observed near the larval habitats, usually in somewhat open habitats such as meadows, along watercourses, forest margins and clearings in woodland areas. Some species occur in major swarms on low vegetation and shrubs, waving their wings. These swarms can persist for over a month and are often accompanied by a certain odour. All this does not account for *Orygma luctuosum* which is a seaside species, to be found under cast up seaweed.

Identification references: Pont 1979 (British Isles); Ozerov 2003 (Russia); Pont & Meier 2002 (Europe); Stackelberg 1989ab (former USSR); CMPD: Meier & Pont 2000; CAT: Zuska & Pont 1984; Ozerov 2005.

Simuliidae (key couplet 17; fig. 50-51)

Systematics: Nematocera; superfamily Chironomoidea; 8 genera and about 230 species in Europe.

Characters: Bloodsucking, minute to medium sized (1.2-6 mm), stout, usually dark coloured Nematocera. Eyes holoptic in the male, dichoptic in the female; ocelli absent; antenna short and stout, tapering toward the end, with 11, rarely 9 segments. Wing short and wide, with a distinct anal lobe; wing clear, in some cases slightly tinged; costa and R-veins stronger than the other veins; vein CuA2 usually curving like an elongate, inverted S. Legs short and stout.

Biology: The larvae are found in several types of running

561 Simulium ornatum Meigen, female; Séguy 1951

water, attached to a substrate like stones, metal objects, plastic bags, plants, branches, etc., at or above the bottom. They live on micro-organisms and small organic particles they filter from the water. Pupation takes place inside a cocoon. The adults feed on nectar and plant juices, females of a number of species feed on the blood of both cold-blooded and warm-blooded vertebrates and are known as 'black flies'. Feeding takes place in daylight, mainly during the morning and evening hours. In the tropics, but also in the tundra, taiga and some East European regions, Simuliidae can occur in such vast numbers as to interfere with, or entirely prevent, human activity.

Identification references: Bass 1998 (larvae and pupae British Isles); Crosskey 1993a (genera feeding on blood); Davies 1968 (British Isles); Jedlička et al 2004 (pupae Central Europe); Jankovsky 2002 (former USSR); Jensen 1997 (genera); Lechthaler & Car 2004 (larvae pupae Central and Western Europe); Rivosecchi 1978 (Italy); Rubtzov 1959-1964 (Lindner), 1989 (former USSR); Yankovskii 2003 (former USSR); CMPD: Jedlička & Stloukalová 1997; CAT: Rubtzov & Yankovsky 1988.

(Siphonellopsidae): part of the Chloropidae.
(Solvidae): name used in the past for Xylomyidae.
(Spaniidae): part of the Rhagionidae.

Sphaeroceridae (key couplet 85, 213; fig. 184-185, 447)

Systematics: Acalyptrate Brachycera; superfamily Sphaeroceroidea; some 40 genera and about 260 species in Europe.

Characters: Minute to medium sized (0.7-5.5 mm), stout, usually dull brown to black coloured flies, characterised by presence of vibrissae in combination with short, swollen first tarsal segment of the hind leg. Arista bare to long pubescent, ocelli virtually always present; Oc-bristles present or absent; P-bristles converging or absent; 1-2 pairs of F-bristles, curving outward; interfrontal bristles present, partially in rows oriented inward; vibrissae present. Wing almost always without markings; costa with breaks at the base, a subcostal break and in some cases a humeral break as well; vein Sc reduced; crossvein BM-Cu present or absent; cell cup open or closed. Tibiae with or without dorsal preapical bristle.

562 Leptocera fontinalis Fallén, female; Papp 1973

Biology: The larvae live in a wide range of rotting organic materials and presumably feed on micro-organisms. They can be found in dung, manure dumps, leaf litter, organic matter washed up on the shore, mushrooms, bracket fungi, rotting vegetable matter, in caves, nests and other places inhabited by animals. The adults are mediocre fliers. They can be found on decaying matter of plant and animal origin, preferably in damp habitats; relatively many species live in caves or animal burrows. A few species occur near human settlements and, in case of mass occurrence in rotting substrates or in manure near human dwellings, they may become a nuisance because they usually also come indoors. Many species living in manure and dung are associated with hoofed animals and often form large populations on pastures and dung heaps near stables. The family shows a high percentage of species with reduced wings, or no wings at all.

Identification references: Nartshuk 1989h (former USSR); Pitkin 1988 (British Isles); CMPD: Roháček 1998c; CAT: Roháček et al. 2002; CAT: Papp 1984i.

Stenomicridae (key couplet 104; fig. 230-231)

Systematics: Acalyptrate Brachycera; superfamily Opomyzoidea; traditionally part of the Aulacigastridae or Periscelididae; in Europe 1 genus, *Stenomicra*, with 3 species.
Characters: Minute (1-2 mm), delicate, yellow to grey flies with inner vertical bristles curving forward. Arista with long rays; ocelli present with the ocellar triangle situated more or less in middle of frons; Oc-bristles absent; P-bristles small and diverging or absent; 2 pairs of F-bristles, lower pair curving inward; interfrontal bristles absent; vibrissae and a series of vibrissa-like bristles present. Wing clear or with tinged spots; costa with a subcostal break; vein Sc incomplete; crossvein BM-Cu absent; crossvein DM-Cu present or absent; cell cup open. Tibiae without dorsal preapical bristle.

563 Stenomicra spec., female
Williams 1939

Biology: The biology of this small family is little known. The larvae probably develop in the water-holding leaf bases of plants, particularly monocotyledons. The adults of some species often occur near water or at moist places and are closely associated with tussocks of large *Carex* species, others also with *Typha*, *Scirpus* or *Cyperus*; the species *Stenomicra soniae* Merz & Roháček may be associated with umbellifers. As in some Chamaemyiidae, the adults are able to walk sideways and backward with the same ease as forward.
Identification references: Collin 1944, Irwin 1982 (British Isles); Merz & Roháček 2005 (species); Stackelberg 1989ao (former USSR); CMPD: Mathis & Papp 1998; CAT: Papp 1984g.

(Stomoxydidae): part of the Muscidae.

Stratiomyidae (key couplet 42, 54, 70; fig. 88, 103, 116, 150)

Systematics: Lower Brachycera; superfamily Stratiomyoidea; some 27 genera and about 140 species in Europe.
Characters: Small to very large (2-25 mm) flies of slender to stout shape, in some cases flattened. Body colour often striking: usually dark with conspicuous white, yellow, or greenish patterns, sometimes with blue to green metallic lustre. Eyes usually holoptic in male; antenna with 7-10 segments, varying between rather uniform to basal segments swollen and terminal segments making up the style; mouthparts often short, in some cases elongate. In most species scutellum with spinelike extensions. Wing usually clear, in some species with markings or partially much darkened; venation characteristic,

564 Sargus flavipes Meigen, male; Rozkošný 1982

e.g. R-veins in front part of the wing with vein R5 almost always ending well before the wing tip; discal cell diamond-shaped, small to very small; cell m3 open; cell cup closed. Legs simple, almost always without bristles, in a few species tibia of the mid leg with a single apical bristle; empodium pulvilliform.
Biology: Contrary to most other lower flies, the larvae of the Stratiomyidae are not predators or parasitoids. They are aquatic to terrestrial and mainly feed on algae and rotting organic matter, in most cases of vegetable origin. The aquatic habitats range from stagnant to running water, moist rock faces, on water vegetation, marshes, saline environments along the seaside or inland. Terrestrial larvae often occur underneath leaves and other rotting vegetable matter, in the topsoil, in some cases in dung, in ants nests and below the bark of trees. The adults feed on nectar and honeydew. They are frequent visitors to flowers and are also found sitting on leaves of herbs, shrubs, or trees. Preferred habitats include woodland, the vicinity of water, dunes or other coastal habitats.
Identification references: Brugge 2002 (Netherlands); Nartshuk 1989e (former USSR); Oldroyd 1969 (British Isles); Rozkošný 1973 (Fennoscandia and Denmark); Rozkošný 1982-1983 (revision), 1997c (genera), 2000 (species); Stubbs & Drake 2001 (British Isles); Woodley 1995 (Beridinae); CMPD: Rozkošný 1997b; CAT: Rozkošný & Nartshuk 1988.

Streblidae (key couplet 169, 209; fig. 175, 372-373, 444)

Systematics: Calyptrate Brachycera; superfamily Hippoboscoidea; in Southern Europe 1 species, *Brachytarsina flavipennis* Macquart.

Characters: Minute to small (0.8-5 mm), yellow to yellowish brown flies. Some non-European species with wings reduced or absent. Head small, not adpressed to the thorax; eyes and ocelli rudimentary or absent. First tarsal segment short; terminal tarsal segment widened.

Biology: The adults are ectoparasites of bats, with a preference for colonial cave-dwelling Microchiroptera, most species being confined to a single species or genus of host. The larvae develop inside the female abdomen of the fly and are deposited one by one, generally in the direct vicinity of bat roosts,

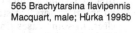

565 *Brachytarsina flavipennis* Macquart, male; Hůrka 1998b

after which they immediately pupate. Adults of fully winged species, although highly mobile, remain most closely associated with the host bat. The flies generally do not leave the host except for deposition of the larvae or when the host is disturbed.

Identification references: Stackelberg 1989bd (former USSR); CMPD: Hůrka 1998b; CAT: Hůrka & Soós 1986b.

Strongylophthalmyiidae (key couplet 168; fig. 332, 368-369)

Systematics: Acalyptrate Brachycera; superfamily Diopsoidea; 2 species in Europe: *Strongylophthalmyia ustulata* (Zetterstedt) and *S. pictipes* Frey.

Characters: Small to medium sized (3.5-5.5 mm), slender, largely black flies with yellowish legs. Arista short pubescent; ocelli present; Oc-bristles present; P-bristles diverging; 2 pairs of F-bristles; interfrontal bristles absent; vibrissae absent. Wing clear with vague markings in the central part or at the wing tip; costa with a subcostal break; vein Sc incomplete; crossvein BM-Cu present; cell cup closed. Tibiae without dorsal preapical bristle.

Biology: The biology of the European species is little known. The larvae live under bark of dying or decaying deciduous trees, apparently with a preference for wet, dead ash wood. The adults

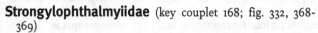

566 *Strongylophthalmyia ustulata* (Zetterstedt), female; Papp 1978

occur on low vegetation in wet deciduous forests near the breeding sites of the larvae.

Identification references: Krivosheina 1981; Shatalkin 1993 (review); Stackelberg 1989v (former USSR); CMPD: Iwasa 1998a; CAT: Krivosheina 1984a.

Synneuridae (key couplet 24; fig. 60)

Systematics: Nematocera; superfamily Scatopsoidea; the families Canthyloscelidae and Synneuridae are often considered one family; in Northern, Central and Eastern Europe 1 species: *Synneuron annulipes* Lundström.

Characters: Small (3-4 mm), stout, black Nematocera with a somewhat laterally compressed thorax and elongate abdomen. Ocelli present; eyes below the antenna not far apart or even touching; antenna with 12 short segments, the terminal segment elongate. Wing clear or slightly tinged; R-veins stronger than the other veins. Legs relatively long and stout.

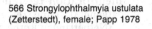

567 Wing Synneuron decipiens Hutson; Peterson & Cook 1981 (North America)

Biology: The biology of the European species of *Synneuron* is little known. The larvae live in moist, decaying wood. Adults are rare, at least in collections; they are presumed to occur especially in old, undisturbed forests.

Identification references: Andersson 1982 (Sweden); Hutson 1977 (revision); CMPD: Haenni 1997b; CAT: Mamaev & Krivosheina 1986a.

Syrphidae (key couplet 64; fig. 135-137)

Systematics: Lower Cyclorrhapha; superfamily Syrphoidea; some 90 genera and about 830 species in Europe.

Characters: Small to very large (3.5-35 mm), slender to stout flies, usually coloured black with white or yellow markings on head, thorax and especially abdomen; sometimes body largely brown, metallic yellow, green, or blue, or with a wide range of colours; in some cases strongly resembling bees, bumblebees, or wasps. Head large, without any distinct bristles, mostly consisting of eyes which are usually holoptic in the male. Third antennal segment usually the largest but sometimes first or second segment longer; third segment carrying

568 Scaeva pyrastri (Linnaeus) female; Oldroyd 1970

an apical or dorsal arista or an apical style. Thorax with very few bristles. Wing usually clear, in some cases tinged or with markings; venation characteristic, virtually always with a spurious vein (*vena spuria*), less distinct or absent in *Eristalinus* and *Psilota*; vein M1 not reaching the wing margin but curving strongly forward and ending in vein R4+5 before the wing margin; cell dm present, cell below cell dm open; cell cup elongate, ending acutely just before the wing margin. Legs usually slender and simple but sometimes, especially in the male, with coxa, trochanter, femur, tibia, or tarsus modified and/or bearing spurs.

Biology: The larvae can be divided into mycophagous and phytophagous species (including a number of species being pests in agriculture, especially bulb cultivation); predators of other insects (such as aphids, ants, larvae of wasps or beetles, etc.), and species feeding on various types of organic detritus (decaying wood, sap exuding from tree wounds, organic detritus underneath tree bark or in nests of birds and other animals, leaf litter, liquid excrement or other organically enriched watery solutions). In some cases larvae of this latter group inhabit semi-aquatic to entirely aquatic habitats. The adults are common, conspicuous flies, to be found in various open to woodland habitats such as gardens, meadows, forest eaves, river banks, etc. They are able to hover in one place and are apparently capable of moving not only forward, but also backward and laterally. The adults feed on nectar and pollen, and several of the frequent visitors to flowers play an important part in pollinating certain plant species.

Identification references: Barendregt 2001 (Netherlands); Dolezil & Rozkošný 1997 (genera); Hippa et al. 2001 (*Eristalis*); Kormann 2002 (species); Reemer 2000 (Netherlands); Rotheray 1993 (larvae); Stackelberg 1989p (former USSR); Speight 1999 (Xylotini); Stubbs & Falk 2002 (British Isles); Van Veen 2004 (Northwestern Europe); Verlinden 1991 (Belgium); CMPD: Thompson & Rotheray 1998; CAT: Peck 1988.

Tabanidae (key couplet 50; fig. 2a, 6b, 20, 92, 110-111)

Systematics: Lower Brachycera; superfamily Tabanoidea; some 13 genera and about 220 species in Europe.

Characters: Stout, medium sized to very large (6-30 mm) flies, devoid of bristles. Body generally brown, black or grey and with a pattern of lighter markings; in some cases body yellowish, greenish or metallic blue. Eyes in living flies often brilliantly reddish or metallic green, sometimes yellowish, often with spots or bands; eyes in male holoptic, in female dichoptic; antenna with 6 or more segments, usually with a large third segment and a tapering style; in most females the mouthparts constitute a powerful, often long piercing proboscis. Wing clear, or tinged, or with transverse bands, or darkened with distinct spots; the costa surrounding the wing entirely; fork of veins R4 and R5 usually strongly diverging with vein R5 ending distinctly beyond the wing tip; cell m3 open; cell cup open or closed. Lower

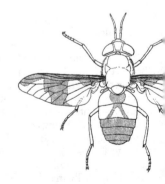

569 Chrysops caecutiens (Linné) female; Séguy 1926

calypter conspicuously large. Legs powerful, tibia of mid leg and in some cases hind leg with apical spurs; empodium pulvilliform.

Biology: The larvae are aquatic, semi-aquatic or terrestrial (living mainly in soil and litter). In most genera they are predators but also able to feed as facultative saprophages. They prey on other invertebrates such as worms, snails and the larvae of other Diptera, including those of their own species. The adults are able fliers and especially active on warm, sunny days. In the majority of species the females feed on blood of mammals including humans; both sexes also feed on nectar and pollen. Some species show territorial behaviour, the males safeguarding their territories by hovering. As a consequence of their blood feeding habits and associated transmission of diseases, Tabanidae can be a serious problem locally for humans as well as cattle.

Identification references: Chvála et al. 1972 (revision); Chvála & Jezek 1997 (genera); Kniepert 2000 (species); Majer 1987 (Central Europe); Oldroyd 1969 (British Isles); Olsufyev 1989 (former USSR); Portillo Rubio 2002 (Iberian Peninsula); Stubbs & Drake 2001 (British Isles); Timmer 1980 (Belgium, Netherlands and Luxembourg); Trojan 1979 (Poland); Zeegers & Haaren 2000 (Netherlands); CMPD: no contribution; CAT: Chvála 1988b.

Tachinidae (key couplet 171-172, 181, 190, 219; fig. 11, 374-375, 386-389, 399, 402, 405-406)

Systematics: Calyptrate Brachycera; superfamily Oestroidea; some 280 genera and about 880 species in Europe.

Characters: Small to large (2-20 mm) flies. Most species are grey to slate black, in some cases with yellow or red spots on the sides of abdomen; some species are metallic green or yellowish grey. Usually many bristles are present, some species may, however, be quite bare or show a bee-like pilosity. Arista usually quite bare, rarely plumose. Wing with vein M1 usually strongly curved forward, sometimes cell r4+5 is petiolate. Lower calypter usually broadly rounded-triangular, in some cases small and elliptical. Meral bristles always present; subscutellum strongly developed (except in *Litophasia*); lappets of posterior spiracle usually distinctly unequal in size, the posterior lappet subcircular, in several genera about equal in size.

570 Mintho rufiventris Fallén, male; Mihályi 1986

Biology: The larvae are endoparasitoids of other insects, sometimes of other arthropods. Eggs are deposited directly on or in the immediate vicinity of the host. The larvae of tachinids which deposit their eggs directly on the host hatch within minutes (ovolarvipary) and penetrate the host. A conspicuous specialisation is the development of microtype eggs by species of the tribe Goniini in the subfamily Exoristinae. Here, tens of thousands of eggs are deposited on leaves. The eggs only develop after a caterpillar has eaten them (along with the leaves they feed on), after which they hatch inside the caterpillar's digestive tract. Hence, most Goniini show a wide range of lepidopterous hosts. Some other groups show host specialisation: the Dexiini are parasitoids of beetles, the Exoristinae and Tachininae are mainly parasitoids of caterpillars and the larvae of the Phasiinae develop inside true bugs (Hemiptera); others parasitise besides caterpillars also larvae of Tipulidae (Tachininae: *Siphona*). Several species are of economic importance, being useful in biological pest control. The adults of the Tachinidae are largely found on walls, leaves, branches and similar spots; only a small number of species regularly visit flowers. Some species are territorial.

Identification references: Andersen 1996 (Siphonini); Belshaw 1993 (British Isles); Mesnil 1944-1980 (Lindner); Mihályi 1986, 1994; Tschorsnig & Herting 1994 (Central Europe); Zimin et al. 1989 (former USSR); CMPD: Tschorsnig & Richter 1998; CAT: Herting & Dely-Draskovits 1993.

(Taeniapteridae): part of the Micropezidae.

Tanypezidae (key couplet 158; fig. 342-343)
Systematics: Acalyptrate Brachycera; superfamily Diopsoidea; in Europe 1 widespread species, *Tanypeza longimana* Fallén.
Characters: The only European species, *Tanypeza longimana*, is a slender, black lustrous, medium sized (5-8 mm) fly with silvery markings on the head and the sides of the thorax, and with long, slender, largely yellow legs. Head rounded; arista pubescent; ocelli present; Oc-bristles present; P-bristles diverging; 2 pairs of F-bristles, curving backward; interfrontal bristles absent; vibrissae absent. Wing unmarked, costa with a weak subcostal break; vein Sc reaching the costa, running parallel to vein R1 and approaching or touching it just before reaching the costa; crossvein BM-Cu present; cell cup closed. Tibiae without dorsal preapical bristle.

571 Tanypeza longimana Fallén, male; Séguy 1934

Biology: The larvae of *T. longimana* feed on rotting vegetable matter, possibly in decaying wood or other plant detritus. The adults prefer low vegetation in moist woodland, in particular near water.
Identification references: Van der Goot & Van Veen 1996 (Northwestern Europe); Stackelberg 1989u (former USSR); CMPD: Roháček 1998a; CAT: Soós 1984b.

(Tendipedidae): name used in the past for Chironomidae.

Tephritidae (key couplet 107, 150; fig. 9, 239, 328-331)
Systematics: Acalyptrate Brachycera; superfamily Tephritoidea; some 70 genera and about 270 species in Europe.
Characters: Generally small to medium sized (2.5-10 mm, extremes from 1-20 mm), usually colourful flies, wing often with characteristic markings and vein Sc abruptly bent forward toward the costa at nearly 90°. Arista bare to short pubescent; ocelli present; Oc-bristles present; P-bristles parallel to diverging; 2-8 pairs of F-bristles, at least 1 but usually several lower pairs curving inward, at least 1 of the upper pairs curving backward, in some cases

572 Ceratitis capitata (Wiedemann), female; Balachowsky & Mesnil 1935

F-bristles inserted on a raised tubercle; interfrontal setulae most often absent or represented by 1-2 tiny setulae near the lunula; vibrissae absent but several genera with strong bristles near the vibrissal angle. Wing with yellow, brown or black markings or dark coloured with lighter markings; in a few species wing clear; costa with a humeral and a subcostal break; apical part of vein Sc usually less distinct or even transparent at about a right angle with respect to the basal part; crossvein BM-Cu present; cell cup closed, nearly always narrowing to an acute angle, closed by a geniculate vein CuA2, vein CuA2 rarely straight or convex. Tibiae without dorsal preapical bristle. Female with oviscape, non retractable basal segment of the ovipositor.
Biology: The larvae of nearly all Tephritidae are phytophagous. The female uses her telescopic ovipositor to deposit eggs in living, healthy plant tissue. The larvae develop in flowers, seeds, fruits, leaves, stems or roots of their host plant, depending on the species. Some species induce gall-forming. Several species are serious pests in agriculture and horticulture, others play a role in the biological control of weeds. An exception to the phytophagous lifestyle are *Euphranta toxoneura* (Loew) and *Chetostoma stackelbergi* Rohdendorf with larvae developing in galls formed by sawflies (Hymenoptera: Symphyta). Adult Tephritidae are good fliers. They are often found on the host plant or while feeding on nectar, pollen, plant juices, rotting plant material or honeydew.
Identification references: Darvas & Papp 2000 (keys to exotic species occurring in Europe); Kabos & Van Aartsen 1984 (Netherlands); Merz 1994 (Northern and Central Europe); Richter 1989f (former USSR); White 1988 (British Isles); White & Elson-Harris 1994 (review of economically important species); CMPD: no contribution; CAT: Foote 1984.

(Tetanoceridae): name used in the past for Sciomyzidae.

Tethinidae (key couplet 116, 118, 168; fig. 251-252, 255, 367)

Systematics: Acalyptrate Brachycera; superfamily Carnoidea; 3 genera and about 35 species in Europe.

Characters: Minute to small (1.5-3.5 mm), yellow to black flies with a grey, yellowish, or brownish pruinosity. Arista bare to short pubescent; ocelli present; Oc-bristles present; P-bristles converging; 1-5 pairs of F-bristles, the upper ones curving outward or backward, the lower pairs, if present, curving outward, forward and/or inward; interfrontal bristles present or absent; vibrissa-like bristles present, in some cases small and inconspicuous; in the subfamily Tethininae a small, lustrous bulge in between the antennae and the anterior vibrissa-like bristle. Wing unmarked, with a pale yellow to brownish tinge; costa with a subcostal break; vein Sc first running parallel to and then merging with R1 just before the costa; crossvein BM-Cu absent or very faded; cell cup present but small. Tibiae without dorsal preapical bristle.

573 Tethina czernyi (Hendel); male; Merz 2002

Biology: Tethinidae are almost exclusively halophilous flies, mainly found along the coast although they occur in saline inland habitats as well; only a few species are also known from habitats that are apparently without increased salinity (forests, meadows, deserts). They are often present in great numbers. Some species are recorded from secondarily man-created habitats, such as meadows polluted by industrial emissions, slaughterhouses and poultry farms, where they live synanthropically in alkaline damp places.

Identification references: Beschovski 1994 (Eastern Europe); Collin 1966 (revision); Munari & Baéz 2000 (Atlantic islands); Munari & Merz 2002 (Mediterranean species); Soós 1981 (Central Europe); Stackelberg 1989au (former USSR); CMPD: Munari 1998; CAT: Soós 1984m; Munari 2002.

Thaumaleidae (key couplet 15; fig. 47-48)

Systematics: Nematocera; superfamily Chironomoidea; 3 genera and about 75 species in Europe.

Characters: Small (3-5 mm), delicate, yellow to orange or dark brown, lustrous Nematocera; upper side of thorax often with a U-shaped colour pattern. Eyes holoptic in both sexes; ocelli absent; antenna short and directed forward, with 12 segments, tapering to its tip. Wing clear, with brown tinge and relatively broad. Legs short and slender, tibiae without apical bristles or spurs.

Biology: The larvae and pupae are found in unpolluted aquatic environments, usually in shaded, hilly or mountain areas. The larvae live in the water film of streams and splash zones of springs, banks, rocks and other objects protruding from the water, holding to the substratum with thoracic and abdominal pseudopodia. They feed on vegetable debris and diatoms. The adults are clumsy fliers and are generally found near the larval habitats.

574 Thaumalea testacea Ruthé, male; Séguy 1951

Identification references: Disney 1999 (British Isles); Stackelberg 1989e (former USSR); Wagner 1997f, 2002 (species); CMPD: Wagner 1997c; CAT: Martinovský & Rozkošný 1988.

Therevidae (key couplet 60; fig. 100, 127)

Systematics: Lower muscomorph Brachycera; superfamily Asiloidea; some 17 genera and about 100 species in Europe.

Characters: Slender to more or less stout, small to large (2.5-15 mm) flies, often strongly pubescent with stronger bristles on thorax and legs; colour varying from pale yellow to black. Frons of female in many species with glossy black markings; eyes usually holoptic in the male; antenna variable, first or third segment often the longest; the latter followed by no, or 1-2 smaller segments and a small apical style. Wing clear or tinged, in some cases with markings or bands; cell m3 strongly narrowing toward hind margin, or closed; cell cup closed. Legs relatively long and slender; tibiae with apical bristles; empodium bristle-like or absent.

575 Thereva nobilitata (Fabric female; Tóth 1977

Biology: The larvae live in sandy soil, underneath tree bark and in dead wood, where they prey on a range of small invertebrates, in particular beetle larvae. The adults are found in various habitats but prefer open, sunny places; in very dry areas they are often found near water. They are able fliers but often sit on the ground or on rocks, leaves, branches, tree trunks, etc. Contrary to many other lower Brachycera the adults are not predatory but feed on nectar, plant or animal juices and other organic matter.

Identification references: Van der Goot 1985 (Northwestern Europe); Haarto & Winqvist 2006 (Finland); Oldroyd 1969; Stubbs & Drake 2001 (British Isles); V.F. Zaitzev 1989a (former USSR); CMPD: Majer 1997b; CAT: Lyneborg 1989.

(Thyreophoridae): part of the Piophilidae.

Tipulidae (key couplet 6, 195; fig. 19, 22-23, 25, 29, 33, 425-427)

Systematics: Nematocera; superfamily Tipuloidea; traditionally this family includes the Cylindrotomidae, Limoniidae and Pediciidae; 10 genera and about 470 species in Europe.

Characters: Medium sized to very large (7-35 mm) Nematocera with long and slender antennae, wings, legs and abdomen. Tipulidae include some of the largest Diptera. Rostrum well-developed and often with nasus; ocelli absent; antenna usually consisting of 13 segments, in some cases more. Thorax with a V-shaped transverse suture. Wing with 2 anal veins and in some species with conspicuous markings; in a few species female with wings short or reduced.

Biology: The larvae are rarely aquatic, generally semi-aquatic to terrestrial; some species live in mosses, liverworts, or in dying and decaying wood of deciduous trees. The larvae feed mainly on plant remains but in a number of species living parts of mosses, liverworts and seed plants are eaten as well; the larvae of *Tipula*, subgenus *Tipula* and some *Nephrotoma* species feed on pasture grasses, tree saplings, young cultivated

576 Tipula trifascingulata Theo female; Séguy 1951

plants, etc., and can cause serious damage to agriculture as they are often present in large numbers. Tipulidae larvae are an important food source especially for birds. The adults occur in a wide range of habitats varying from dark moist forests to dry semi-deserts or high alpine regions. They do not feed; at most, they rarely drink from dew-drops or take some nectar from flowers and similar liquids.

Identification references: Hofsvang 1997 (genera); Mannheims 1951-1968; Oosterbroek et al 2006 (Ctenophorinae); Theowald 1973-1980 (Lindner); Savchenko 1989b (former USSR); Stubbs 1972-1996 (British Isles); CMPD: no contribution; CAT: Oosterbroek & Theowald 1992; Oosterbroek 2006.

(Trepidariidae): name used in the past for Micropezidae: Calobatinae.

Trichoceridae (key couplet 4; fig. 27-28)

Systematics: Nematocera; superfamily Trichoceroidea; 2 genera, *Diazosma* and *Trichocera*, and about 50 species in Europe.

Characters: Small to medium sized (3-9 mm), generally light coloured Nematocera, with long slender antennae, wings, legs and abdomen. Ocelli present; antenna with 18 segments. Wing with a short, curved vein A2, the longest in *Diazosma*. Tibiae with terminal bristles or spurs. Female of *Trichocera* with a more or less downward curving ovipositor.

Biology: The larvae are terrestrial and saprophagous, living in moist earth, rotting vegetable matter, under bark of fallen coniferous trees, mushrooms and bracket fungi, carrion (including human corpses), excrement, etc. The adults of *Trichocera* are active from autumn to spring, throughout the winter ("winter gnats"), with especially the females found walking on snow. Adults of *Diazosma* occur mainly in summer and early autumn. The males have a characteristic bouncing flight and often form swarms, in particular on sunny days.

Identification references: Dahl 1966, 1967 (Sweden); Stackelberg 1989a (former USSR); Starý & Martinovský 1993 (*Diazosma*); CMPD: Dahl & Krzeminska 1997; CAT: Dahl 1992.

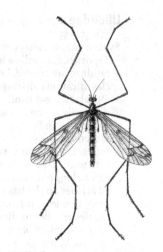

577 Trichocera maculipennis Meigen, male; Zilahi-Sebess 1960

(Trichomyiidae): part of the Psychodidae.

Trixoscelididae (key couplet 93; fig. 192, 205-206)

Systematics: Acalyptrate Brachycera; superfamily Sphaeroceroidea; traditionally part of the Heleomyzidae; in Europe 1 genus, *Trixoscelis*, with about 25 species.

Characters: Small (2-4 mm), mostly grey flies. Arista with microscopic hairs to short plumose; ocelli present; Oc-bristles present, situated next to the ocellar triangle at the same level as the anterior ocellus or lower; P-bristles converging; 2 pairs of F-bristles, curving backward; interfrontal setulae absent; vibrissae present. Wing usually patterned or partly tinged, in only a few species wing clear; costa with a subcostal break; vein Sc complete but apical part thin and especially near costa close to vein R1; crossvein BM-Cu present; cell cup closed. Tibiae with dorsal preapical bristle

Biology: The larvae of *Trixoscelis* are unknown; some species have been bred from nests of birds. Adults are thermophilous and prefer open well insolated habitats, like dunes, beaches, heather, sandy steppes and forest steppes.

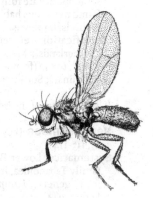

578 Trixoscelis frontalis (Fallén), male; Papp 1988i

Identification references: Hackman 1970 (Europe); Soós 1981 (Central Europe); Stackelberg 1989am (former USSR); CMPD: Papp 1998i; CAT: Soós 1984j.

(Trypetidae): name used in the past for Tephritidae.
(Tylidae): name used in the past for Micropezidae.

Ulidiidae (key couplet 139, 141, 161, 163; fig. 301, 349-351, 356-357)

Systematics: Acalyptrate Brachycera; superfamily Tephritoidea; following recent classifications the Otitinae are treated here as a subfamily; traditional classifications distinguish between Ulidiidae and Otitidae as two families or include the Ulidiinae as a subfamily of the Otitidae; some 17 genera and about 105 species in Europe.

Characters: Small to large (2.5-11 mm), moderately robust flies, colour varying from partially yellow and dark to all black, in some cases grey or metallic. Arista bare to long pubescent; ocelli present; in a few species all bristles on the head are reduced or absent, but the usual bristle configuration is: Oc-bristles present, P-bristles parallel to diverging, 1-2 pairs of F-bristles, scattered interfrontal setulae present, and vibrissae absent. Wing usually with a pattern of spots or

579 Physiphora alceae (Preyssler), male; Stackelberg 1956

stripes or with a spot at the wing tip, rarely all clear; costa with or without a subcostal and/or a humeral break; vein Sc complete; crossvein BM-Cu present; cell cup closed. Tibiae usually without dorsal preapical bristle. Female with oviscape, non retractable basal segment of the ovipositor.

Biology: The biology of most species is little known. Larvae are found in rotting vegetable matter, leaf litter, fruit, dung, sap exuding from tree wounds and underneath tree bark; some species are fully phytophagous, feeding on sugar beet, onions, maize, etc. Adults are found in various habitats ranging from saline biotopes to sandy areas or moist and marshy grounds, on flowers, tree trunks or on excrement.

Identification references: Clements 1990 (British Isles); Kabos & Van Aartsen 1984 (Netherlands); Merz 1996a, 1996c (Switzerland); Soós 1980 (Central Europe); Richter 1989d, 1989e (former USSR); CMPD: Greve 1998 (Otitinae only); CAT: V.F. Zaitzev 1984 (Ulidiinae); Soós 1984f (Otitinae).

(Usiidae): part of the Bombyliidae.

Vermileonidae (key couplet 41, 52; fig. 87, 113-114)

Systematics: Lower Brachycera; superfamily Tabanoidea; in Europe 9 species in 2 genera, *Lampromyia* (5) and *Vermileo* (4).

Characters: Medium sized to large (5-18 mm), slender, pale to dark brown flies, devoid of bristles. Eyes dichoptic; apical segments of the antenna tapering, making up a style; mouthparts conspicuously elongate in *Lampromyia*, short in *Vermileo*. Wing long and slender, without alula; clear, tinged or with markings along the veins or across the wing; cell m3 and cell cup both either

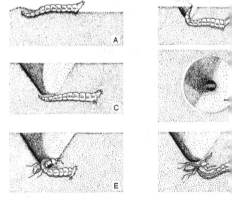

580 Vermileonid larva digging a funnel-trap in dry sand; Wheeler 1931

open or closed. Legs long, tibiae with apical bristles or spurs; empodium pulvilliform, in *Lampromyia* pulvilli and/or empodium sometimes extremely reduced.

Biology: The larvae of the Vermileonidae prey on small insects and other invertebrates, using a funnel trap they dig out in dry and sandy areas, similar to the traps of ant lions (Neuroptera: Myrmeleontidae). The adults visit flowers.

Identification references: Nagatomi et al. 1999 (World genera); Stuckenberg 1965 (*Vermileo*), 1998 (*Lampromyia*); CMPD: Nagatomi 1997b; CAT: Majer 1988f.

Xenasteiidae (key couplet 101; fig. 226-227)
Systematics: Acalyptrate Brachycera; superfamily Opomyzoidea; in Southern Europe 2 species, *Tunisimyia convergens* Ventura & Carles-Tolrá (Balearic Islands) and *Xenasteia excellens* (Papp) (Italy, Malta).
Characters: Minute (1.3-1.8 mm), brown to black lustrous flies; legs and parts of the head yellowish. Arista micropubescent; ocelli present; Oc-bristles present; small, thin pseudopostocellar bristles present, curving forward or converging; 3 pairs of F-bristles, the lower pair curving inward, the others backward or the upper pair obliquely backward; interfrontal bristles present; vibrissae present. Wing unmarked; alula with long hair-like bristles; venation reduced; costa with humeral and subcostal breaks; vein Sc complete up to the costa, but apical part pallid; vein M1 largely pallid, not reaching the wing margin; crossveins BM-Cu and DM-Cu both absent; cell cup closed; anal vein absent. Tibiae without dorsal preapical bristle.

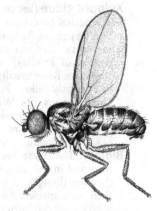

581 Xenasteia excellans (Papp), male; Papp 1980

Biology: The larvae of this small family are unknown. The adults are found in saline habitats, both coastal and inland and in deserts. They are found on flowering plants and trees.
Identification references: Vanin 2003 (Italy); Ventura & Carles-Tolrá 2002 (Balearic Islands); CMPD: Papp 1998f; CAT: Papp 1984l.

Xylomyidae (key couplet 41, 53; fig. 86, 115)
Systematics: Lower Brachycera; superfamily Stratiomyoidea; in Europe 8 species in 2 genera, *Solva* (5) and *Xylomya* (3).
Characters: Medium sized to large (6-20 mm), slender flies, without bristles on thorax and abdomen. Body dark, with limited to extensive lighter markings. Eyes dichoptic; antenna with acute tip, with 10 segments. Wing clear or tinged; both the cell m3 and cell cup closed. Tibia of mid and hind leg with apical bristles or spurs; empodium pulvilliform.
Biology: Larvae, often several together, living underneath the bark of deciduous trees, feeding on injured and dead insect larvae and other small invertebrates. The adults are mainly found near the larval habitats, often sitting on trees.

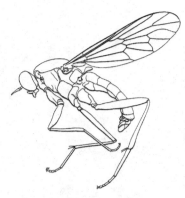

582 Xylomya maculata (Meigen), male; Séguy 1926

Identification references: Brugge 2002 (Netherlands); Krivosheina 1999a, 1999b; Nartshuk 1989e (former USSR); Oldroyd 1969; Stubbs & Drake 2001 (British Isles); Rozkošný 1973 (Fennoscandia and Denmark); CMPD: Nagatomi & Rozkošný 1997a; CAT: Krivosheina 1988b.

Xylophagidae (key couplet 43, 52; fig. 90, 112)

Systematics: Lower Brachycera; superfamily Xylophagoidea; in Europe 1 genus, *Xylophagus*, with 5 species.

Characters: Medium sized to large (5-11 mm) flies, largely devoid of bristles, with long, slender wings, legs and abdomen. Body usually dark with yellow markings, in some cases largely yellow. Eyes dichoptic; antenna simple, with 12, rarely 11 segments. Wing without alula and with markings along the veins or across the wing; cell m3 open; cell cup open or closed. Legs in some cases banded; tibiae with apical bristles or spurs; empodium pulvilliform.

Biology: The larvae live underneath tree bark, in rot holes of trees and in decaying deciduous and coniferous trees where they are thought to prey on other invertebrates, including their own species. Adults are mainly found near the larval habitats, especially in more rich and humid forests, often sitting on tree trunks.

583 Xylophagus ater Meigen, male; Verrall 1909

Identification references: Van der Goot 1985 (Northwestern Europe); Haenni 1997d (Switzerland); Krivosheina & Mamaev 1972 (species); Nartshuk 1989d (former USSR); Oldroyd 1969; Stubbs & Drake 2001 (British Isles); CMPD: Nagatomi & Rozkošný 1997b; CAT: Krivosheina & Mamaev 1988d.

Andorra: Carles-Tolrá Hjorth-Andersen, M. (ed.), 2002. Catalogo de los Diptera de España, Portugal y Andorra. – Monografias Sociedad Entomologica Aragonesa 8: 1-323.

Belgium: Grootaert, P., Bruyn, L. de & Meyer, M. de (eds.), 1992. Catalogue of the Diptera of Belgium. – Studiedocumenten van het K.B I.N. 70: 1–338.

Bulgaria: Lavciev, V., 2003. Diptera: Fanniidae, Muscidae, Stomoxydidae. - Catalogus Faunae Bulgaricae 5: 1-77 (the other volumes in this series are not about Diptera).

Czech Republic: Chvála, M. (ed.), 1997. Checklist of Diptera (Insecta) of the Czech and Slovak Republics. Karolinum-Charles University Press, Prague: 1–130.

Denmark: Petersen, F.T. & Meier, R. (eds.), 2001. A preliminary list of the Diptera of Denmark. – Steenstrupia 26: 119–276.

Finland: Hackmann, W., 1980. A check list of the Finnish Diptera. - Notulae Entomologicae 60: 17-48 (Part I. Nematocera and Brachycera (s. str.)); 117-162 (Part II. Cyclorrhapha).

Germany: Schumann, H., Bährmann, R. & Stark, A. (eds.), 1999. Checkliste der Dipteren Deutschlands. – Studia Dipterologica, Supplement 2: –354. (Additions in Studia Dipterologica 9 (2003): 437–445, and 11 (2005): 619-630.)

Great Britain: Chandler, P.J., (ed.), 1998. Checklists of insects of the British Isles (New Series) part 1: Diptera. – Handbooks for the Identification of British Insects 12: i–vx, 1–234. (Additions and corrections in subsequent issues of Dipterists Digest.)

Hungary: Papp, L. (ed.), 2001. Checklist of the Diptera of Hungary. Hungarian Natural History Museum, Budapest: 1–550.

Ireland: see Great Britain.

Italy: Minelli, A., Ruffo, S. & Posta, S. la (eds.), 1995. Checklist delle specie della fauna Italiana, vols. 63-78 (each volume with separate page numbers).

Lithuania: Pakalniskis, S., Rimsaite, J., Spragnauskaite-Bernotiene, R., Butautaite, R. & Podenas, S., 2000. Checklist of Lithuanian Diptera. Acta Zoologica Lituanica 10(1): 3–57. (Additions: 2000: Acta Zoologica Lituanica 10 (3): 20–26).

Netherlands: Beuk, P.L.T. (ed.), 2002. Checklist of the Diptera of the Netherlands. KNNV Uitgeverij, Utrecht: 1–448

Poland: Razowski, J. (ed.), 1991. Checklist of animals of Poland, Vol. 2, Polska Akademia Nauk: 1 – 342 (Diptera 77–269).

Portugal: see Andorra

Slovakia: see Czech Republic

Spain: see Andorra.

Switzerland: Merz, B., Bächli, G, Haenni, J.- P. & Gonseth, Y. (eds.), 1998. Diptera – Checklist Fauna Helvetica 1: 1-369. (Additions: 2002: Mitteilungen der Entomologische Gesellschaft Basel 51: 110-140.)

For a systematic survey of all European animals at family, genus, (sub)species and/or country level, including distribution maps for the (sub)species, see the Fauna European Web Service website at http://www.faunaeur.org

REFERENCES

Albrecht, A., 1990. Revision, phylogeny and classification of the genus Dorylomorpha (Diptera: Pipunculidae). - Acta Zoologica Fennica 188: 1-240.

Alexander, C.P. & Byers, G.W., 1981. Tipulidae. - In: McAlpine, J.F. et al. (eds), Manual of Nearctic Diptera, Volume 1. - Research Branch, Agriculture Canada, Monograph 27: 153-190.

Andersen, S., 1996. The Siphonini (Diptera: Tachinidae) of Europe. - Fauna Entomologica Scandinavica 33: 1-148.

Andersson, H., 1971. The Swedish species of Chyromyidae (Diptera) with lectotype designations. - Entomologisk Tidskrift 92: 95-99.

Andersson, H., 1976. Revision of the Anthomyza species of Northwest Europe (Diptera: Anthomyzidae) I. The gracilis group. - Entomologica Scandinavica 7: 41-52.

Andersson, H., 1977. Taxonomic and phylogenetic studies on Chloropidae (Diptera) with special reference to Old World genera. - Entomologica Scandinavica, Supplement 8: 1-200.

Andersson, H., 1982. The Swedish species of the families Synneuridae, Canthyloscelidae and Scatopsidae (Diptera). - Entomologisk Tidskrift 103: 5-11.

Andersson, H., 1984a. Revision of the Anthomyza species of Northwest Europe (Diptera: Anthomyzidae) II. The pallida group. - Entomologica Scandinavica 15: 15-24.

Andersson, H., 1984b. Family Anthomyzidae. - In: Soós, Á. & Papp, L. (eds), Catalogue of Palaearctic Diptera 10: 50-53

Andersson, H., 1989. Taxonomic notes on Fennoscandian Micropezidae (Diptera). - Notulae Entomologicae 69: 153-162.

Andersson, H., 1990. De svenska prickflugorna (Diptera, Pallopteridae), med typdesigneringar och nya synonymer. - Entomologisk Tidskrift 111: 123-131.

Andersson, H., 1991. Family Lonchopteridae. - In: Soós, Á. & Papp, L. (eds), Catalogue of Palaearctic Diptera 7: 139-142.

Andersson, H., 1997a. Diptera Ptychopteridae, Phantom crane flies. In: Nilsson, A. (ed.), Aquatic insects of North Europe. A taxonomic handbook. Volume 2. Odonata - Diptera. Apollo Books, Stenstrup: 193-207.

Andersson, H., 1997b. Diptera Scatophagidae, Dung flies. Same as 1997a: 401-410.

Ashe, P. & Cranston, P.S., 1990. Family Chironomidae. - In: Soós, Á. & Papp, L. (eds), Catalogue of Palaearctic Diptera 2: 113-355.

Bächli, G., 1997. Die Arten der Tanypezidae, Dryomyzidae, Periscelidae, Aulacigastridae und Stenomicridae in der Schweiz (Diptera). - Mitteilungen der Entomologischen Gesellschaft Basel 47: 29-34.

Bächli, G., 1998. Family Drosophilidae. - In: Papp, L. & Darvas, B. (eds), Contributions to a Manual of Palaearctic Diptera, Volume 3. Science Herald, Budapest: 503-513.

Bächli, G. & Burla, H., 1985. Diptera Drosophilidae. - Insecta Helvetica, Serie A 7: 1-116.

Bächli, G. & Teresa Rocha Pité, M., 1984. Family Drosophilidae. - In: Soós, Á. & Papp, L. (eds), Catalogue of Palaearctic Diptera 10: 186-220.

Bächli, G., Vilela, C.R., Andersson Escher, S. & Saura, A., 2004. The Drosophilidae (Diptera) of Fennoscandia and Denmark. - Fauna Entomologica Scandinavica 39: 1-362.

Bährmann, R. & Bellstedt, R., 1988. Beobachtungen und Untersuchungen zum Vorkommen der Lonchopteriden

auf dem Gebiet der DDR, mit einer Bestimmungstabelle der Arten. - Deutsche Entomologische Zeitschrift (NF) 35: 265-279.

Balachowsky, A. & Mesnil, L., 1935. Les insectes nuisibles aux plantes cultivées, leurs moeurs, leur destruction: traité d' entomologie agricole concernant la France, la Corse, l'Afrique du Nord et les régions limitrophes 1: i-xvi, 1-1137.

Bankowska, R., 1979. Conopidae. - Fauna Polski 7: 1-133.

Barendregt, A., 2001. Zweefvliegentabel. Ninth edition. Jeugdbondsuitgeverij, Utrecht: 1-96.

Barták, M., 1982. The Czechoslovak species of Rhamphomyia (Diptera, Empididae) with description of a new species from Central Europe. - Acta Universitatis Carolinae (Biologica) 1880: 381-461.

Barták, M., 1986. The Czechoslovak species of Lonchopteridae (Diptera). In: Olejnicek, J. & Spitzer, K. (eds), Diptera Bohemoslovaca 4. - Jihosec Muzeum, Ceske Budejovice 5: 61-69.

Barták, M., 1998. Family Lonchopteridae. - In: Papp, L. & Darvas, B. (eds), Contributions to a Manual of Palaearctic Diptera, Volume 3. Science Herald, Budapest: 13-16.

Bass, J., 1998. Last-instar larvae and pupae of the Simuliidae of Britain and Ireland. A key with brief ecological notes. - Freshwater Biological Association, Scientific Publication 55: 1-101.

Becker, N. (ed.), 2003. Mosquitoes and their control. - Kluwer Academic Press/Plenum Publishers, New York: i-xxi, 1-498.

Belshaw, R., 1993. Tachinid flies. Diptera: Tachinidae. - Handbooks for the Identification of British Insects 10(4ai): 1-170.

Berest, Z.L. & Mamaev, B.M., 1989. A key to females of the Palaearctic genera of gall midges of the subfamily Lestremiinae (Diptera, Cecidomyiidae). Kiev: 1-13 (In Russian).

Bernardi, N., 1973. The genera of the Nemestrinidae (Diptera: Brachycera). - Arqivos Zoologicos de Sao Paulo 24: 211-318,

Beschovski, V.L., 1994. Contribution to the study of West Palaearctic Tethinidae (Diptera). - Acta Zoologica Bulgarica 47: 16-29.

Beuk, P.L.Th. (ed.), 2002. Checklist of the Diptera of the Netherlands. - KNNV Publishing, Utrecht: 1-448.

Beuk, P.L.Th. & Jong, H. de, 1994. De Nederlandse soorten van de Camillidae (Diptera). - Entomologische Berichten, Amsterdam 54: 1-6.

Boorman, J., 1993. Biting midges (Ceratopogonidae). - In: Lane, R.P. & Crosskey, R.W. (eds), Medical Insects and Arachnids. Chapmann & Hall, London etc.: 288-309.

Boorman, J., 1997. Family Ceratopogonidae. - In: Papp, L. & Darvas, B. (eds), Contributions to a Manual of Palaearctic Diptera, Volume 2. Science Herald, Budapest: 349-368.

Borkent, A. & Wirth, W.W., 1999. World species of biting midges (Diptera: Ceratopogonidae). - Bulletin of the American Museum of Natural History 233: 1-257.

Borowiec, L., 1984. Hippoboscidae - Klucze do Oznaczania owadów Polski 131: 1-40.

Brake, I., 2000. Phylogenetic systematics of the Milichiidae (Diptera, Schizophora). - Entomologica Scandinavica, Supplement 57: 1-120.

Brinkmann, R., 1997. Diptera Cylindrotomidae. - In: Nilsson, A. (ed.), Aquatic insects of North Europe. A taxonomic handbook. Volume 2. Odonata - Diptera. Apollo Books, Stenstrup: 99-104.

Brugge, B., 2002. Wapenvliegentabel. Second edition. Jeugdbondsuitgeverij, Utrecht: 1-94.

Brunel, E., 1998. Family Opomyzidae. - In: Papp, L. & Darvas, B. (eds), Contributions to a Manual of Palaearctic Diptera, Volume 3. Science Herald, Budapest: 259-266.

Burnet, B., 1960. The European species of the genus Coelopa (Dipt., Coelopidae). - Entomologists Monthly Magazine 96: 8-13.

Büttiker, W., 1994. Die Lausfliegen der Schweiz (Diptera, Hippoboscidae) mit Bestimmungsschlüssel. - Documenta Faunistica Helvetiae 15: 1-118.

Byers, G.W., 1983. The crane fly genus Chionea in North America. - Kansas University Science Bulletin 52: 59-195.

Canzoneri, S. & Meneghini, D., 1983. Ephydridae e Canaceidae. - Fauna d'Italia 20: i-xii, 1-337.

Carles-Tolrá, M., 1993. A new species of Geomyza, with new acalyptrate records to the Iberian Peninsula (Diptera, Acalyptratae). - Historia Animalium 2: 49-55.

Carles-Tolrá, M., 1994. Three new species of Opomyzidae (Diptera) from Spain. - Entomological Review 73: 91-95 2: 49-55.

Carles-Tolrá, M., 2006. Syllegomydas algericus (Gerstaecker): Genero y especie nuevos para Europa (Diptera: Mydidae). - Boletin de la Sociedad Entomologica Aragonesa 38: 76.

Chandler, P.J., 1975. Notes on the status of three unusual Acalyptrate flies (Diptera) (Micropezidae, Megamerinidae, Tanypezidae). - Proceedings of the British Entomological and Natural History Society 8: 66-72.

Chandler, P.J., 1978. A revision of the British Asteiidae (Diptera) including two additions to the British fauna. - Proceedings of the British Entomological and Natural History Society 11: 23-34.

Chandler, P.J., 1981. Nemedina alamirabilis sp.n., a new genus and species of Diptera Eremoneura of uncertain affinities, from Hungary. - Acta Zoologica Academiae Scientiarum Hungaricae 27: 103-113.

Chandler, P.J., 1983. Pseudopomyza atrimana (Meigen) (Diptera: Pseudopomyzidae), a fly of an acalyptrate family new to the British list. - Proceedings and Transactions of the British Entomological and Natural History Society 16: 87-91.

Chandler, P.J., 1986. The British species of Diastata Meigen and Campichoeta Macquart (Diptera, Drosophiloidea). - Proceedings and Transactions of the British Entomological and Natural History Society 19: 9-16.

Chandler, P.J., 1987. The families Diastatidae and Campichoetidae (Diptera, Drosophiloidea) with a revision of Palaearctic and Nepalese species of Diastata Meigen. - Entomologica Scandinavica 18: 1-50.

Chandler, P.J., 1991a. Family Opetiidae. - In: Soós, Á. & Papp, L. (eds), Catalogue of Palaearctic Diptera 7: 204-205.

Chandler, P.J., 1991b. Family Platypezidae. - Same as 1991a: 205-217.

Chandler, P.J., 1998a. Family Opetiidae. - In: Papp, L. & Darvas, B. (eds), Contributions to a Manual of Palaearctic Diptera, Volume 3. Science Herald, Budapest: 17-25.

Chandler, P.J., 1998b. Family Campichoetidae. - Same as 1998a: 515-522.

Chandler, P.J., 1998c. Family Diastatidae. - Same as 1998a: 523-530.

Chandler, P.J., 2001. The flat-footed flies (Diptera: Opetiidae and Platypezidae) of Europe. - Fauna Entomologica Scandinavica 36: 1-276.

Chandler, P.J., 2002. Heterotricha Loew and allied genera (Diptera: Sciaroidea): offshoots of the stem group of the Mycetophilidae and/or Sciaridae? - Annales de la Societé Entomologique de France (N.S.) 38: 101-144.

Chandler, P.J. & Ribeiro, E., 1995. The Sciaroidea (Diptera) (excluding Sciaridae) of the Atlantic islands (Canary Islands, Madeira and the Azores). - Boletim do Museu Municipal do Funchal (História Natural), Supplement 3: 1-170.

Chandler, P.J. & Shatalkin, A.I., 1998. Family Platypezidae. - In: Papp, L. & Darvas, B. (eds), Contributions to a Manual of Palaearctic Diptera, Volume 3. Science Herald, Budapest: 27-49.

Chvála, M., 1975. The Tachydromiinae (Dipt. Empididae) of Fennoscandia and Denmark. - Fauna Entomologica Scandinavica 3: 1-336.

Chvála, M., 1980. Acroceridae (Diptera) of Czechoslovakia. - Acta Universitatis Carolinae, Biologica 12: 253-267.

Chvála, M., 1983. The Empidoidea (Diptera) of Fennoscandia and Denmark II. General part. The families Hybotidae, Atelestidae and Microphoridae. - Fauna Entomologica Scandinavica 12: 1-279.

Chvála, M., 1986. Revision of Palaearctic Microphoridae (Diptera). 1. Microphor Macq. - Acta Entomologica Bohemoslovaca 83: 432-454.

Chvála, M., 1987. Revision of Palaearctic Microphoridae (Diptera). 2. Schistostoma Beck. - Acta Entomologica Bohemoslovaca 84: 133-155.

Chvála, M., 1988a. Revision of Palaearctic Microphoridae (Diptera). 3. Parathalassiinae (Parathalassius Mik and Microphorella Becker). - Acta Entomologica Bohemoslovaca 85: 352-372.

Chvála, M., 1988b. Family Tabanidae. - In: Soós, Á. & Papp, L. (eds), Catalogue of Palaearctic Diptera 5: 97-171.

Chvála, M., 1989a. Monograph of Northern and Central European species of Platypalpus (Diptera, Hybotidae), with data on the occurrence in Czechoslovakia. - Acta Universitatis Carolinae, Biologica 32: 209-376.

Chvála, M., 1989b. Family Atelestidae. - In: Soós, Á. & Papp, L. (eds), Catalogue of Palaearctic Diptera 6: 169-170, 336.

Chvála, M., 1989c. Family Microphoridae. - Same as 1989b: 171-174.

Chvála, M., 1994. The Empidoidea (Diptera) of Fennoscandia and Denmark. III. Genus Empis. - Fauna Entomologica Scandinavica 29: 1-192.

Chvála, M., 2005. The Empidoidea (Diptera) of Fennoscandia and Denmark. IV. Genus Hilara. - Fauna Entomologica Scandinavica 40: 1-233.

Chvála, M. & Jezek, J., 1997. Diptera Tabanidae, Horse flies. - In: Nilsson, A. (ed.), Aquatic insects of North Europe. A taxonomic handbook. Volume 2. Odonata - Diptera. Apollo Books, Stenstrup: 295-309.

Chvála, M. & Kovalev, V.G., 1989. Family Hybotidae. - In: Soós, Á. & Papp, L. (eds), Catalogue of Palaearctic Diptera 6: 174-227.

Chvála, M., Lyneborg, L. & Moucha, J., 1972. The Horse Flies of Europe (Diptera, Tabanidae). - Entomological Society of Copenhagen: 1-499, pl. 1-8.

Chvála, M. & Smith, K.G.V., 1988. Family Conopidae. - In: Soós, Á. & Papp, L. (eds), Catalogue of Palaearctic Diptera 8: 245-272.

Chvála, M. & Wagner, R., 1989. Family Empididae. - In:

Soós, Á. & Papp, L. (eds), Catalogue of Palaearctic Diptera 6: 228-336.

Clements, D.K., 1990. Provisional key to the Otitidae and Platystomatidae of the British Isles. - Dipterists Digest 6: 32-41.

Coe, R.L., 1966. Diptera Pipunculidae. - Handbooks for the Identification of British Insects 10(2c): 1-83.

Cogan, B.H., 1984a. Family Canacidae. - In: Soós, Á. & Papp, L. (eds), Catalogue of Palaearctic Diptera 10: 124-126.

Cogan, B.H., 1984b. Family Ephydridae. - Same as 1984a: 126-176.

Cole, J., 1996. A second British site for Prosopantrum flavifrons (Tonnoir & Malloch) (Diptera, Heleomyzidae). - Entomologists Monthly Magazine 132: 310.

Collin, J.E., 1944. The British species of Anthomyzidae (Diptera). - Entomologists Monthly Magazine 80: 265-272.

Collin, J.E., 1945. The British genera and species of Oscinellinae. - Transactions of the Royal Entomological Society 97: 117-148.

Collin, J.E., 1961. Empididae. - British Flies 6: v-viii, 1-782.

Collin, J.E., 1966. A revision of the palaearctic species of Tethina and Rhicnoessa. - Bolletino del Museo Civico di Storia Naturale di Venezia 16: 19-32.

Collis, D.H. & McAlpine, D.K., 1991. Diptera. - In: Naumann, I.D. et al. (eds), The Insects of Australia (2nd ed.). Vol. 2. Melbourne University Press, Melbourne: 717-786.

Colwell, D.D., Hall, M.J.R. & Scholl, P.J., 2004. The oestrid flies. - CABI Publishing, Wallingford: i-x, 1-359.

Colyer, C.N. & Hammond, C.O., 1968. Flies of the British Isles (2nd ed.). Warne & Co., London, New York: 1-384.

Cook, E.F., 1969. A synopsis of the Scatopsidae of the Palaearctic Part I. Rhegmoclematini. - Journal of Natural History 3: 393-407.

Cook, E.F., 1972. A synopsis of the Scatopsidae of the Palaearctic Part II. Swammerdamellini. - Journal of Natural History 6: 625-634.

Cook, E.F., 1974. A synopsis of the Scatopsidae of the Palaearctic Part III. The Scatopsini. - Journal of Natural History 8: 61-100.

Couri, M.C. & Pont, A.C., 1999. A key to the world genera of the Coenosiini (Diptera, Muscidae, Coenosiinae). - Studia Dipterologica 6: 93-102.

Courtney, G.W., 2000. Family Blephariceridae. - In: Papp, L. & Darvas, B. (eds), Contributions to a Manual of Palaearctic Diptera, Volume 2. Science Herald, Budapest: 9-30.

Courtney, G.W., Sinclair, B.J. & Meier, R., 2000. Morphology and terminology of Diptera larvae. - In: Papp, L. & Darvas, B. (eds), Contributions to a Manual of Palaearctic Diptera, Volume 2. Science Herald, Budapest: 85-163.

Cranston, P.S., Ramsdale, C.D., Snow, L.R. & White, G.B., 1987. Keys to the adults, male hypopygia, fourth-instar larvae and pupae of the British mosquitoes (Culicidae) with notes on their ecology and medical importance. - Freshwater Biological Association, Scientific Publication 48: 1-152.

Crosskey, R.W., 1977. A review of the Rhinophoridae (Diptera) and a revision of the Afrotropical species. - Bulletin of the British Museum of Natural History (Entomology) 36: 1-66.

Crosskey, R.W., 1993a. Blackflies (Simuliidae). - In: Lane, R.P. & Crosskey, R.W. (eds), Medical Insects and Arachnids. Chapmann & Hall, London etc.: 241-287.

Crosskey, R.W., 1993b. Stable-flies and horn-flies (bloodsucking Muscidae). – Same as 1993a: 389-402.

Crosskey, R.W. & Lane, R.P., 1993. House-flies, blow-flies and their allies (calyptrate Diptera). - In: Lane, R.P. & Crosskey, R.W. (eds), Medical Insects and Arachnids. Chapman & Hall, London etc.: 403-428.

Czerny, L., 1928. 54. Opomyzidae. - In: Lindner, E. (ed.), Die Fliegen der paläarktischen Region. Schweizerbart, Stuttgart 6(54c): 1-15.

Czerny, L., 1930. 38a, b. Dryomyzidae und Neottiophilidae. - In: Lindner, E. (ed.), Die Fliegen der paläarktischen Region. Schweizerbart, Stuttgart 5(38a, b): 1-15.

Czerny, L., 1936. 51. Chamaemyiidae (Ochthiphilidae). - In: Lindner, E. (ed.), Die Fliegen der paläarktischen Region. Schweizerbart, Stuttgart 5(51): 1-15.

Dahl, C., 1957. Die Gattung Trichocera in Spitzbergen, Bäreninsel und Jan Mayen (Dipt.). - Opuscula Entomologica 22: 227-237.

Dahl, C., 1966. Notes on the taxonomy and distribution of Swedish Trichoceridae. - Opuscula Entomologica 31: 93-118.

Dahl, C., 1967. Additional notes on the taxonomy and distribution of Swedish Trichoceridae. - Opuscula Entomologica 32: 188-200.

Dahl, C., 1992. Family Trichoceridae. - In: Soós, Á., Papp, L. & Oosterbroek, P. (eds), Catalogue of Palaearctic Diptera 1: 31-37.

Dahl, C., 1997. Diptera Culicidae, Mosquitoes. - In: Nilsson, A. (ed.), Aquatic insects of North Europe. A taxonomic handbook. Volume 2. Odonata - Diptera. Apollo Books, Stenstrup: 163-186.

Dahl, C. & Krzeminska, E., 1997. Family Trichoceridae. - In: Papp, L. & Darvas, B. (eds), Contributions to a Manual of Palaearctic Diptera, Volume 2. Science Herald, Budapest: 227-237.

Darvas B. & Papp, L., 2000. Exotic dipteran pests in Europe. - In: Papp, L. & Darvas, B. (eds), Contributions to a Manual of Palaearctic Diptera, Volume 1. Science Herald, Budapest: 693-750.

D'Assis Fonseca, E.C.M., 1968. Diptera Cyclorrhapha Calyptrata section (b) Muscidae. - Handbooks for the Identification of British Insects 10(4b): 1-119.

D'Assis Fonseca, E.C.M., 1978. Diptera Orthorrhapha Brachycera Dolichopodidae. - Handbooks for the Identification of British Insects 9(5): i-iv, 1-90.

Davies, L., 1968. A key to the British species of Simuliidae (Diptera) in the larval, pupal and adult stages. - Freshwater Biological Association, Scientific Publication 24: 1-125.

Dely-Draskovits, Á., 1978. Chloropidae. - Fauna Hungariae 133: 61-194.

Dely-Draskovits, Á., 1981. Scathophagidae. - Fauna Hungariae 145: 1-53.

Dely-Draskovits, Á., 1983. Ptychopteridae. - Fauna Hungariae 156: 1-88.

Dely-Draskovits, Á., 1993. Family Anthomyiidae. - In: Soós, Á. & Papp, L. (eds), Catalogue of Palaearctic Diptera 13: 11-102.

Dempewolf M., 2004 Arthropods of economic importance: Agromyzidae of the world. - World Biodiversity Database CD-ROM series (Mac/Win).

Dienske, J.W., 1987. An illustrated key to the genera and subgenera of the western palaearctic Limoniidae (Insecta, Diptera) including a description of the external morphology. - Stuttgarter Beiträge zur Entomologie, Serie A 409: 1-52.

Disney, R.H.L., 1983. Scuttle flies. Diptera, Phoridae (except Megaselia). - Handbooks for the Identification of British Insects 10(6): 1-81.

Disney, R.H.L., 1989. Scuttle flies. Diptera, Phoridae. Genus Megaselia. - Handbooks for the Identification of

British Insects 10(8): 1-155.

Disney, R.H.L., 1991. Family Phoridae. - In: Soós, Á. & Papp, L. (eds), Catalogue of Palaearctic Diptera 7: 143-204.

Disney, R.H.L., 1994. Scuttle flies: the Phoridae. Chapman & Hall, London: i-xii, 1-467.

Disney, R.H.L., 1998. Family Phoridae. - In: Papp, L. & Darvas, B. (eds), Contributions to a Manual of Palaearctic Diptera, Volume 3. Science Herald, Budapest: 51-79.

Disney, R.H.L., 1999. British Dixidae (meniscus midges) and Thaumaleidae (trickle midges): Keys with ecological notes. - Freshwater Biological Association Scientific Publications 56: 1-129.

Dobson, J.R., 1999. A 'bee-louse' Braula schmitzi Örösi Pál (Diptera: Braulidae) new to the British Isles, and the status of Braula spp. in England and Wales. - British Journal of Entomology and Natural History 11: 139-148.

Dolezil, Z. & Rozkošný, R., 1997. Diptera Syrphidae, Hover flies. - In: Nilsson, A. (ed.), Aquatic insects of North Europe. A taxonomic handbook. Volume 2. Odonata - Diptera. Apollo Books, Stenstrup: 347-361.

Draber Monko, A., 1989. Rhinophoridae. - Klucze do Oznaczania owadow Polski 141: 1-60.

Drake, C.M., 1992. Two new species of Geomyza with notes on the combinata group (Diptera: Opomyzidae). - British Journal of Entomology and Natural History 5: 143-153.

Drake, C.M., 1993. A review of the British Opomyzidae (Diptera). - British Journal of Entomology and Natural History 6: 159-176.

Duda, O., 1930. 2. Bibionidae. - In: Lindner, E. (ed.), Die Fliegen der paläarktischen Region. Schweizerbart, Stuttgart 2(1)4: 1-18.

Dunk, K. von der, 1996. Bestimmungsschlüssel für Raubfliegen (Dipt., Asilidae). - Galathea 12: 131-146.

Dunk, K. von der, 1998. Key to Central European species of Pipunculidae (Diptera). - Studia Dipterologica 4: 289-335.

Ebejer, M.J., 1998a. A review of the Palaearctic species of Aphaniosoma Becker (Diptera, Chyromyidae) with descriptions of new species and a key for the identification of adults. - Deutsche Entomologische Zeitschrift 45: 191-230.

Ebejer, M.J., 1998b. A new species of Gymnochiromyia Hendel (Diptera: Chyromyidae) from the Mediterranean, with notes, lectotype designations and a key to the species from the West Palaearctic. - Studia Dipterologica 5: 19-29.

Ebejer, M.J., 2005. A contribution to the knowledge of the Chyromyidae (Diptera) of Italy with description of a new species of Aphaniosoma Becker. - Revue Suisse de Zoologie 112: 859-867.

Edwards, F.W., 1932. A note on the genus Vermitigris Wheeler (Diptera, Rhagionidae). - Stylops 1: 217-220,

Elberg, K.Y., 1989. Family Anthomyiidae. - In: Bei-Bienko, G.Y. & Steyskal, G.C. (eds), Keys to the insects of the European Part of the USSR. Volume V. Diptera and Siphonaptera, Part 2. Brill, Leiden, etc.: 760-838 (originally published in Russian in 1970).

Enderlein, G., 1942. Klassifikation der Pyrgotiden. - Sitzungsbericht der Gesellschaft der Naturforschende Freunde in Berlin 1941: 98-134.

Engel, E.O., 1932-1937. 25. Bombyliidae. - In: Lindner, E. (ed.), Die Fliegen der paläarktischen Region. Schweizerbart, Stuttgart 4(3)25: 1-619.

Erzinclioglu, Z., 1996. Blowflies. - Naturalist's Handbook 23: 1-71.

Evenhuis, N.L., 2002. Catalog of the Mythicomyiidae of the world (Insecta: Diptera). - Bishop Museum Bulletin in Entomology 10: 1-85.

Evenhuis, N.L. & Greathead, D.J., 1999. World catalog of bee flies (Diptera: Bombyliidae). - Backhuys, Leiden: i-xlviii, 1-756.

Falcoz, L, 1926. Diptères Pupipares. - Faune de France 14: 1-64.

Falk, S.J., 2005. The identification and status of Dryomyza decrepita Zetterstedt (Diptera, Dryomyzidae). - Dipterists Digest 12: 7-12.

Ferrar, P., 1987. A guide to the breeding habits and immature stages of Diptera Cyclorrhapha. Entomonograph 8(1 & 2): 1-907.

Fitzgerald, S.J. & Werner, D., 2004. A key to the Penthetria Meigen (Diptera, Bibionidae) of Europe and the first record of Penthetria heteroptera (Say) from the Palaearctic Region. - Studia Dipterologica 11: 207-210.

Földvári, M. & De Meyer, M., 2000. Revision of the Central and West European Tomosvaryella Aczél (Diptera, Pipunculidae). - Acta Zoologica Academiae Scientiarum Hungaricae 45: 299-334.

Foote, R.H., 1984. Family Tephritidae. - In: Soós, Á. & Papp, L. (eds), Catalogue of Palaearctic Diptera 9: 66-149.

Freeman, P., 1983. Sciarid flies. Diptera, Sciaridae. - Handbooks for the Identification of British Insects 9(6): 1-68.

Freeman, P. & Lane, R.P., 1985. Bibionid and Scatopsid flies. Diptera: Bibionidae & Scatopsidae. - Handbooks for the Identification of British Insects 9(7): 1-74.

Freidberg, A., 2005. A new Palaearctic species of Xenasteia Hardy (Diptera: Xenasteiidae). - Israel Journal of Entomology 28: 133-137.

Frey, R., 1952. Über Chiropteromyza n.gen. und Pseudopomyza Strobl (Diptera, Haplostomata). - Notulae Entomologicae 32: 5-8.

Frutiger, A. & Jolidon, C., 2000. Bestimmungsschlüssel für die Larven und Puppen der in der Schweiz, in Österreich und in Deutschland vorkommenden Netzflügelmücken (Diptera: Blephariceridae), mit Hinweisen zu ihrer Verbreitung und Phänologie. - Mitteilungen der Schweizerischen Entomologischen Gesellschaft 73: 93-108.

Gagné, R.J., 2004. Catalog of the Cecidomyiidae (Diptera) of the World. - Memoirs of the Entomological Society of Washington 25: 1-408.

Gatt, P., 2003. New species and records of Microphorella Becker (Diptera: Empidoidea, Dolichopodidae) from the Mediterranean region. - Revue Suisse de Zoologie 110(4): 669-684.

Geiger, W., 1986. Diptera. Limoniidae 1: Limoniinae. - Diptera Helvetica, Fauna 8: 1-131.

Geller-Grimm, F., 2003. Fotoatlas und Bestimmungsschlüssel der Raubfliegen Deutschlands - CD-ROM.

Gerbachevskaya, A.A., 1986. Family Sciaridae. - In: Soós, Á. & Papp, L. (eds), Catalogue of Palaearctic Diptera 4: 11-72.

Gerbachevskaya, A.A., 1989. Family Sciaridae (Lycoriidae). - In: Bei-Bienko, G.Y. & Steyskal, G.C. (eds), Keys to the insects of the European Part of the USSR. Volume V. Diptera and Siphonaptera, Part 1. Brill, Leiden, etc.: 488-538 (originally published in Russian in 1969).

Gill, G.D. & Peterson, B.V., 1987. Heleomyzidae. - In: McAlpine, J.F. et al. (eds), Manual of Nearctic Diptera, Volume 2. - Research Branch, Agriculture Canada, Monograph 28: 973-980.

Goetghebuer. M. & Lenz, F., 1936-1960. 13. Tendipedidae. -

In: Lindner, E. (ed.), Die Fliegen der paläarktischen Region. Schweizerbart, Stuttgart 3(1)13: 1-260.

Goot, V.S. van der, 1963. Tweevleugelige insecten - Diptera. VII. De snavelvliegen (Rhagionidae), viltvliegen (Therevidae), mugvliegen (Cyrtidae) en wolzwevers (Bombyliidae) van Nederland. - Wetenschappelijke Mededelingen Koninklijke Nederlandse Natuurhistorische Vereniging 46: 1-19.

Goot, V.S. van der, 1985. De snavelvliegen (Rhagionidae), roofvliegen (Asilidae) en aanverwante families van Noordwest-Europa. - Wetenschappelijke Mededelingen Koninklijke Nederlandse Natuurhistorische Vereniging 171: 1-66.

Goot, V.S. van der, 1990. Dansvliegen, determineertabel voor de wat grotere soorten van het geslacht Empis en alle soorten van het geslacht Hybos in de Benelux. Jeugdbondsuitgeverij, Utrecht: 1-34.

Goot, V.S. van der, Aartsen, B. van & Chvála, M., 2000. The Dutch species of the dance fly genus Hilara (Diptera: Empididae). - Nederlandse Faunistische Mededelingen 12: 121-149.

Goot, V.S. van der & Veen, M.P. van, 1996. De spillebeenvliegen, wortelvliegen en wolzwevers van Noordwest-Europa. Second edition. Jeugdbondsuitgeverij, Utrecht: 1-57.

Gornostaev, N.G., 2001. A key to the drosophilid flies (Diptera, Drosophilidae) of European Russia and neighbouring countries. - Entomologicheskoe Obozrenie 80(4): 908-915 (in Russian; English translation in Entomological Review 81(6): 681-686).

Gorodkov, K.B., 1984a. Family Helcomyzidae. - In: Soós, Á. & Papp, L. (eds), Catalogue of Palaearctic Diptera 9: 149-150.

Gorodkov, K.B., 1984b. Family Coelopidae. - same as 1984a: 151-152.

Gorodkov, K.B., 1984c. Family Heleomyzidae. - In: Soós, Á. & Papp, L. (eds), Catalogue of Palaearctic Diptera 10: 15-45.

Gorodkov, K.B., 1986. Family Scathophagidae. - In: Soós, Á. & Papp, L. (eds), Catalogue of Palaearctic Diptera 11: 11-41.

Gorodkov, K.B., 1989a. Family Thyreophoridae. - In: Bei-Bienko, G.Y. & Steyskal, G.C. (eds), Keys to the insects of the European Part of the USSR. Volume V. Diptera and Siphonaptera, Part 2. Brill, Leiden, etc.: 351 (originally published in Russian in 1970).

Gorodkov, K.B., 1989b. Family Helomyzidae (Heleomyzidae). - Same as 1989a: 510-537.

Gorodkov, K.B., 1989c. Family Scatophagidae (Cordyluridae, Scatomyzidae, Scopeumatidae). - Same as 1989a: 732-759.

Gorodkov, K.B. & Kovalev, V.G., 1989. Family Empididae. - In: Bei-Bienko, G.Y. & Steyskal, G.C. (eds), Keys to the insects of the European Part of the USSR. Volume V. Diptera and Siphonaptera, Part 1. Brill, Leiden, etc.: 218-239 (originally published in Russian in 1969).

Greathead, D.J. & Evenhuis, N.L., 1997. Family Bombyliidae. - In: Papp, L. & Darvas, B. (eds), Contributions to a Manual of Palaearctic Diptera, Volume 2. Science Herald, Budapest: 487-512.

Greathead, D.J. & Evenhuis, N.L., 2001. Annotated keys to the genera of African Bombylioidea (Diptera: Bombyliidae; Mythicomyiidae). - African Invertebrates, 42: 105-224.

Gregor, F. & Rozkošný, R., 1995. A key to the identification of Central European Fanniidae (Diptera). - Entomological Problems, Supplement 1: 1-72.

Gregor, F., Rozkošný, R., Barták, M. & Vanhara, J., 2002.

The Muscidae (Diptera) of Central Europe. - Folia Facultatis Scientiarum Naturalium Universitatis Masarykianae Brunensis, Biologia 107: 1-280.

Greve, L., 1993. Family Pallopteridae (Diptera) in Norway. - Fauna Norvegiae, Series B, 40: 37-44.

Greve, L., 1998. Family Otitidae. - In: Papp, L. & Darvas, B. (eds), Contributions to a Manual of Palaearctic Diptera, Volume 3. Science Herald, Budapest: 185-192.

Greve, L. & Nielsen, T.R., 1991. A survey of the family Micropezidae in Norway. - Fauna Norvegiae, Series B, 38: 77-87.

Grichanov, I.Y., 2004. Keys to Swedish genera and species of Dolichopodidae (Diptera). - International Journal of Dipterological Research 15: 123-161.

Grimaldi, D., 1997. The bird flies, genus Carnus: species revision, generic relationships, and a fossil Meoneura in amber (Diptera: Carnidae). - American Museum Novitates 3190: 1-30.

Grunin, K.Y., 1964-1969. 64b. Hypodermatidae. - In: Lindner, E. (ed.), Die Fliegen der paläarktischen Region. Schweizerbart, Stuttgart 8(1)64b: 1-160.

Grunin, K.Y., 1966-1969. 64a. Oestridae. - Same as 1964-1969, 8(2)64a: 1-97.

Grunin, K.Y., 1969. 64a. Gasterophilidae. - Same as 1964-1969, 8(3)64a: 1-61.

Grunin, K.Y., 1989a. Family Gasterophilidae. - In: Bei-Bienko, G.Y. & Steyskal, G.C. (eds), Keys to the insects of the European Part of the USSR. Volume V. Diptera and Siphonaptera, Part 2. Brill, Leiden, etc.: 976-978 (originally published in Russian in 1970).

Grunin, K.Y., 1989b. Family Hippoboscidae. - Same as 1989a: 979-988.

Grunin, K.Y., 1989c. Family Calliphoridae. - Same as 1989a: 995-1020.

Grunin, K.Y., 1989d. Family Oestridae. - Same as 1989a: 1104-1107.

Grunin, K.Y., 1989e. Family Hypodermatidae. - Same as 1989a: 1108-1110.

Gullan, P.J. & Cranston, P.S., 2005. The inescts: an outline of entomology (3rd ed.). - Blackwell Publishing, Malden etc.: i-xviii, 1-505.

Gutsevich, A.V., 1989. Family Culicidae. - In: Bei-Bienko, G.Y. & Steyskal, G.C. (eds), Keys to the insects of the European Part of the USSR. Volume V. Diptera and Siphonaptera, Part 1. Brill, Leiden, etc.: 218-239 (originally published in Russian in 1969).

Haarto, A. & Winqvist, K., 2006. Finnish flies of the family Therevidae. - Entomologica Fennica 17: 46-55.

Hackman, W., 1970. Trixoscelidae (Diptera) from southern Spain and descriptions of a new Trixoscelis species from northern Europe. - Entomologica Scandinavica 1: 127-134.

Hackman, W., Laštovka, P., Matile, L. & Väisänen, R., 1988. Family Mycetophilidae. - In: Soós, Á. & Papp, L. (eds), Catalogue of Palaearctic Diptera 3: 220-327.

Haenni, J.-P., 1982. Révision des espèces de groupe de Dilophus febrilis (L.) avec description d'une espèce nouvelle (Diptera, Bibionidae). - Revue Suisse de Zoologie 89: 337-354.

Haenni, J.-P., 1988. Note sur quelques Diptères associés à un gîte de chauves-souris arboricoles. - Bulletin de la Société Neuchâteloise des Sciences Naturelles 111: 49-53.

Haenni, J.-P., 1992. Family Scatopsidae. - In: Papp, L. & Darvas, B. (eds), Contributions to a Manual of Palaearctic Diptera, Volume 2. Science Herald, Budapest: 255-272.

Haenni, J.-P., 1997b. Family Canthyloscelidae. - Same as 1997a: 273-279.

Haenni, J.-P., 1997c. Anisopodidae (Diptera) de la faune de Suisse, avec la description d'une espèce nouvelle. - Mitteilungen der Schweizerischen Entomologischen Gesellschaft 70: 177-186.

Haenni, J.-P., 1997d. Xylophagidae (Diptera) de la faune Suisse. - Bulletin de la Société Neuchâteloise des Sciences Naturelles 120: 125-129.

Haenni, J.-P., 2004. Mouches. Histoire naturelles des insectes diptères et leurs relations avec l'homme. Muséum d'histoire naturelle de Neuchâtel: 1-168.

Hall, J.C., 1981. Bombyliidae. - In: McAlpine, J.F. et al. (eds), Manual of Nearctic Diptera, Volume 1. - Research Branch, Agriculture Canada, Monograph 27: 589-602.

Hardy, D.E., 1981. Bibionidae. - In: McAlpine, J.F. et al. (eds), Manual of Nearctic Diptera, Volume 1. - Research Branch, Agriculture Canada, Monograph 27: 217-222.

Haren, J.C.M. van & Verdonschot, P.F.M., 1995. Proeftabel Nederlandse Culicidae. - Instituut voor Bos- en Natuuronderzoek Rapport 173: 1-106.

Hendel, F., 1928. Zweiflügler oder Diptera. II. Allgemeiner Teil. - In: Dahl, F. (ed.), Die Tierwelt Deutschlands, Jena, 11.

Hendel, F., 1931-1936. 59. Agromyzidae. - In: Lindner, E. (ed.), Die Fliegen der paläarktischen Region. Schweizerbart, Stuttgart 6(2)59: 1-570.

Hendel, F., 1933. 36. Pyrgotidae. - In: Lindner, E. (ed.), Die Fliegen der paläarktischen Region. Schweizerbart, Stuttgart 5(1)36: 1-15.

Hennig, W., 1937a. 44. Tanypezidae. - In: Lindner, E. (ed.), Die Fliegen der paläarktischen Region. Schweizerbart, Stuttgart 5(1)44: 1-6.

Hennig, W., 1937b. 60a. Milichiidae et Carnidae. - In: Lindner, E. (ed.), Die Fliegen der paläarktischen Region. Schweizerbart, Stuttgart 6(1)60a: 1-91.

Hennig, W., 1938. 60c. Braulidae. - In: Lindner, E. (ed.), Die Fliegen der paläarktischen Region. Schweizerbart, Stuttgart 6(1)60c: 1-14.

Hennig, W., 1940. 45. Ulidiidae. - In: Lindner, E. (ed.), Die Fliegen der paläarktischen Region. Schweizerbart, Stuttgart 5(1)45: 1-34.

Hennig, W., 1955-1964. 63b. Muscidae. - In: Lindner, E. (ed.), Die Fliegen der paläarktischen Region. Schweizerbart, Stuttgart 7(2)63b: 1-1110.

Hennig, W., 1966-1976. 63b. Anthomyiidae. - In: Lindner, E. (ed.), Die Fliegen der paläarktischen Region. Schweizerbart, Stuttgart 7(1)63a: i-lxx, 1-974, i-cxiv.

Hennig, W., 1958. Die Familien der Diptera Schizophora und ihre phylogenetischen Verwandtschaftsbeziehungen. - Beiträge zur Entomologie 8: 505-688.

Hennig, W., 1972. Beiträge zur Kenntnis der rezenten und fossilen Carnidae, mit besonderer Berücksichtigung einer neuen Gattung aus Chile (Diptera: Cyclorrhapha). - Stuttgarter Beiträge zur Naturkunde 240: 1-20.

Hennig, W., 1973. Diptera (Zweiflügler). - Handbuch der Zoologie. IV Band: Arthropoda. 2 Hälfte: Insecta. Zweite Auflage. 2: 1-337.

Hering, E.M., 1957. Bestimmungstabellen der Blattminen von Europa, 1-3. Junk, Den Haag: 1-1185 + 221.

Herting, B., 1961. 64e. Rhinophoridae. - In: Lindner, E. (ed.), Die Fliegen der paläarktischen Region. Schweizerbart, Stuttgart 9(64e): 1-36.

Herting, B., 1993. Family Rhinophoridae. - In: Soós, Á. & Papp, L. (eds), Catalogue of Palaearctic Diptera 13: 102-117.

Herting, B. & Dely-Draskovits, Á., 1993. Family Tachinidae. - In: Soós, Á. & Papp, L. (eds), Catalogue of Palaearctic Diptera 13: 118-458.

Hilger, S., 2000. Family Diopsidae. - In: Papp, L. & Darvas,

B. (eds), Contributions to a Manual of Palaearctic Diptera, Volume 4. Science Herald, Budapest: 335-343.

Hippa, H., Nielsen, T.R. & Steenis, J. van., 2001. The West Palaearctic species of the genus Eristalis Latreille (Diptera, Syrphidae). - Norwegian Journal of Entomology 48: 289-327.

Hofsvang, T., 1997. Diptera Tipulidae, Crane Flies. - In: Nilsson, A. (ed.), Aquatic insects of North Europe. A taxonomic handbook. Volume 2. Odonata - Diptera. Apollo Books, Stenstrup: 93-98.

Hollmann-Schirrmacher, V., 1988. Phylogeny of the subfamily Ilythinae (Diptera, Ephydridae), with special reference to the genus Philygria. - Studia Dipterologica, Supplment 5: 1-144.

Huang, Y.-M., 2002. A pictorial key to the mosquito genera of the world, including subgenera of Aedes and Ochlerotatus (Diptera: Culicidae).- Insecta Koreana 19: 1-132.

Huckett, H.C. & Vockeroth, J.R., 1987. Muscidae. - In: McAlpine, J.F. et al. (eds), Manual of Nearctic Diptera, Volume 2. - Research Branch, Agriculture Canada, Monograph 28: 1115-1131.

Huijbregts, H., 2002. Nederlandse bromvliegen (Diptera: Calliphoridae), inclusief zeven soorten nieuw voor Nederland. - Entomologische Berichten, Amsterdam 62: 82-89.

Hull, F.M., 1962. Robber flies of the world. - Bulletin of the United States National Museum 224: 1-907.

Hůrka, K., 1998a. Family Nycteribiidae. - In: Papp, L. & Darvas, B. (eds), Contributions to a Manual of Palaearctic Diptera, Volume 3. Science Herald, Budapest: 829-838.

Hůrka, K., 1998b. Family Streblidae. - Same as 1998a: 839-848.

Hůrka, K. & Soós, Á., 1986a. Family Nycteribiidae. - In: Soós, Á. & Papp, L. (eds), Catalogue of Palaearctic Diptera 11: 226-234.

Hůrka, K & Soós, Á., 1986b. Family Streblidae. - Same as 1986a: 234-236.

Hutson, A.M., 1977. A revision of the families Synneuridae and Canthyloscelidae (Diptera). - Bulletin of the British Museum of Natural History (Entomology) 35: 67-100.

Hutson, A.M., 1984. Keds, flat-flies and bat-flies. Diptera, Hippoboscidae and Nycteribiidae. - Handbooks for the Identification of British Insects 10(7): 1-40.

Hutson, A.M., Ackland, D.M. & Kidd, L.N., 1980. Mycetophilidae (Bolitophilinae, Ditomyiinae, Diadocidiinae, Keroplatinae, Sciophilinae and Manotinae) Diptera, Nematocera. - Handbooks for the Identification of British Insects 9(3): 1-111.

Irwin, A.G., 1982. A new species of Stenomicra Coquillett (Diptera, Aulacigastridae) from Anglesey, North Wales. - Entomologists Monthly Magazine 118: 235-238.

Ismay, J.W., 1999. The British and Irish genera of Chloropinae (Dipt., Chloropidae). - Entomologists Monthly Magazine 135: 1-37.

Ismay, J.W. & Nartshuk, E.P., 2000. Family Chloropidae. - In: Papp, L. & Darvas, B. (eds), Contributions to a Manual of Palaearctic Diptera, Volume 4. Science Herald, Budapest: 387-429.

Ismay, J.W. & Smith, D., 1994. Prosopantrum flavifrons (Tonnoir and Malloch) (Diptera, Heleomyzidae) new to Britain and the Northern Hemisphere. - Dipterists Digest, Second Series 1: 1-5.

Iwasa, M., 1998a. Family Strongylophthalmyiidae. - In: Papp, L. & Darvas, B. (eds), Contributions to a Manual of Palaearctic Diptera, Volume 3. Science Herald, Budapest: 173-175.

Iwasa, M., 1998b. Family Psilidae. - Same as 1998a: 177-183.
James, M.T., 1981. Xylophagidae. - In: McAlpine, J.F. et al. (eds), Manual of Nearctic Diptera, Volume 1. - Research Branch, Agriculture Canada, Monograph 27: 489-492.
James, M.T. & Turner, W.J., 1981. Rhagionidae. - In: McAlpine, J.F. et al. (eds), Manual of Nearctic Diptera, Volume 1. - Research Branch, Agriculture Canada, Monograph 27: 483-488.
Jankovsky, A.V., 2002. Key for the identification of blackflies (Diptera: Simuliidae) of Russia and adjacent territories (former USSR). - Opredeliteli po Fauna Rossii 170: 1-570 (in Russian).
Jaschhof, M., 1998. Revision der "Lestremiinae" (Diptera, Cecidomyiidae) der Holarktis. - Studia Dipterologica, Supplement 4: 1-552.
Jaschhof, M., Jaschhof, C., Viklund, B. & Kallweit, U., 2006. On the morphology and systematic position of Sciarosoma borealis Chandler, based on new material from Fennoscandia (Diptera: Sciaroidea). - Studia Dipterologica 12: 231-241.
Jedlička, L., Kúdela, M. & Stloukalová, V., 2004. Key to the identification of blackfly pupae (Diptera: Simuliidae) of Central Europe. - Biologia, Bratislava 59, Supplement 15: 157-178.
Jedlička, L. & Stloukalová, V., 1997. Family Simuliidae. - In: Papp, L. & Darvas, B. (eds), Contributions to a Manual of Palaearctic Diptera, Volume 2. Science Herald, Budapest: 331-347.
Jensen, F., 1997. Diptera Simuliidae, Blackflies. - In: Nilsson, A. (ed.), Aquatic insects of North Europe. A taxonomic handbook. Volume 2. Odonata - Diptera. Apollo Books, Stenstrup: 209-241.
Jervis, M.A., 1992. A taxonomic revision of the pipunculid fly genus Chalarus Walker, with particular reference to the European fauna. - Zoological Journal of the Linnean Society 105: 243-352.
Jong, H. de, 2000. Family Scathophagidae. - In: Papp, L. & Darvas, B. (eds), Contributions to a Manual of Palaearctic Diptera, Volume 4. Science Herald, Budapest: 431-445.
Jong, H. de, Noordam, A.P. & Zeegers, Th., 2000. The Acroceridae (Diptera) of The Netherlands. - Entomologische Berichten, Amsterdam 60: 171-179.
Jung, H.F., 1958. 9a. Psychodidae-Bruchomyiinae. - In: Lindner, E. (ed.), Die Fliegen der paläarktischen Region. Schweizerbart, Stuttgart 3(1)9a: 6-10.
Kabos, W.J., 1964. Tweevleugelige insekten. Diptera. De Nederlandse vliegen (Muscidae). - Wetenschappelijke Mededelingen Koninklijke Nederlandse Natuurhistorische Vereniging 33: 1-32.
Kabos, W.J. & Aartsen, B. van, 1984. De Nederlandse boorvliegen (Tephritidae) en prachtvliegen (Otitidae) van Noordwest-Europa. - Wetenschappelijke Mededelingen Koninklijke Nederlandse Natuurhistorische Vereniging 163: 1-64.
Kalweit, U. & Jaschhof, M., 2004. Sciarosoma borealis Chandler, 2002: a remarkable addition to the German fauna of Sciaroidea (Diptera: Bibionomorpha). - Studia Dipterologica 11: 127-128.
Kassebeer, C.F., 2001. Die einheimischen Arten der Gattung Aulacigaster Macquart, 1835 (Diptera, Aulacigastridae). - Dipteron 4: 23-32.
Kehlmaier, C., 2005. Taxonomic revision of European Eudorylini (Insecta, Diptera, Pipunculidae). - Verhandlungen des Naturwissenschaftlichen Vereins zu Hamburg (NF) 41: 45-353.
Kelsey, L.P., 1969. A revision of the Scenopinidae (Diptera) of the world. - Bulletin of the United States National Museum 227: 1-336.
Kelsey, L.P. & Soós, Á., 1989. Family Scenopinidae. - In: Soós, Á. & Papp, L. (eds), Catalogue of Palaearctic Diptera 6: 35-43.
Kelsey, L.P., 1981. Scenopinidae. - In: McAlpine, J.F. et al. (eds), Manual of Nearctic Diptera, Volume 1. - Research Branch, Agriculture Canada, Monograph 27: 525-528.
Klink, A.G. & Moller Pillot H.K.M., 2003. Chironomidae larvae: key to the higher taxa and species of the lowlands of northwestern Europe. - World Biodiversity Database CD-ROM series (Mac/Win).
Kniepert, F.-W., 2000. Insecta: Diptera: Tabanidae. - Süßwasserfauna von Mitteleuropa 21(19): 111-204.
Knutson, L.V. & Lyneborg, L., 1965. Danish Acalyptrate flies, 3, Sciomyzidae. - Entomologiske Meddelelser 34: 61-101.
Koenig, D. & Young, C., 2007. First observation of parasitic relations between big-headed flies, Nephrocerus Zetterstedt (Diptera, Pipunculidae) and crane flies, Tipula Linnaeus (Diptera: Tipulidae, Tipulinae), with larval and puparial descriptions for the genus Nephrocerus. - Proceedings of the Entomologcal Society of Washington (in press).
Kormann, K., 2002. Schwebfliegen und Blasenkopffliegen Mitteleuropas. - Fauna Naturführer, Band 1: 1-272, 220 colour plates.
Korneyev, V.A., 2001. A key to genera of Palaearctic Platystomatidae (Diptera), with descriptions of a new genus and a new species. - Entomological Problems 32: 1-16.
Korneyev, V.A., 2004. Genera of Palaearctic Pyrgotidae (Diptera, Acalyptrata), with nomenclatural notes and a key. - Vestnik Zoologii 38: 19-46.
Kovalev, V.G. & Morge, G., 1984. Family Lonchaeidae. - In: Soós, Á. & Papp, L. (eds), Catalogue of Palaearctic Diptera 9: 247-259.
Kozánek, M., De Meyer, M. & Albrecht, A., 1998. Family Pipunculidae. - In: Papp, L. & Darvas, B. (eds), Contributions to a Manual of Palaearctic Diptera, Volume 3. Science Herald, Budapest: 141-150.
Krivosheina, M.G., 2000. Family Axymyiidae. - In: Papp, L. & Darvas, B. (eds), Contributions to a Manual of Palaearctic Diptera, Volume 4. Science Herald, Budapest: 31-39.
Krivosheina, N.P., 1981. New Palaearctic species of the genus Strongylophthalmyia (Diptera, Strongylophthalmyiidae). - Entomologicheskoe Obozrenie 60: 183-186 (in Russian, English translation in: Entomological Review, Washington 60: 162-165).
Krivosheina, N.P., 1984a. Family Strongylophthalmyiidae. - In: Soós, Á. & Papp, L. (eds), Catalogue of Palaearctic Diptera 9: 27-28.
Krivosheina, N.P., 1984b. Family Odiniidae. - Same as 1984a: 260-262.
Krivosheina, N.P., 1984c. Family Pseudopomyzidae. - In: Soós, Á. & Papp, L. (eds), Catalogue of Palaearctic Diptera 10: 49.
Krivosheina, N.P., 1986a. Family Pleciidae. - In: Soós, Á. & Papp, L., Catalogue of Palaearctic Diptera 4: 313-316.
Krivosheina, N.P., 1986b. Family Bibionidae. - Same as 1986a: 319-330.
Krivosheina, N.P., 1986c. Family Anisopodidae. - Same as 1986a: 330-332.
Krivosheina, N.P., 1988a. Family Diadociidae. - In: Soós, Á. & Papp, L. (eds), Catalogue of Palaearctic Diptera 3: 210-211.
Krivosheina, N.P., 1988b. Family Xylomyidae. - In: Soós, Á.

& Papp, L. (eds), Catalogue of Palaearctic Diptera 5: 38-42.

Krivosheina, N.P., 1989a. Family Scatopsidae. - In: Bei-Bienko, G.Y. & Steyskal, G.C. (eds), Keys to the insects of the European Part of the USSR. Volume V. Diptera and Siphonaptera, Part 1. Brill, Leiden, etc.: 646-662 (originally published in Russian in 1969).

Krivosheina, N.P., 1989b. Family Hyperoscelididae. - Same as 1989a: 663.

Krivosheina, N.P., 1989c. Family Bibionidae. - Same as 1989a: 667-680.

Krivosheina, N.P., 1997a. Family Pachyneuridae. - In: Papp, L. & Darvas, B. (eds), Contributions to a Manual of Palaearctic Diptera, Volume 2. Science Herald, Budapest: 29-34.

Krivosheina, N.P., 1997b. Family Hesperinidae. - Same as 1997a: 35-39.

Krivosheina, N.P., 1997c. Family Anisopodidae. - Same as 1997a: 239-248.

Krivosheina, N.P., 1997d. Family Mycetobiidae. - Same as 1997a: 249-254.

Krivosheina, N.P., 1997e. Family Scenopinidae. - Same as 1997a: 531-538.

Krivosheina, N.P., 1999a. Xylophilous flies of the genus Solva Walker (Diptera, Xylomyidae) of the fauna of Russia and adjacent countries. - Entomologicheskoe Obozrenie 78(1): 196-206.

Krivosheina, N.P., 1999b. Xylophilous flies of the genera Macroceromys and Xylomya (Diptera, Xylomyidae) of the fauna of Russia and adjacent countries. - Zoologicheskii Zhurnal 78(2): 202-216.

Krivosheina, N.P. & Haenni, J.-P., 1986. Family Scatopsidae. - In: Soós, Á. & Papp, L. (eds), Catalogue of Palaearctic Diptera 4: 297-310.

Krivosheina, N.P. & Mamaev, B.M., 1972. A review of the Palaearctic species of the genus Xylophagus Meig. (Diptera, Xylophagidae). - Entomologicheskoe Obozrenie 51: 430-444 (in Russian; English translation in: Entomological Review, Washington 51: 258-267).

Krivosheina, N.P. & Mamaev, B.M., 1986. Family Hesperinidae. - In: Soós, Á. & Papp, L. (eds), Catalogue of Palaearctic Diptera 4: 318-319.

Krivosheina, N.P. & Mamaev, B.M., 1988a. Family Pachyneuridae. - In: Soós, Á. & Papp, L. (eds), Catalogue of Palaearctic Diptera 3: 192-193.

Krivosheina, N.P. & Mamaev, B.M., 1988b. Families Keroplatidae and Macroceridae. - Same as 1988a: 199-210; 212-217.

Krivosheina, N.P. & Mamaev, B.M., 1988c. Family Manotidae. - Same as 1988a: 218.

Krivosheina, N.P. & Mamaev, B.M., 1988d. Family Xylophagidae. - In: Soós, Á. & Papp, L. (eds), Catalogue of Palaearctic Diptera 5: 35-38.

Krivosheina, N.P. & Menzel, F., 1998. The Palaearctic species of the genus Sylvicola Harris, 1776 (Diptera, Anisopodidae). - Beiträge zur Entomologie 48: 201-217.

Krzeminski, W., 1986. Ptychopteridae of Poland (Diptera, Nematocera). - Polskie Pismo Entomologiczne 56: 105-131.

Kuchlein, J.H. & Vos, R. de, 1999. Geannoteerde naamlijst van de Nederlandse vlinders. Backhuys, Leiden: 1-302.

Kurina, O. & Schacht, W., 2003. Synopsis of literature for determination of European fungus gnats (Diptera Sciaroidea: Ditomyiidae, Bolitophilidae, Keroplatidae, Diadocidiidae, Mycetophilidae, Lygistorrhinidae). - Entomofauna 24: 177-200.

Lane, R.P., 1993. Sandflies (Phlebotominae). - In: Lane, R.P. & Crosskey, R.W. (eds), Medical Insects and Arachnids.

Chapmann & Hall, London etc.: 78-119.

Lane, R.P. & Crosskey, R.W. (eds), 1993. Medical Insects and Arachnids. Chapmann & Hall, London etc.: i-xv, 1-723.

Langton, P.H. & Visser, H., 2003. Chironomidae exuviae: a key to pupal exuviae of the West Palaearctic Region. - World Biodiversity Database CD-ROM series (Mac/Win).

Lantsov, V.I. & Chernov, Y.I., 1983. Tipuloid crane-flies in the tundra zone. Moscow: 1-175 (In Russian).

Larsen, M.N. & Meier, R., 2004. Species diversity, distribution and conservation status of Asilidae (Insecta: Diptera) in Denmark. - Steenstrupia 28: 177-241.

Laštovka, P. & Matile, L., 1972. Révision des Diadocidia holarctiques (Dipt. Mycetophilidae). - Annales de la Societé Entomologique de France (N.S.) 8: 205-223.

Lechthaler, W. & Car, M., 2004. Simuliidae. Key to larvae and pupae from Central and Western Europe. - CD-ROM, Wolfgang Lechthaler, Vienna.

Lehr, P.A., 1988. Family Asilidae. - In: Soós, Á. & Papp, L. (eds), Catalogue of Palaearctic Diptera 5: 197-326.

Lehr, P.A., 1996. Robber flies of the subfamily Asilinae (Diptera, Asilidae) of the Palaearctic. - Dalnauka, Vladivostok: 1-184.

Lindegaard, C., 1997. Diptera Chironomidae, Non-biting midges. - In: Nilsson, A. (ed.), Aquatic insects of North Europe. A taxonomic handbook. Volume 2. Odonata - Diptera. Apollo Books, Stenstrup: 265-294.

Lindner, E., 1925. 20. Rhagionidae. - In: Lindner, E. (ed.), Die Fliegen der paläarktischen Region. Schweizerbart, Stuttgart 4(1)20: 1-49.

Lindner, E., 1936-1938. 18. Stratiomyiidae. - In: Lindner, E. (ed.), Die Fliegen der paläarktischen Region. Schweizerbart, Stuttgart 4(1)18: 1-218.

Lindner, E., 1948. Handbuch. - In: Lindner, E. (ed.), Die Fliegen der paläarktischen Region. Schweizerbart, Stuttgart 1: i-xii, 1-422.

Lyneborg, L., 1964. Danske Acalyptrate fluer, 2, Psilidae, Platystomatidae og Otitidae. - Entomologiske Meddelelser 32: 367-388.

Lyneborg, L., 1989. Family Therevidae. - In: Soós, Á. & Papp, L. (eds), Catalogue of Palaearctic Diptera 6: 11-35.

Maa, T.C., 1963. Genera and species of Hippoboscidae (Diptera): types, synonymy, habitats and natural groupings. - Pacific Insects Monograph 6: 1-186.

MacGowan, I. & Rotheray, G.E., 2000. New species, additions and possible deletions to British Lonchaea Fallén (Diptera, Lonchaeidae). - Dipterists Digest, Second Series 7: 37-49.

Majer, J., 1987. Tabanidae. - Fauna Hungariae 161: 1-425.

Majer, J., 1988a. Family Athericidae. - In: Soós, Á. & Papp, L. (eds), Catalogue of Palaearctic Diptera 5: 11-13.

Majer, J., 1988b. Family Rhagionidae. - Same as 1988a: 14-28.

Majer, J., 1988c. Family Rachiceridae. - Same as 1988a: 29-30.

Majer, J., 1988d. Family Hilarimorphidae. - Same as 1988a: 30-31.

Majer, J., 1988e. Family Coenomyiidae. - Same as 1988a: 31-34.

Majer, J., 1988f. Family Vermileonidae. - Same as 1988a: 34-35.

Majer, J., 1997a. Family Rhagionidae. - In: Papp, L. & Darvas, B. (eds), Contributions to a Manual of Palaearctic Diptera, Volume 2. Science Herald, Budapest: 433-438.

Majer, J., 1997b. Family Therevidae. - Same as 1997a: 519-529.

Majer, J., 1997c. European Asilidae. - Same as 1997a: 549-567.

Mamaev, B.M., 1989a. Family Mycetobiidae. - In: Bei-Bienko, G.Y. & Steyskal, G.C. (eds), Keys to the insects of the European Part of the USSR. Volume V. Diptera and Siphonaptera, Part 1. Brill, Leiden, etc.: 402-403 (originally published in Russian in 1969).

Mamaev, B.M., 1989b. Family Cecidomyiidae (Itonididae). - Same as 1989a: 539-645.

Mamaev, B.M., 1989c. Family Axymyiidae. - Same as 1989a: 664-665.

Mamaev, B.M., 1990. New taxa of gall midges of the subfamily Porricondylinae (Diptera, Cecidomyiidae) and a key to the Palaearctic genera of Porricondylinae - Acta Zoologica Bulgarica 40: 12-28 (In Russian).

Mamaev, B.M., 2001. The tribe Holoneurini in the Palaearctic (Diptera, Cecidomyiidae). - All-Russian Institute of Continuous Education in Forestry 17: 1-11 (In Russian).

Mamaev, B.M. & Krivosheina, N.P., 1986a. Family Synneuridae. - In: Soós, Á. & Papp, L. (eds), Catalogue of Palaearctic Diptera 4: 311.

Mamaev, B.M. & Krivosheina, N.P., 1986b. Family Synneuridae. - Same as 1986a: 311-312.

Mamaev, B.M. & Krivosheina, N.P., 1986c. Family Axymyiidae. - In: Soós, Á. & Papp, L. (eds), Catalogue of Palaearctic Diptera 4: 317-318.

Mamaev, B.M. & Krivosheina, N.P., 1988a. Family Ditomyiidae. - In: Soós, Á. & Papp, L. (eds), Catalogue of Palaearctic Diptera 3: 197-199.

Mamaev, B.M. & Krivosheina, N.P., 1988b. Family Mycetobiidae. - Same as 1988a: 218-219.

Mannheims, B., 1951-1968. 15. Tipulidae. - In: Lindner, E. (ed.), Die Fliegen der paläarktischen Region. Schweizerbart, Stuttgart 3(5)1: 1-321.

Martinek, V., 1978. The female of Opomyza thalhammeri and a new species of the genus Geomyza (Diptera, Opomyzidae). - Acta Entomologica Bohemoslovaca 75: 336-343.

Martinovský, J. & Rozkošný, R., 1988. Family Thaumaleidae. - In: Soós, Á. & Papp, L. (eds), Catalogue of Palaearctic Diptera 3: 186-192.

Mathis, W.N., 1982. Canacidae of Israel, with a review of the Palaearctic species of the genus Canace Haliday (Diptera). - Entomologica Scandinavica 13: 57-66.

Mathis, W.N., 1998. Family Canacidae. - In: Papp, L. & Darvas, B. (eds), Contributions to a Manual of Palaearctic Diptera, Volume 3. Science Herald, Budapest: 251-257.

Mathis, W.N. & Freidberg, A., 1982. New beach flies of the genus Xanthocanace Hendel, with a review of the species from the western Palaearctic (Diptera: Canacidae). - Memoirs of the Entomological Society of Washington 10: 97-104.

Mathis, W.N. & Papp, L., 1998. Family Periscelididae. - In: Papp, L. & Darvas, B. (eds), Contributions to a Manual of Palaearctic Diptera, Volume 3. Science Herald, Budapest: 285-294.

Mathis, W.N. & Zatwarnicki, T., 1995. World catalog of shore flies (Diptera: Ephydridae). - Memoirs on Entomology, International 4: 1-430.

Mathis, W.N. & Zatwarnicki, T., 1998. Family Ephydridae. - In: Papp, L. & Darvas, B. (eds), Contributions to a Manual of Palaearctic Diptera, Volume 3. Science Herald, Budapest: 537-570.

Matile, L., 1990. Recherches sur la systématique et l'evolution des Keroplatidae (Diptera, Mycetophiloidea). - Mémoires du Museum National d'Histoire Naturelle,

Zoologie 148: 1-682.

Matile, L., 1993. Les Diptères d'Europe Occidentale. I. Introduction, Techniques d'étude et Morphologie. Nématocères, Brachycères Orthorrhaphes et Aschizes. - Boubee, Paris: 1-439.

Matile, L., 1995. Les Diptères d'Europe Occidentale. II. Biologie, Brachycères Schizophores. - Boubee, Paris: 1-381.

McAlpine, D.K., 1998a. Family Platystomatidae. - In: Papp, L. & Darvas, B. (eds), Contributions to a Manual of Palaearctic Diptera, Volume 3. Science Herald, Budapest: 193-199.

McAlpine, D.K., 1998b. Family Coelopidae. - Same as 1998a: 335-340.

McAlpine, D.K., 1998c. Family Helcomyzidae. - Same as 1998a: 341-344.

McAlpine, D.K., 1998d. Family Heterocheilidae. - Same as 1998a: 345-347.

McAlpine, D.K. & Shatalkin, A.I., 1998. Family Pseudopomyzidae. - In: Papp, L. & Darvas, B. (eds), Contributions to a Manual of Palaearctic Diptera, Volume 3. Science Herald, Budapest: 155-163.

McAlpine, J.F., 1964. Descriptions of new Lonchaeidae (Diptera). I & II. - Canadian Entomologist 96: 661-700, 701-757.

McAlpine, J.F., 1975. Identities of lance flies (Diptera: Lonchaeidae) described by de Meijere, with notes on related species. - Canadian Entomologist 107: 989-1007.

McAlpine, J.F., 1977. A revised classification of the Piophilidae, including 'Neottiophilidae' and 'Thyreophoridae' (Diptera: Schizophora). - Memoirs of the Entomological Society of Canada 103: 2 + 1-66.

McAlpine, J.F., 1981a. Morphology and terminology – Adults. - In: McAlpine, J.F. et al. (eds), Manual of Nearctic Diptera, Volume 1. - Research Branch, Agriculture Canada, Monograph 27: 9-63.

McAlpine, J.F., 1981b. Key to families – Adults. - In: McAlpine, J.F. et al. (eds), Manual of Nearctic Diptera, Volume 1. - Research Branch, Agriculture Canada, Monograph 27: 89-124.

McAlpine, J.F., 1981c. Morgea freidbergi new species, a living sister-species of the fossil M. mcalpinei, and a key to the world genera of Pallopteridae. - Canadian Entomologist 113: 81-91.

McAlpine, J.F., 1987a. Acartophthalmidae. - In: McAlpine, J.F. et al. (eds), Manual of Nearctic Diptera, Volume 2. - Research Branch, Agriculture Canada, Monograph 28: 859-861.

McAlpine, J.F., 1987b. Periscelididae. - Same as 1987a: 895-898.

McAlpine, J.F., 1987c. Chyromyidae. - Same as 1987a: 985-988.

McAlpine, J.F., 1987d. Curtonotidae. - Same as 1987a: 1007-1010.

McAlpine, J.F., et al. (eds), 1981-1989. Manual of Nearctic Diptera. Volume 1 (1981), Research Branch, Agriculture Canada, Monograph 27: i-vi, 1-674; Volume 2 (1987) ditto, Monograph 28: i-vi, 675-1332; Volume 3 (1989) ditto, Monograph 32: i-vi, 1333-1581.

McAlpine, J.F. & Steyskal, G.C., 1982. A revision of Neosilba McAlpine with a key to the world genera of Lonchaeidae (Diptera). - Canadian Entomologist 114: 105-137.

McLean, I.F.G., 1998a. Family Chamaemyiidae. - In: Papp, L. & Darvas, B. (eds), Contributions to a Manual of Palaearctic Diptera, Volume 3. Science Herald, Budapest: 415-423.

McLean, I.F.G., 1998b. Leucopis psyllidiphaga sp.n., a new

species of silverfly (Diptera, Chamaemyiidae) from Britain. - Dipterists Digest 5: 49-54.

Meier, R. & Pont, A.C., 2000. Family Sepsidae. - In: Papp, L. & Darvas, B. (eds), Contributions to a Manual of Palaearctic Diptera, Volume 4. Science Herald, Budapest: 367-386.

Meijere, J.C.H. de, 1906. Die Lonchopteren des palaearktischen Gebietes. - Tijdschrift voor Entomologie 49: 44-98.

Meijere, J.C.H. de, 1913. Zur Kenntnis von Carnus hemapterus Nitzsch. - Schriften der Physikalischen-ökonomischen Gesellschaft zu Königsberg 53: 1-18.

Menzel, F., 2000. Die Trauermücken-Fauna der Bundesrepublik Deutschland. - Beiträge zur Entomologie 50: 317-355.

Menzel, F. & Mohrig, W., 1997. Family Sciaridae. - In: Papp, L. & Darvas, B. (eds), Contributions to a Manual of Palaearctic Diptera, Volume 2. Science Herald, Budapest: 51-69.

Menzel, F. & Mohrig, W., 2000. Revision der paläarktischen Trauermücken (Diptera, Sciaridae). - Studia Dipterologica, Supplement 9 (1999): 3-761.

Menzel, F., Smith, J.E. & Chandler, P.J., 2006. The sciarid fauna of the British Isles (Diptera: Sciaridae) including the descriptions of six new species. - Zoological Journal of the Linnean Society 46: 1-147.

Merz, B., 1994. Diptera, Tephritidae. - Insecta Helvetica 10: 1-186.

Merz, B., 1996a. Systematik und Faunistik der Gattung Herina (Diptera, Otitidae) der Schweiz. - Mitteilungen der Schweizerischen Entomologischen Gesellschaft 69: 329-344.

Merz, B., 1996b. Die Piophilidae (Diptera) der Schweiz mit Beschreibung einer neuen Art. - Same as 1996a: 345-360.

Merz, B., 1996c. Zur Faunistik der Pyrgotidae, Platystomatidae und Ulidiidae (= Otitidae) (Diptera, Tephritoidea) der Schweiz mit spezieller Berücksichtigung von Otites Latreille. - Same as 1996a: 405-416.

Merz, B., 1996d. Die Asteiidae (Diptera) der Schweiz. - Revue Suisse de Zoologie 103: 893-904.

Merz, B., 1997. Die Micropezidae (Diptera) der Schweiz. - Mitteilungen der Schweizerischen Entomologischen Gesellschaft 70: 93-100.

Merz, B., 1998a. Die Megamerinidae, Strongylophthalmyiidae, Pseudopomyzidae, Chyromyidae und Camillidae der Schweiz (Diptera, Acalyptrata). - Mitteilungen der Entomologischen Gesellschaft Basel 47: 130-138.

Merz, B., 1998b. Family Pallopteridae. - In: Papp, L. & Darvas, B. (eds), Contributions to a Manual of Palaearctic Diptera, Volume 3. Science Herald, Budapest: 201-210.

Merz, B., 2002. Einführung in die Familie Lauxaniidae (Diptera, Acalyptrata) mit Angaben zur Fauna der Schweiz. - Mitteilungen der Entomologischen Gesellschaft Basel 52: 29-128.

Merz, B., Bächli, G. & Haenni, J.-P., 2001. Erster Nachtrag zur Checkliste der Diptera der Schweiz. - Mitteilungen der Entomologischen Gesellschaft Basel 51: 110-140.

Merz, B. & Haenni, J.-P., 2000. Morphology and terminology of adult Diptera (other than terminalia). - In: Papp, L. & Darvas, B. (eds), Contributions to a Manual of Palaearctic Diptera, Volume 1. Science Herald, Budapest: 21-51.

Merz, B. & Roháček, J., 2005. The Western Palaearctic species of Stenomicra Coquillett (Diptera, Periscelididae, Stenomicrinae), with description of a new species of the subgenus Podocera Czerny. - Revue Suisse de Zoologie 112: 519-539.

Mesnil, L.P., 1944-1975. 64g. Larvaevorinae (Tachininae). - In: Lindner, E. (ed.), Die Fliegen der paläarktischen Region. Schweizerbart, Stuttgart 10(1-3)64g: 1-1435.

Mesnil, L.P., 1980. 64f. Dexiinae. - In: Lindner, E. (ed.), Die Fliegen der paläarktischen Region. Schweizerbart, Stuttgart 9(64f): 1-52.

Meuffels, H., 2001. Tabel tot de in Nederland voorkomende genera van de familie Dolichopodidae (Diptera). - Vliegenmepper 10(2): 3-15.

Meuffels, H. & Grootaert, P., 1990. The identity of Sciapus contristans (Wiedemann, 1817) (Diptera: Dolichopodidae), and a revision of the species group of its relatives. - Bulletin de l'Institut Royal des Sciences Naturelles de Belgique, Entomologie 60: 161-178.

Meyer, M. De, 1969. The West-Palaearctic species of the pipunculid genera Cephalops and Beckerias (Diptera): classification, phylogeny and geographical distribution. - Journal of Natural History 23: 725-765.

Meyer, M. De, 1996. World catalogue of Pipunculidae (Diptera). - Studiedocumenten Koninklijk Belgisch Instituut voor Natuurwetenschappen 86: 1-127.

Michelsen, V., 1999. Wood gnats of the genus Sylvicola (Diptera, Anisopodidae): taxonomic status, family assignment, and review of nominal species described by J.C. Fabricius. - Tijdschrift voor Entomologie 142: 69-75.

Mihályi, F., 1960. Tephritidae. - Fauna Hungariae 56: 1-80.

Mihályi, F., 1972. Diptera. - Fauna Hungariae 107: 1-76.

Mihályi, F., 1979. Calliphoridae - Sarcophagidae. - Fauna Hungariae 135: 1-152.

Mihályi, F., 1986. Tachinidae - Rhinophoridae. - Fauna Hungariae 161: 1-425.

Mihályi, F., 1994. Die Tachiniden Ungarns und des Karpathenbeckens. – Translated from Hungarian by the author; originally published in Fauna Hungariae 15: 1-425.

Mikolajczyk, W., 1977. Bibionidae. - Klucze do Oznaczania owadow Polski 96: 1-20.

Minář, J., 1980. Hypodermatidae, Oestridae, Gasterophilidae. - Fauna CSSR 22: 391-446.

Minář, J., 1990. Family Culicidae. - In: Soós, Á. & Papp, L. (eds), Catalogue of Palaearctic Diptera 2: 74-113.

Minář, J., 2000a. Family Culicidae. - In: Papp, L. & Darvas, B. (eds), Contributions to a Manual of Palaearctic Diptera, Volume 4. Science Herald, Budapest: 93-111.

Minář, J., 2000b. Family Gasterophilidae. - Same as 2000a: 455-466.

Minář, J., 2000c. Family Oestridae. - Same as 2000a: 467-478.

Minář, J., 2000d. Family Hypodermatidae. - Same as 2000a: 479-494.

Möhn, E., 1966-1971. 6. Cecidomyiidae. - In: Lindner, E. (ed.), Die Fliegen der paläarktischen Region. Schweizerbart, Stuttgart 2(2)6: 1-248.

Mohrig, W., 1969. Die Culiciden Deutschlands. Untersuchungen zur Taxonomie, Biologie und Ökologie der einheimischen Stechmücken. - Parasitologische Schriftenreihe 18: 1-260.

Morge, G., 1959-1962. Monographie der Palaearktischen Lonchaeidae. - Beiträge zur Entomologie, 9: 1-92, 323-371, 909-945; 12: 381-434.

Morge, G., 1963. Die Lonchaeidae und Pallopteridae Österreichs und der angrenzenden Gebiete, 1, Die Lonchaeidae. - Naturkundliche Jahrbücher der Stadt Linz, 9: 123-312.

Morge, G., 1967. Die Lonchaeidae und Pallopteridae Öster-

reichs und der angrenzenden Gebiete, 2, Die
Pallopteridae. - Naturkundliche Jahrbücher der Stadt
Linz, 13: 141-212.

Morge, G., 1974. Die Lonchaeidae und Pallopteridae Österreichs und der angrenzenden Gebiete, 3. -
Naturkundliche Jahrbücher der Stadt Linz, 1973: 11-88.

Morge, G., 1984. Family Pallopteridae. - In: Soós, Á. &
Papp, L. (eds), Catalogue of Palaearctic Diptera 9: 242-
246.

Munari, L., 1998. Family Tethinidae. - In: Papp, L. &
Darvas, B. (eds), Contributions to a Manual of
Palaearctic Diptera, Volume 3. Science Herald,
Budapest: 243-250.

Munari, L., 2002. Beach flies (Diptera: Tethinidae) of the
Palaearctic Region: An annotated checklist, including
world distribution. - Societa Veneziani di Scienze
Naturali Lavori 27: 17-25.

Munari, L. & Baéz, M., 2000. The Tethinidae of
Macaronesia: A faunal revision, with descriptions of
two new species (Diptera). - Bolletino del Museo Civico
di Storia Naturale di Venezia 50: 3-30.

Munari, L. & Merz, B., 2002. Contribution to the knowledge of the fauna and taxonomy of Mediterranean Beach
Flies (Diptera, Tethinidae). - Mitteilungen der
Schweizerischen Entomologischen Gesellschaft 76: 221-
233.

Nagatomi, A., 1970. Rachiceridae (Diptera) from the
Oriental and Pacific regions. - Pacific Insects 12: 417-
466.

Nagatomi, A., 1984. Notes on Athericidae (Diptera). -
Memoirs of the Kagoshima University Research Centre
for the South Pacific 5: 87-106.

Nagatomi, A., 1997a. Family Rachiceridae. - In: Papp, L. &
Darvas, B. (eds), Contributions to a Manual of
Palaearctic Diptera, Volume 2. Science Herald,
Budapest: 379-386.

Nagatomi, A., 1997b. Family Vermileonidae. - Same as
1997a: 447-458.

Nagatomi, A., 1997c. Family Hilarimorphidae. - Same as
1997a: 513-517.

Nagatomi, A. & Rozkošný, R., 1997a. Family Xylomyidae. -
In: Papp, L. & Darvas, B. (eds), Contributions to a
Manual of Palaearctic Diptera, Volume 2. Science
Herald, Budapest: 369-378.

Nagatomi, A. & Rozkošný, R., 1997b. Family Xylophagidae.
- Same as 1997a: 413-420.

Nagatomi, A., Yang, C.-K. & Yang, D., 1999. The Chinese
species of and the World genera of Vermileonidae
(Diptera). - Tropics, Japanese Society of Tropical
Ecology Monograph Series 1: 1-154.

Nagatomi, A. & Yang, D., 1998. A review of extinct meso-
zoic genera and families of Brachycera (Insecta, Diptera,
Orthorrhapha). - Entomologists Monthly Magazine 134:
92-192.

Nartshuk, E.P., 1984a. Family Megamerinidae. - In: Soós, Á.
& Papp, L. (eds), Catalogue of Palaearctic Diptera 9: 25-
26.

Nartshuk, E.P., 1984b. Family Cryptochetidae. - In: Soós, Á.
& Papp, L. (eds), Catalogue of Palaearctic Diptera 10: 67-
68.

Nartshuk, E.P., 1984c. Family Chloropidae. - Same as
1984b: 222-298.

Nartshuk, E.P., 1987. Grassflies (Diptera: Chloropidae),
their system, evolution and association with plants.
Nauka, Leningrad: 1-279 (In Russian).

Nartshuk, E.P., 1988. Family Acroceridae. - In: Soós, Á. &
Papp, L. (eds), Catalogue of Palaearctic Diptera 5: 186-
196.

Nartshuk, E.P., 1989a. Family Blepharoceridae. - In: Bei-
Bienko, G.Y. & Steyskal, G.C. (eds), Keys to the insects
of the European Part of the USSR. Volume V. Diptera
and Siphonaptera, Part 1. Brill, Leiden, etc.: 205-208
(originally published in Russian in 1969).

Nartshuk, E.P., 1989b. Family Rhagionidae. - Same as
1989a: 683-696.

Nartshuk, E.P., 1989c. Family Hilarimorphidae. - Same as
1989a: 697.

Nartshuk, E.P., 1989d. Family Xylophagidae. - Same as
1989a: 698-700.

Nartshuk, E.P., 1989e. Family Stratiomyidae. - Same as
1989a: 701-738.

Nartshuk, E.P., 1989f. Family Acroceridae. - Same as 1989a:
773-777.

Nartshuk, E.P., 1989g. Family Scenopinidae. - Same as
1989a: 838-842.

Nartshuk, E.P., 1989h. Family Sphaeroceridae (Borboridae,
Cypselidae). - In: Bei-Bienko, G.Y. & Steyskal, G.C.
(eds), Keys to the insects of the European Part of the
USSR. Volume V. Diptera and Siphonaptera, Part 2.
Brill, Leiden, etc.: 559-590 (originally published in
Russian in 1970).

Nartshuk, E.P., 1989i. Family Ephydridae. - Same as 1989h:
605-646.

Nartshuk, E.P., 1997. Family Acroceridae. - In: Papp, L. &
Darvas, B. (eds), Contributions to a Manual of
Palaearctic Diptera, Volume 2. Science Herald,
Budapest: 469-485.

Nartshuk, E.P., 2000. Family Cryptochetidae. - In: Papp, L.
& Darvas, B. (eds), Contributions to a Manual of
Palaearctic Diptera, Volume 4. Science Herald,
Budapest: 345-353.

Nartshuk, E.P., 2003. Key to the families of Diptera
(Insecta) of the fauna of Russia and adjacent countries. -
Proceedings of the Zoological Institue 294: 1-250.

Nartshuk, E.P., Smirnov, E.S. & Fedoseeva, L.I., 1989.
Family Chloropidae. - In: Bei-Bienko, G.Y. & Steyskal,
G.C. (eds), Keys to the insects of the European Part of
the USSR. Volume V. Diptera and Siphonaptera, Part 2.
Brill, Leiden, etc.: 667-731 (originally published in
Russian in 1970).

Negrobov, O.P., 1991. Family Dolichopodidae. - In: Soós, Á.
& Papp, L. (eds), Catalogue of Palaearctic Diptera 7: 11-
139.

Negrobov, O.P. & Stackelberg, A.A., 1989. Family
Dolichopodidae. - In: Bei-Bienko, G.Y. & Steyskal, G.C.
(eds), Keys to the insects of the European Part of the
USSR. Volume V. Diptera and Siphonaptera, Part 1.
Brill, Leiden, etc.: 1026-1152 (originally published in
Russian in 1969).

Niesiolowski, S., 1992. Empididae Aquatica. - Fauna Polski
14: 1-129.

Nijveldt, W., 1969. Gall midges of economic importance.
Volume 8. Gall midges miscellaneous. Crosby
Lockwood & Son, London: 1-221.

Oka, H., 1930. Morphologie und Oekologie von Clunio
pacificus Edwards (Diptera, Chironomidae). -
Zoologische Jahrbücher, Abteilung für Systematik 59:
253-280.

Olafsson, E., 1991. Taxonomic revision of western
Palaearctic species of the genera Scatella R.D. and
Lamproscatella Hendel, and studies on their phyloge-
netic positions within the subfamily Ephrydinae
(Diptera, Ephydridae). - Entomologica Scandinavica,
Supplement 37: 1-100.

Oldroyd, H., 1964. The natural history of flies. Weidenfeld
& Nicolson, London: i-xiv, 1-324.

Oldroyd, H., 1969. Diptera Brachycera Section (a) Tabanoidea and Asiloidea. - Handbooks for the Identification of British Insects 9(4a): 1-132.

Oldroyd, H., 1970. Diptera. I. Introduction and key to families. - Handbooks for the Identification of British Insects 9(1): 1-180.

Olsufyev, N.G., 1989. Family Tabanidae. - In: Bei-Bienko, G.Y. & Steyskal, G.C. (eds), Keys to the insects of the European Part of the USSR. Volume V. Diptera and Siphonaptera, Part 1. Brill, Leiden, etc.: 739-769 (originally published in Russian in 1969).

Örösi Pál, Z., 1966. Die Bienenlaus-Arten. - Angewandte Entomologie 7: 138-171.

Oosterbroek, P., 1981. De Europese Diptera. Determineertabel, biologie en literatuuroverzicht van de families van de muggen en vliegen. - Wetenschappelijke Mededelingen Koninklijke Nederlandse Natuurhistorische Vereniging 148: 1-81.

Oosterbroek, P., 1998. The families of Diptera of the Malay Archipelago. - Fauna Malesiana Handbook 1, Brill, Leiden etc.: i-xii, 1-227.

Oosterbroek, P., 2006. Catalogue of the craneflies of the world (Insecta, Diptera, Nematocera, Tipuloidea). - Available online at: http://www.science.uva.nl/zma/

Oosterbroek, P., Bygebjerg, R. & Munk, T., 2006. The Western Palaearctic species of Ctenophorinae (Diptera, Tipulidae); key, distribution and references. - Entomologische Berichten, Amsterdam 66 (in press).

Oosterbroek, P. & Jong, H. de, 2005. Eginia ocypterata (Diptera: Muscidae) in The Netherlands. - Entomologische Berichten, Amsterdam 65: 142-144.

Oosterbroek, P., Jong, H. de, Sijstermans, L., 2005. De Europese families van de muggen en de vliegen (Diptera). Determinatie, diagnose, biologie. - KNNV Publishing, Utrecht: 1-205.

Oosterbroek, P. & Theowald, Br., 1992. Family Tipulidae. - In: Soós, Á., Papp, L. & Oosterbroek, P. (eds), Catalogue of Palaearctic Diptera 1: 56-178.

Ostroverkhova, G.P. & Stackelberg, A.A., 1989. Family Mycetophilidae. - In: Bei-Bienko, G.Y. & Steyskal, G.C. (eds), Keys to the insects of the European Part of the USSR. Volume V. Diptera and Siphonaptera, Part 1. Brill, Leiden, etc.: 404-487 (originally published in Russian in 1969).

Ozerov, A.L., 1986. Review of the family Acartophthalmidae (Diptera) with description of a new species. - Zoologicheskii Zhurnal 65(5): 807-809. [In Russian]

Ozerov, A.L., 1991. On the taxonomy of flies of the subfamily Calobatinae (Diptera, Micropezidae). - Zoologicheskii Zhurnal 70(11): 63-72. [In Russian]

Ozerov, A.L., 1998. Family Dryomyzidae. - In: Papp, L. & Darvas, B. (eds), Contributions to a Manual of Palaearctic Diptera, Volume 3. Science Herald, Budapest: 349-355.

Ozerov, A.L., 2000. Family Piophilidae. - In: Papp, L. & Darvas, B. (eds), Contributions to a Manual of Palaearctic Diptera, Volume 4. Science Herald, Budapest: 355-365.

Ozerov, A.L., 2003. The Sepsidae (Diptera) of Russia - Sbornik Trudov Zoologicheskogo Muzeja MGU 45: 1-182 (in Russian).

Ozerov, A.L., 2005. World catalogue of the family Sepsidae (Insecta: Diptera). - ZoologicheskieIssledovanija (Zoological Studies) 8: 1-76.

Pape, T., 1987. The Sarcophagidae (Diptera) of Fennoscandia and Denmark. - Fauna Entomologica Scandinavica 19: 1-203.

Pape, T., 1996. A catalogue of world Sarcophagidae (Diptera). - Memoirs on Entomology, International 8: 1-558.

Pape, T., 1998a. Family Sarcophagidae. - In: Papp, L. & Darvas, B. (eds), Contributions to a Manual of Palaearctic Diptera, Volume 3. Science Herald, Budapest: 649-678.

Pape, T., 1998b. Family Rhinophoridae. - Same as 1998a: 679-689.

Pape, T. & Carlberg. U., 2001 (onwards). A pictorial guide to the Sarcophagidae of the world (Insecta: Diptera). - Available online at: http://www2.nrm.se/en/sarcophagidae/welcome.html.

Papp, L., 1973. Sphaeroceridae - Drosophilidae. - Fauna Hungariae 112: 1-146.

Papp, L., 1975. Ephydridae. - Fauna Hungariae 120: 1-128.

Papp, L., 1978. Odiniidae - Chloropidae. - Fauna Hungariae 133: 1-202.

Papp, L., 1980. New taxa of acalyptrate flies (Diptera: Tunisimyiidae fam. n., Risidae, Ephydridae: Nannodastiinae subfam. n.). - Acta Zoologica Hungarica 26: 415-431.

Papp, L., 1984a. Family Lauxaniidae. - In: Soós, Á. & Papp, L. (eds), Catalogue of Palaearctic Diptera 9: 193-217.

Papp, L., 1984b. Family Periscelididae. - same as 1984a: 233-234.

Papp, L., 1984c. Family Thyreophoridae. - same as 1984a: 240-241.

Papp, L., 1984d. Family Agromyzidae. - same as 1984a: 263-343.

Papp, L., 1984e. Family Acartophthalmidae. - In: Soós, Á. & Papp, L. (eds), Catalogue of Palaearctic Diptera 10: 14-15.

Papp, L., 1984f. Family Aulacigastridae. - Same as 1984e: 60-61.

Papp, L., 1984g. Family Stenomicridae. - Same as 1984e: 61-62.

Papp, L., 1984h. Family Asteiidae. - Same as 1984e: 63-66.

Papp, L., 1984i. Family Sphaeroceridae. - Same as 1984e: 68-107.

Papp, L., 1984j. Family Milichiidae. - Same as 1984e: 110-118.

Papp, L., 1984k. Family Carnidae. - Same as 1984e: 118-124.

Papp, L., 1984l. Family Xenasteiidae. - Same as 1984e: 176-177.

Papp, L., 1984m. Family Risidae. - Same as 1984e: 177-178.

Papp, L., 1984n. Family Braulidae. - Same as 1984e: 178-180.

Papp, L., 1984o. Family Camillidae. - Same as 1984e: 180-182.

Papp, L., 1984p. Family Diastatidae. - Same as 1984e: 182-185.

Papp, L., 1984q. Family Curtonotidae. - Same as 1984e: 221-222.

Papp, L., 1985. Key to the world species of Camillidae (Diptera). - Acta Zoologica Hungarica 31: 217-227.

Papp, L., 1994. A new Cremifania species from Hungary (Diptera, Chamaemyiidae). - Annales Historico-Naturales Musei Nationalis Hungarici 86: 105-107.

Papp, L., 1998a. The palaearctic species of Aulacigaster Macquart (Diptera: Aulacigastridae). - Acta Zoologica Academiae Scientiarium Hungaricae 43: 225-234.

Papp, L., 1998b. Family Carnidae. - In: Papp, L. & Darvas, B. (eds), Contributions to a Manual of Palaearctic Diptera, Volume 3. Science Herald, Budapest: 211-217.

Papp, L., 1998c. Family Odiniidae. - Same as 1998b: 233-242.

Papp, L., 1998d. Family Aulacigastridae. - Same as 1998b: 279-284.

Papp, L., 1998e. Family Asteiidae. - Same as 1998b: 295-303.

Papp, L., 1998f. Family Xenasteiidae. - Same as 1998b: 305-308.

Papp, L., 1998g. Family Braulidae. - Same as 1998b: 325-330.

Papp, L., 1998h. Family Cremifaniidae. - Same as 1998b: 409-414.

Papp, L., 1998i. Families of Heleomyzoidea. - Same as 1998b: 425-455.

Papp, L., 1998j. Family Curtonotidae. - Same as 1998b: 497-502.

Papp, L., 1998k. Family Camillidae. - Same as 1998b: 531-535.

Papp, L., 1998l. Nidomyiini, a new tribe, genus and species of Borboropsidae (Diptera), with the redefinition of the family. - Acta Zoologica Academiae Scientiarium Hungaricae 44: 297-310.

Papp, L. & Darvas, B. (eds), 1997-2000. Contributions to a Manual of Palaearctic Diptera. Science Herald, Budapest. Volume 1 (2000), General and applied dipterology: 1-978; Volume 2 (1997), Nematocera and Lower Brachycera: 1-592; Volume 3 (1998), Higher Brachycera: 1-880; Volume 4 (2000), Appendix: 1-604.

Papp, L. & Földvári, M., 2001. A new genus and three new species of Hybotidae with new records of the Hungarian Empidoidea (Diptera). - Acta Zoologica Academiae Scientiarum Hungaricae 47: 349-361.

Papp, L., Földvári, M. & Paulovics, P., 1997. Sphyracephala europaea sp. n. (Diptera: Diopsidae) from Hungary represents a family new to Europe. - Folia Entomologica Hungarica 58: 137-146.

Papp, L. & Mathis, W.N., 1998. Family Nannodastiidae. - In: Papp, L. & Darvas, B. (eds), Contributions to a Manual of Palaearctic Diptera, Volume 3. Science Herald, Budapest: 309-314.

Papp, L. & Mathis, W.N., 2001. A review of the family Nannodastiidae (Diptera). - Proceedings of the Entomological Society of Washington 103: 337-348.

Papp, L. & Ozerov, A.L., 1998. Family Acartophthalmidae. - In: Papp, L. & Darvas, B. (eds), Contributions to a Manual of Palaearctic Diptera, Volume 3. Science Herald, Budapest: 227-232.

Papp, L. & Schumann, H., 2000. Key to families, adults. - In: Papp, L. & Darvas, B. (eds), Contributions to a Manual of Palaearctic Diptera, Volume 1. Science Herald, Budapest: 163-200.

Papp, L. & Shatalkin, A.I., 1998. Family Lauxaniidae. - In: Papp, L. & Darvas, B. (eds), Contributions to a Manual of Palaearctic Diptera, Volume 3. Science Herald, Budapest: 383-400.

Papp, L. & Wheeler, T.A., 1998. Family Milichiidae. - In: Papp, L. & Darvas, B. (eds), Contributions to a Manual of Palaearctic Diptera, Volume 3. Science Herald, Budapest: 315-324.

Paramonov, S.J., 1945. Bestimmungstabelle der palaearktischen Nemestrinus-Arten (Nemestrinidae, Diptera). - Eos 21: 279-295.

Paramonov, S.J., 1951. Bestimmungstabelle der palaearktischen Arten der Gattung Rhynchocephalus (Nemestrinidae, Diptera). - Zoologischer Anzeiger 146(5/6): 118-127.

Paramonov, S.J., 1956. Übersicht der paläarktischen Arten der Gattung Hirmoneura (Nemestrinidae, Diptera). - Zoologischer Anzeiger 156(9/10): 234-242.

Parent, O., 1938. Diptères Dolichopodidae. - Faune de France 72: 1-720.

Pasini, M. & Ferrarese, U., 1998. I Chironomidi delle risaie dell'Italia Nord-orientale: chiavi per il riconoscimento di larve, pupe e adulti. - Bolletino di Zoologia Agraria e

Bachicoltura, Serie II, 30: 79-113.

Peck, L.V., 1988. Family Syrphidae. - In: Soós, Á. & Papp, L. (eds), Catalogue of Palaearctic Diptera 8: 11-230.

Pedersen, B.V., 1971. Studies of the Mycetobia Meig. in Scandinavia, with records of two species new to the region. - Entomologiske Meddelelser 39: 63-67.

Perfilyev, P.P., 1989. Family Phlebotomidae. - In: Bei-Bienko, G.Y. & Steyskal, G.C. (eds), Keys to the insects of the European Part of the USSR. Volume V. Diptera and Siphonaptera, Part 1. Brill, Leiden, etc.: 167-198 (originally published in Russian in 1969).

Peters, T.M., 1981. Dixidae. - In: McAlpine, J.F. et al. (eds), Manual of Nearctic Diptera, Volume 1. - Research Branch, Agriculture Canada, Monograph 27: 329-333.

Peterson, B.V., 1981. Simuliidae. - In: McAlpine, J.F. et al. (eds), Manual of Nearctic Diptera, Volume 1. - Research Branch, Agriculture Canada, Monograph 27: 355-391.

Peterson, B.V., 1987. Lonchopteridae. - In: McAlpine, J.F. et al. (eds), Manual of Nearctic Diptera, Volume 2. - Research Branch, Agriculture Canada, Monograph 28: 675-680.

Peterson, B.V. & Cook, E.F., 1981. Synneuridae. - In: McAlpine, J.F. et al. (eds), Manual of Nearctic Diptera, Volume 1. - Research Branch, Agriculture Canada, Monograph 27: 321-324.

Peus, F., 1952., 17. Cylindrotomidae. - In: Lindner, E. (ed.), Die Fliegen der paläarktischen Region. Schweizerbart, Stuttgart 3(5)3: 1-80.

Peus, F., 1958., 10b. Liriopeidae. - In: Lindner, E. (ed.), Die Fliegen der paläarktischen Region. Schweizerbart, Stuttgart 3(1)10b: 10-44.

Pinder, L.C.V., 1978a. A key to adult males of British Chironomidae. Vol. 1. - Freshwater Biological Association Scientific Publications 37: 1-169.

Pinder, L.C.V., 1978b. A key to adult males of British Chironomidae. Vol. 2. Illustrations of the hypopygia. - Freshwater Biological Association Scientific Publications 37: figures 77-189.

Pitkin, B.R., 1988. Lesser dung flies. Diptera Sphaeroceridae. - Handbooks for the Identification of British Insects 10(5e): 1-175.

Plassmann, E., 1988. Family Bolitophilidae. - In: Soós, Á. & Papp, L. (eds), Catalogue of Palaearctic Diptera 3: 193-196.

Pont, A.C., 1979. Sepsidae (Diptera Cyclorrhapha, Acalyptrata). - Handbooks for the Identification of British Insects 10(5c): 1-35.

Pont, A.C., 1986a. Family Fanniidae. - In: Soós, Á. & Papp, L. (eds), Catalogue of Palaearctic Diptera 11: 41-57.

Pont, A.C., 1986b. Family Muscidae. - Same as 1986a: 57-215.

Pont, A.C., 2000. Family Fanniidae - In: Papp, L. & Darvas, B. (eds), Contributions to a Manual of Palaearctic Diptera, Volume 4. Science Herald, Budapest: 447-454.

Pont, A.C. & Meier, R., 2002. The Sepsidae (Diptera) of Europe. -Fauna Entomologica Scandinavica 37: 1-221.

Portillo Rubio, M., 2002. Diptera Tabanidae. - Fauna Iberica 18: 1-309.

Povolny, D. & Verves, Y.G., 1997. The fleshflies of Central Europe (Insecta, Diptera, Sarcophagidae). - Spixiana Supplement 24: 1-260.

Reemer, M., 2000. Zweefvliegen veldgids (Diptera, Syrphidae). Jeugdbondsuitgeverij, Utrecht: 1-64, 16 plts.

Remm, H., 1988. Families Ceratopoginidae and Leptoconopidae. - In: Soós, Á. & Papp, L. (eds), Catalogue of Palaearctic Diptera 3: 11-114.

Remm, H., 1989. Family Ceratopogonidae (Heleidae). - In: Bei-Bienko, G.Y. & Steyskal, G.C. (eds), Keys to the insects of the European Part of the USSR. Volume V. Diptera and Siphonaptera, Part 1. Brill, Leiden, etc.: 300-330 (originally published in Russian in 1969).

Reusch, H. & Oosterbroek, P., 1997. Diptera Limoniidae and Pediciidae, Short-palped crane flies. - In: Nilsson, A. (ed.), Aquatic insects of North Europe. A taxonomic handbook. Volume 2. Odonata - Diptera. Apollo Books, Stenstrup: 105-132.

Revier, J.M. & Goot, V.S. van der, 1989. Slakkendodende vliegen (Sciomyzidae) van Noordwest-Europa. - Wetenschappelijke Mededelingen Koninklijke Nederlandse Natuurhistorische Vereniging 191: 1-64.

Richter, V.A., 1988. Family Nemestrinidae. - In: Soós, Á. & Papp, L. (eds), Catalogue of Palaearctic Diptera 5: 171-181.

Richter, V.A., 1989a. Family Nemestrinidae. - In: Bei-Bienko, G.Y. & Steyskal, G.C. (eds), Keys to the insects of the European Part of the USSR. Volume V. Diptera and Siphonaptera, Part 1. Brill, Leiden, etc.: 770-771 (originally published in Russian in 1969).

Richter, V.A., 1989b. Family Asilidae. - Same as 1989a: 778-820.

Richter, V.A., 1989c. Family Platystomatidae. - In: Bei-Bienko, G.Y. & Steyskal, G.C. (eds), Keys to the insects of the European Part of the USSR. Volume V. Diptera and Siphonaptera, Part 2. Brill, Leiden, etc.: 194-196 (originally published in Russian in 1970).

Richter, V.A., 1989d. Family Otitidae. - Same as 1989c: 197-207.

Richter, V.A., 1989e. Family Ulidiidae. - Same as 1989c: 208-211.

Richter, V.A., 1989f. Family Tephritidae (Trypetidae). - Same as 1989c: 212-276.

Richter, V.A., 1997a. Family Nemestrinidae. - In: Papp, L. & Darvas, B. (eds), Contributions to a Manual of Palaearctic Diptera, Volume 2. Science Herald, Budapest: 459-468.

Richter, V.A., 1997b. Family Mydidae. - Same as 1997a: 539-547.

Richter, V.A. & Zaitzev, V.F., 1988. Family Mydiidae. - In: Soós, Á. & Papp, L. (eds), Catalogue of Palaearctic Diptera 5: 181-186.

Rivosecchi, L., 1978. Simuliidae. - Fauna d'Italia 13: i-viii, 1-533.

Rivosecchi, L., 1992. Sciomyzidae. - Fauna d'Italia 30: i-xi, 1-270.

Rivosecchi, L., 1996. Chiavi analitiche illustrate sui Conopidae (Diptera) della fauna Italiana. - Bolletino del Museo Civico di Storia Naturale di Verona 20: 135-151.

Robinson, H. & VockerothJ.R., 1981. Dolichopodidae. - In: McAlpine, J.F. et al. (eds), Manual of Nearctic Diptera, Volume 1. - Research Branch, Agriculture Canada, Monograph 27: 625-639.

Rognes, K., 1991. Blowflies (Diptera, Calliphoridae) of Fennoscandia and Denmark. - Fauna Entomologica Scandinavica 24: 1-272.

Rognes, K., 1998. Family Calliphoridae. - In: Papp, L. & Darvas, B. (eds), Contributions to a Manual of Palaearctic Diptera, Volume 3. Science Herald, Budapest: 617-648.

Rognes, K., 2002. Blowflies (Diptera, Calliphoridae) of Israel and adjacent areas, including a new species from Tunisia. - Fauna Entomologica Scandinavica, Supplementum 58: 1-148.

Roháček, J., 1992. Typhamyza gen. n. for Anthomyza bifasciata Wood, with description of immature stages (Diptera, Anthomyzidae). - Bollettino del Museo Regionale di Scienze Naturali Torino 10: 187-207.

Roháček, J., 1998a. Family Tanypezidae. - In: Papp, L. &

Darvas, B. (eds), Contributions to a Manual of Palaearctic Diptera, Volume 3. Science Herald, Budapest: 165-171.

Roháček, J., 1998b. Family Anthomyzidae. - Same as 1998a: 267-278.

Roháček, J., 1998c. Family Sphaeroceridae. - Same as 1998a: 463-496.

Roháček, J., 1999. Taxonomy and distribution of West Palaearctic Anthomyzidae (Diptera), with special regards to the Mediterranean and Macaronesian faunas. - Bollettino del Museo Regionale di Scienze Naturali, Torino 16: 189-224.

Roháček, J., 2006. A monograph of Palaearctic Anthomyzidae (Diptera). Part 1. - Casopis Slezského Zemského Muzea Opava (A) 55, Supplement 1: 1-328.

Roháček, J. & Barták, M., 1990. Micropezidae (Diptera) of Czechoslovakia. - Casopis Slezského Zemského Muzea Opava (A) 39: 97-111.

Roháček. J., Marshall, S.A., Norrbom, A.L., Buck, M., Quiros, D.I. & Smith, I., 2002. World catalog of Sphaeroceridae (Diptera). - Slézske Zemske Muzeum, Opava: 1-415.

Rohdendorf, B.B., 1930-1982. 64h. Sarcophaginae. - In: Lindner, E. (ed.), Die Fliegen der paläarktischen Region. Schweizerbart, Stuttgart 11(64h): 1-235.

Rohdendorf, B.B., 1937. Fam. Sarcophagidae. Part 1. - Fauna USSR 19: i-xv, 1-500.

Rohdendorf, B.B., 1989. Family Sarcophagidae. - In: Bei-Bienko, G.Y. & Steyskal, G.C. (eds), Keys to the insects of the European Part of the USSR. Volume V. Diptera and Siphonaptera, Part 2. Brill, Leiden, etc.: 1021-1098 (originally published in Russian in 1970).

Rohdendorf, E.B., 1989. Family Agromyzidae. - In: Bei-Bienko, G.Y. & Steyskal, G.C. (eds), Keys to the insects of the European Part of the USSR. Volume V. Diptera and Siphonaptera, Part 2. Brill, Leiden, etc.: 377-504 (originally published in Russian in 1970).

Rotheray, G.E., 1993. Colour guide to hoverfly larvae (Diptera, Syrphidae) in Britain and Europe. - Dipterists Digest 9: 1-155.

Rozkošný, R., 1973. The Stratiomyidae (Diptera) of Fennoscandia and Denmark. - Fauna Entomologica Scandinavica 1: 1-140.

Rozkošný, R., 1982-1983. A biosystematic study of the European Stratiomyidae (Diptera). Junk, Den Haag, etc.: Volume 1: 1-401, Volume 2: 1-431.

Rozkošný, R., 1984. The Sciomyzidae (Diptera) of Fennoscandia and Denmark. - Fauna Entomologica Scandinavica 14: 1-224.

Rozkošný, R., 1987. A review of the Palaearctic Sciomyzidae (Diptera). - Folia Facultatis Scientiarum Naturalium Universitatis Purkynianae Brunensis, Biologia 86: 1-100.

Rozkošný, R., 1990. Family Dixidae. - In: Soós, Á. & Papp, L. (eds), Catalogue of Palaearctic Diptera 2: 66-71.

Rozkošný, R., 1992. Family Ptychopteridae. - In: Soós, Á., Papp, L. & Oosterbroek, P. (eds), Catalogue of Palaearctic Diptera 1: 370-373.

Rozkošný, R., 1997a. Family Ptychopteridae. - In: Papp, L. & Darvas, B. (eds), Contributions to a Manual of Palaearctic Diptera, Volume 2. Science Herald, Budapest: 291-297.

Rozkošný, R., 1997b. Family Stratiomyidae. - Same as 1997a: 387-411.

Rozkošný, R., 1997c. Diptera Stratiomyidae, Soldier flies. - In: Nilsson, A. (ed.), Aquatic insects of North Europe. A taxonomic handbook. Volume 2. Odonata - Diptera. Apollo Books, Stenstrup: 321-332.

Rozkošný, R., 1997d. Diptera Sciomyzidae, Snail killing flies. - Same as 1997c: 363-381.

Rozkošný, R., 1998a. Family Sciomyzidae. - In: Papp, L. & Darvas, B. (eds), Contributions to a Manual of Palaearctic Diptera, Volume 3. Science Herald, Budapest: 357-376.

Rozkošný, R., 1998b. Family Phaeomyiidae. - Same as 1998a: 377-382.

Rozkošný, R., 2000. Insecta: Diptera: Stratiomyidae. - Süßwasserfauna von Mitteleuropa 21(18): 1-110.

Rozkošný, R., 2002. Insecta: Diptera: Sciomyzidae. - Süßwasserfauna von Mitteleuropa 21(23): 15-122.

Rozkošný, R. & Elberg, K., 1984. Family Sciomyzidae. - In: Soós, Á. & Papp, L. (eds), Catalogue of Palaearctic Diptera 9: 167-193.

Rozkošný, R. & Gregor, F., 1997. Diptera Muscidae, Muscid flies. - In: Nilsson, A. (ed.), Aquatic insects of North Europe. A taxonomic handbook. Volume 2. Odonata - Diptera. Apollo Books, Stenstrup: 411-425.

Rozkošný, R. & Gregor, F., 2004. Insecta: Diptera: Muscidae. - Süßwasserfauna von Mitteleuropa 21(29): i-viii, 1-111.

Rozkošný, R., Gregor, F. & Pont, A.C., 1997. The European Fanniidae (Diptera). - Acta Scientiarum Naturalium Academiae Bohemicae, Brno 31(2): 1-80.

Rozkošný, R. & Nagatomi, A., 1997a. Family Coenomyiidae. - In: Papp, L. & Darvas, B. (eds), Contributions to a Manual of Palaearctic Diptera, Volume 2. Science Herald, Budapest: 421-426.

Rozkošný, R. & Nagatomi, A., 1997b. Family Athericidae. - Same as 1997a: 439-446.

Rozkošný, R. & Nartshuk, E.P., 1988. Family Stratiomyidae. - In: Soós, Á. & Papp, L. (eds), Catalogue of Palaearctic Diptera 5: 42-96.

Rubtzov, I.A., 1959-1964. 14. Simuliidae. - In: Lindner, E. (ed.), Die Fliegen der paläarktischen Region. Schweizerbart, Stuttgart 3(4)14: 1-689.

Rubtzov, I.A., 1989. Family Simuliidae. - In: Bei-Bienko, G.Y. & Steyskal, G.C. (eds), Keys to the insects of the European Part of the USSR. Volume V. Diptera and Siphonaptera, Part 1. Brill, Leiden, etc.: 331-369 (originally published in Russian in 1969).

Rubtzov, I.A. & Yankovsky, A.V., 1988. Family Simuliidae. - In: Soós, Á. & Papp, L. (eds), Catalogue of Palaearctic Diptera 3: 114-186.

Sabrosky, C.W., 1983. A synopsis of the world species of Desmometopa Loew (Diptera: Milichiidae). - Contributions of the American Entomological Institute 19: 1-69.

Sabrosky, C.W., 1987. Carnidae. - In: McAlpine, J.F. et al. (eds), Manual of Nearctic Diptera, Volume 2. - Research Branch, Agriculture Canada, Monograph 28: 909-912.

Sack, P, 1933. 22. Nemestrinidae. - In: Lindner, E. (ed.), Die Fliegen der paläarktischen Region. Schweizerbart, Stuttgart 4(1)22: 1-42.

Sack, P, 1936. 21. Cyrtidae (Acroceridae). - In: Lindner, E. (ed.), Die Fliegen der paläarktischen Region. Schweizerbart, Stuttgart 4(1)21: 1-36.

Saether, O.A., 1997a. Diptera Chaoboridae, Phantom midges. - In: Nilsson, A. (ed.), Aquatic insects of North Europe. A taxonomic handbook. Volume 2. Odonata - Diptera. Apollo Books, Stenstrup: 149-161.

Saether, O.A., 1997b. Family Chaoboridae. - In: Papp, L. & Darvas, B. (eds), Contributions to a Manual of Palaearctic Diptera, Volume 2. Science Herald, Budapest: 305-317.

Saether, O.A., 2002. Insecta: Diptera: Chaoboridae. - Süßwasserfauna von Mitteleuropa 21(10): 1-38.

Saether, O.A., Ashe, P. & Murray, D.A., 2000. Family Chironomidae. - In: Papp, L. & Darvas, B. (eds), Contributions to a Manual of Palaearctic Diptera, Volume 4. Science Herald, Budapest: 113-334.

Sasakawa, M., 1998. Family Clusiidae. - In: Papp, L. & Darvas, B. (eds), Contributions to a Manual of Palaearctic Diptera, Volume 3. Science Herald, Budapest: 219-225.

Savchenko, E.N., 1989a. Family Cylindrotomidae. - In: Bei-Bienko, G.Y. & Steyskal, G.C. (eds), Keys to the insects of the European Part of the USSR. Volume V. Diptera and Siphonaptera, Part 1. Brill, Leiden, etc.: 72-74 (originally published in Russian in 1969).

Savchenko, E.N., 1989b. Family Tipulidae. - Same as 1989a: 75-118.

Savchenko, E.N., 1989c. Family Limoniidae (Limnobiidae). - Same as 1989a: 119-163.

Savchenko, E.N., 1989d. Limoniidae fauna of the USSR. - Akademia Nauk Ukrainian SSR, Kiev: 1-377 (In Russian).

Savchenko, E.N., Oosterbroek, P. & Starý, J., 1992. Family Limoniidae. - In: Soós, Á., Papp, L. & Oosterbroek, P. (eds), Catalogue of Palaearctic Diptera 1: 183-369.

Schaffner, F., Angel, G, Geoffroy, B., Hervy, J.-P., Rhaiem, A. & Brunhes, J., 2001. Les Moustiques d'Europe. The Mosquitoes of Europe. Montpellier, France: CD-ROM (Windows).

Schauff, M.E., 2002. Collecting and preserving insects and mites: techniques and tools. - Available online at: http://www.sel.barc.usda.gov/selhome/collpres/collpres.htm

Schumann, H., 1986. Family Calliphoridae. - In: Soós, Á. & Papp, L. (eds), Catalogue of Palaearctic Diptera 12: 11-58.

Schuurmans Stekhoven Jr, J.H. & Broek, E. van den, 1969. Luisvliegen Nycteribiidae en Hippoboscidae. Second edition. - Wetenschappelijke Mededelingen Koninklijke Nederlandse Natuurhistorische Vereniging 16: 1-19.

Séguy, E., 1926. Diptères (Brachycères) (Stratiomyidae - Omphralidae). - Faune de France 13: 1-308.

Séguy, E., 1934. Diptères (Brachycères) (Muscidae Acalypterae et Scatophagidae). - Faune de France 28: 1-832.

Séguy, E., 1940. Diptères Nématocères. - Faune de France 36: 1-68.

Séguy, E., 1949. Un Cyrtosia nouveau et synopsis des espèces méditerranéennes (Dipt. Bombyliidae). - Revue Française d'Entomologie 16: 83-85

Séguy, E., 1950. La biologie des Diptères. - Encyclopédie Entomologique (A) 26: 1-609.

Séguy, E., 1951. Ordre des Diptères (Diptera Linné, 1758). - In: Grassé, P.-G. (ed.), Traité de Zoologie 10: 449-744.

Service, M.W., 1993. Mosquitoes (Culicidae). - In: Lane, R.P. & Crosskey, R.W. (eds), Medical Insects and Arachnids. Chapman & Hall, London etc.: 120-240.

Shatalkin, A.I., 1993. Taxonomy of flies of the family Strongylophthalmyiidae (Diptera). - Zoologicheskii Zhurnal 72(10): 1124-131 (in Russian, English translation in: Entomological Review, Washington 73(6) (1994): 155-161).

Shatalkin, A.I., 2000. Keys to the Palaearctic flies of the family Lauxaniidae (Diptera). - Zoologicheskie Issledovania 5: 1-102 (in Russian).

Shilova A.I., 1989. Family Chironomidae (Tendipedidae). - In: Bei-Bienko, G.Y. & Steyskal, G.C. (eds), Keys to the insects of the European Part of the USSR. Volume V. Diptera and Siphonaptera, Part 1. Brill, Leiden, etc.: 240-299 (originally published in Russian in 1969).

Shorrocks, B., 1972. Invertebrate Types. Drosophila. Ginn, London: 1-144.

Sifner, F., 2003. The family Scathophagidae (Diptera) of the Czech and Slovak Republics (with notes on selected Palaearctic taxa). - Sbornik Narodniho Muzea v Praze 59: 1-90.

Simova-Tosic, D. & Stojanovic, A., 1999. Second finding of Sphyracephala europaea Papp et Földvári in Europe (Diptera: Diopsidae). - Acta Entomologica Serbica 4: 149-157.

Sinclair, B.J. & Cumming, J.M., 2006. The morphology, higher-level phylogeny and classification of the Empidoidea (Diptera). - Zootaxa 1180: 1-172.

Sinclair, B.J. & Papp, L., 2004. The rediscovery of Nemedina alamirabilis Chandler from Hungary (Diptera: Empidoidea), and first description of the male. - Bonner Zoologische Beiträge 52: 155-158.

Sinclair, B.J. & Shamshev, I.V., 2003. A new species of Nemedina Chandler from mid-Asia (Diptera: Empidoidea) resolves the phylogenetic position of this enigmatic genus. - Journal of Natural History 37: 2949-2958.

Skartveit, J., 1997. Family Bibionidae. - In: Papp, L. & Darvas, B. (eds), Contributions to a Manual of Palaearctic Diptera, Volume 2. Science Herald, Budapest: 41-50.

Skierska, B., 1977. Culicidae. - Klucze do Oznaczania owadow Polski 99: 1-120.

Skuhravá, M., 1986. Family Cecidomyiidae. - In: Soós, Á. & Papp, L. (eds), Catalogue of Palaearctic Diptera 4: 72-297.

Skuhravá, M., 1997. Family Cecidomyiidae. - In: Papp, L. & Darvas, B. (eds), Contributions to a Manual of Palaearctic Diptera, Volume 2. Science Herald, Budapest: 71-204.

Skuhravá, M., Skuhravý, V. & Jørgensen, J., 2006. Gall midges (Diptera: Cecidomyiidae) of Denmark. - Entomologiske Meddelelser 74: 1-94.

Smith, K.G.V., 1969a. Diptera Lonchopteridae. - Handbooks for the Identification of British Insects 10(2ai): 1-9.

Smith, K.G.V., 1969b. Diptera Conopidae. - Handbooks for the Identification of British Insects 10(3a): 1-18.

Smith, K.G.V., 1989. An introduction to the immature stages of British flies. Diptera larvae, with notes on eggs, puparia and pupae. - Handbooks for the Identification of British Insects 10(14): 1-280.

Smithers, C. N., 1982. Handbook of insect collecting, collection, preparation, preservation and storage. - David & Charles, Newton Abbot etc.: 1-120.

Snow, K.R., 1990. Mosquitoes. - Naturalist's Handbook 14: 1-66.

Søli, G.E.E., 1992. Norwegian species of Sylvicola Harris, 1776 (Diptera: Anisopodidae). - Fauna Norvegiae, Series B, 39: 49-54.

Søli, G.E.E., Vockeroth, J.R. & Matile, L., 2000. Families of Sciaroidea. - In: Papp, L. & Darvas, B. (eds), Contributions to a Manual of Palaearctic Diptera, Volume 4. Science Herald, Budapest: 49-92.

Soós, Á., 1955. Muscidae Pupiparae. - Fauna Hungariae 1: 1-20.

Soós, Á., 1959. Muscidae Acalyptratae 1. - Fauna Hungariae 48: 1-88.

Soós, Á., 1980. Psilidae - Platystomatidae. - Fauna Hungariae 143: 1-100.

Soós, Á., 1981. Heleomyzidae - Tethinidae. - Fauna Hungariae 149: 1-137.

Soós, Á., 1984a. Family Micropezidae. - In: Soós, Á. & Papp, L. (eds), Catalogue of Palaearctic Diptera 9: 19-24.

Soós, Á., 1984b. Family Tanypezidae. - Same as 1984a: 26-27.

Soós, Á., 1984c. Family Psilidae. - Same as 1984a: 28-35.

Soós, Á., 1984d. Family Pyrgotidae. - Same as 1984a: 36-38.

Soós, Á., 1984e. Family Platystomatidae. - Same as 1984a: 38-45.

Soós, Á., 1984f. Family Otitidae. - Same as 1984a: 45-59.

Soós, Á., 1984g. Family Dryomyzidae. - Same as 1984a: 152-154.

Soós, Á., 1984h. Family Neottiophilidae. - Same as 1984a: 241-242.

Soós, Á., 1984i. Family Clusiidae. - In: Soós, Á. & Papp, L. (eds), Catalogue of Palaearctic Diptera 10: 11-14.

Soós, Á., 1984j. Family Trixoscelididae. - Same as 1984i: 45-48.

Soós, Á., 1984k. Family Opomyzidae. - Same as 1984i: 53-56.

Soós, Á., 1984l. Family Chyromyidae. - Same as 1984i: 56-60.

Soós, Á., 1984m. Family Tethinidae. - Same as 1984i: 107-110.

Soós, Á., 1987. Clusiidae. - In: McAlpine, J.F. et al. (eds), Manual of Nearctic Diptera, Volume 2. - Research Branch, Agriculture Canada, Monograph 28: 853-857.

Soós, Á. & Húrka, K., 1986. Family Hippoboscidae. - In: Soós, Á. & Papp, L. (eds), Catalogue of Palaearctic Diptera 11: 215-226.

Soós, Á. & Minář, J., 1986a. Family Gasterophilidae. - In: Soós, Á. & Papp, L. (eds), Catalogue of Palaearctic Diptera 11: 237-240.

Soós, Á. & Minář, J., 1986b. Family Oestridae. - Same as 1986a: 240-244.

Soós, Á. & Minář, J., 1986c. Family Hypodermatidae. - Same as 1986a: 244-251.

Soós, Á. & Oosterbroek, P., 1992. Family Cylindrotomidae. - In: Soós, Á., Papp, L. & Oosterbroek, P. (eds), Catalogue of Palaearctic Diptera 1: 179-182.

Soós, Á. & Papp, L., 1984. Genera of uncertain family position. - In: Soós, Á. & Papp, L. (eds), Catalogue of Palaearctic Diptera 10: 229.

Speight, M.C.D., 1999. A key to the European Xylotini (Dip: Syrphidae). - Entomologist's Records and Journal of Variation 111: 211-217.

Spencer, K.A., 1972. Diptera, Agromyzidae. - Handbooks for the Identification of British Insects 10(5g): 1-136.

Spencer, K.A., 1976. The Agromyzidae (Diptera) of Fennoscandia and Denmark. - Fauna Entomologica Scandinavica 5: 3-304, 305-606.

Spencer, K.A., 1987. Agromyzidae. - In: McAlpine, J.F. et al. (eds), Manual of Nearctic Diptera, Volume 2. - Research Branch, Agriculture Canada, Monograph 28: 869-879.

Spencer, K.A. & Martinez, M., 1987. Additions and corrections to the Agromyzidae section of the Catalogue of Palaearctic Diptera (Papp 1984). - Annales de la Societé Entomologique de France (N.S.) 23: 253-271.

Stackelberg, A.A., 1956. Synantropic Diptera in the fauna of the USSR. - Opredeliteli po fauna SSSR 60: 1-164. (In Russian).

Stackelberg, A.A., 1989a. Family Trichoceridae. - In: Bei-Bienko, G.Y. & Steyskal, G.C. (eds), Keys to the insects of the European Part of the USSR. Volume V. Diptera and Siphonaptera, Part 1. Brill, Leiden, etc.: 68-71 (originally published in Russian in 1969).

Stackelberg, A.A., 1989b. Family Ptychopteridae. - Same as 1989a: 164-166.

Stackelberg, A.A., 1989c. Family Dixidae. - Same as 1989a: 209-215.

Stackelberg, A.A., 1989d. Family Chaoboridae. - Same as 1989a: 216-217.

Stackelberg, A.A., 1989e. Family Thaumaleidae. - Same as

1989a: 370-372.
Stackelberg, A.A., 1989f. Family Pachyneuridae. - Same as 1989a: 373
Stackelberg, A.A., 1989g. Family Bolitophilidae. - Same as 1989a: 374-385.
Stackelberg, A.A., 1989h. Family Ditomyiidae. - Same as 1989a: 386-388.
Stackelberg, A.A., 1989i. Family Ceroplatidae (Keroplatidae). - Same as 1989a: 389-394.
Stackelberg, A.A., 1989j. Family Diadocidiidae. - Same as 1989a: 395.
Stackelberg, A.A., 1989k. Family Macroceridae. - Same as 1989a: 396-400.
Stackelberg, A.A., 1989l. Family Manotidae. - Same as 1989a: 401.
Stackelberg, A.A., 1989m. Family Hesperinidae. - Same as 1989a: 666.
Stackelberg, A.A., 1989n. Family Anisopodidae. - Same as 1989a: 681-682.
Stackelberg, A.A., 1989o. Family Lonchopteridae. - Same as 1989a: 1153-1155.
Stackelberg, A.A., 1989p. Family Syrphidae. - In: Bei-Bienko, G.Y. & Steyskal, G.C. (eds), Keys to the insects of the European Part of the USSR. Volume V. Diptera and Siphonaptera, Part 2. Brill, Leiden, etc.: 10-148 (originally published in Russian in 1970).
Stackelberg, A.A., 1989r. Family Calobatidae (Trepidariidae). - Same as 1989p: 179.
Stackelberg, A.A., 1989s. Family Micropezidae (Tylidae, in part). - Same as 1989p: 176-178.
Stackelberg, A.A., 1989t. Family Megamerinidae. - Same as 1989p: 180-181.
Stackelberg, A.A., 1989u. Family Tanypezidae. - Same as 1989p: 182.
Stackelberg, A.A., 1989v. Family Strongylophthalmyiidae. - Same as 1989p: 183.
Stackelberg, A.A., 1989w. Family Psilidae. - Same as 1989p: 184-191.
Stackelberg, A.A., 1989x. Family Pyrgotidae. - Same as 1989p: 192-193.
Stackelberg, A.A., 1989y. Family Helcomyzidae. - Same as 1989p: 277.
Stackelberg, A.A., 1989z. Family Coelopidae. - Same as 1989p: 278-279.
Stackelberg, A.A., 1989aa. Family Dryomyzidae. - Same as 1989p: 280-281.
Stackelberg, A.A., 1989ab. Family Sepsidae. - Same as 1989p: 282-289.
Stackelberg, A.A., 1989ac. Family Sciomyzidae. - Same as 1989p: 290-317.
Stackelberg, A.A., 1989ad. Family Lauxaniidae. - Same as 1989p: 318-330.
Stackelberg, A.A., 1989ae. Family Periscelididae. - Same as 1989p: 346.
Stackelberg, A.A., 1989af. Family Piophilidae. - Same as 1989p: 347-350.
Stackelberg, A.A., 1989ag. Family Neottiophilidae. - Same as 1989p: 352-353.
Stackelberg, A.A., 1989ah. Family Pallopteridae. - Same as 1989p: 354-357.
Stackelberg, A.A., 1989ai. Family Lonchaeidae. - Same as 1989p: 358-373.
Stackelberg, A.A., 1989aj. Family Odiniidae. - Same as 1989p: 374-376.
Stackelberg, A.A., 1989ak. Family Clusiidae. - Same as 1989p: 505-508.
Stackelberg, A.A., 1989al. Family Acartophthalmidae. - Same as 1989p: 509.

Stackelberg, A.A., 1989am. Family Trixoscelididae. - Same as 1989p: 538.
Stackelberg, A.A., 1989an. Family Pseudopomyzidae. - Same as 1989p: 539.
Stackelberg, A.A., 1989ao. Family Anthomyzidae. - Same as 1989p: 540-545.
Stackelberg, A.A., 1989ap. Family Opomyzidae. - Same as 1989p: 546-550.
Stackelberg, A.A., 1989aq. Family Chyromyidae. - Same as 1989p: 551-552.
Stackelberg, A.A., 1989ar. Family Aulacigastridae. - Same as 1989p: 553-554.
Stackelberg, A.A., 1989as. Family Asteiidae. - Same as 1989p: 555-556.
Stackelberg, A.A., 1989at. Family Cryptochaetidae. - Same as 1989p: 557-556.
Stackelberg, A.A., 1989au. Family Tethinidae. - Same as 1989p: 591-592.
Stackelberg, A.A., 1989av. Family Milichiidae. - Same as 1989p: 593-601.
Stackelberg, A.A., 1989aw. Family Canaceidae. - Same as 1989p: 602-603.
Stackelberg, A.A., 1989ax. Family Braulidae. - Same as 1989p: 604.
Stackelberg, A.A., 1989ay. Family Camillidae. - Same as 1989p: 647.
Stackelberg, A.A., 1989az. Family Diastatidae. - Same as 1989p: 648-650.
Stackelberg, A.A., 1989ba. Family Drosophilidae. - Same as 1989p: 651-665.
Stackelberg, A.A., 1989bb. Family Curtonotidae. - Same as 1989p: 666.
Stackelberg, A.A., 1989bc. Family Eginiidae. - Same as 1989p: 975.
Stackelberg, A.A., 1989bd. Family Streblidae. - Same as 1989p: 987-988.
Stackelberg, A.A., 1989be. Family Nycteribiidae. - Same as 1989p: 989-994.
Stackelberg, A.A., 1989bf. Family Rhinophoridae. - Same as 1989p: 1097-1102.
Stackelberg, A.A. & Negrobov, O.P., 1930-1979. 29. Dolichopodidae. - In: Lindner, E. (ed.), Die Fliegen der paläarktischen Region. Schweizerbart, Stuttgart 4(5)29: 1-530.
Starý, J. & Martinovský, J., 1993. A review of the genus Diazosma (Diptera: Trichoceridae). - European Journal of Entomology 90: 70-85. BEIDE auteurs overal met streepje op y
Steffan, W.A., 1981. Sciaridae. - In: McAlpine, J.F. et al. (eds), Manual of Nearctic Diptera, Volume 1. - Research Branch, Agriculture Canada, Monograph 27: 247-255.
Steyskal, G.C., 1987. Sepsidae. - In: McAlpine, J.F. et al. (eds), Manual of Nearctic Diptera, Volume 2. - Research Branch, Agriculture Canada, Monograph 28: 945-950.
Steyskal, G.C. & Knutson, L.V., 1981. Empididae. - In: McAlpine, J.F. et al. (eds), Manual of Nearctic Diptera, Volume 1. - Research Branch, Agriculture Canada, Monograph 27: 607-624.
Stojanovich, C.J. & Scott, H.G., 1995a. Illustrated key to the mosquitoes of Fennoscandia, Finland, Sweden, Denmark, Norway. Printed in the USA by the authors: 1-132.
Stojanovich, C.J. & Scott, H.G., 1995b. Mosquitoes of European Russia. Printed in the USA by the authors: 1-106.
Stubbs, A.E., 1972-1974. Introduction to craneflies. - Bulletin of the Amateur Entomologist's Society 31: 46-54, 83-93; 32: 14-23, 58-63, 101-107; 33: 18-23, 142-145.

Stubbs, A.E., 1982. An identification guide to the British Clusiidae. - Proceedings and Transactions of the British Entomological and Natural History Society 15: 89-93.

Stubbs, A.E., 1994. Test keys to Families of Nematocera; Genera of Tipulidae; Species of Cylindrotomidae, Pediciidae, Molophilus, Tasiocera. - Cranefly Recording Scheme Newsletter 7 (test keys included in this newsletter).

Stubbs, A.E., 1996. Test keys for Tipula and long-palped craneflies other than Tipula; Corrections and addenda. - Bulletin of the Dipterists Forum 41 (test keys included in this bulletin).

Stubbs, A.E., 1997. Test key for subfamily Limnophilinae. - Bulletin of the Dipterists Forum 43 (test keys included in this bulletin).

Stubbs, A.E., 1998. Test key for subfamily Limoniinae. - Bulletin of the Dipterists Forum 45 (test keys included in this bulletin).

Stubbs, A.E., 1999. Amendments to test keys. - Bulletin of the Dipterists Forum 47 (test keys included in this bulletin).

Stubbs, A.E. & Chandler, P.J., 1978. A dipterists handbook. - Amateur Entomologist 15: i-ix, 1-155.

Stubbs, A.E. & Drake, M., 2001. British soldierflies and their allies. British Entomology and Natural History Society: i-ix, 1-512, 20 col pls.

Stubbs, A.E. & Falk, S.J., 2002. British hoverflies (2nd edition). – British Entomological and Natural History Society, Reading: i-x; 1-469.

Stuckenberg, B.R., 1965. Notes on the Palaearctic species of Vermileo, with the description of a new species from Crete (Diptera, Vermileonidae). - Annals and Magazine of Natural History (13) 8: 495-500.

Stuckenberg, B.R., 1971. A review of the Old World genera of Lauxaniidae (Diptera). - Annals of the Natal Museum 20: 499-610.

Stuckenberg, B.R., 1973. The Athericidae, a new family in the lower Brachycera. - Annals of the Natal Museum 21: 649-673.

Stuckenberg, B.R., 1998. A revision of the Palaearctic species of Lampromyia Macquart (Diptera, Vermileonidae), with the description of a new Iberian species and a cladogram for the genus. - Bonner Zoologische Beiträge 48: 67-98.

Stuke, J.-H. & Merz, B., 2005. Prosopantrum flavifrons (Tonnoir & Mallach, 1927) in Mitteleuropa nachgewiesen (Diptera: Heleomyzoidea s.l., Cnemostapthidae). - Studia Dipterologica 11: 358.

Suwa, M. & Darvas, B., 1998. Family Anthomyiidae. - In: Papp, L. & Darvas, B. (eds), Contributions to a Manual of Palaearctic Diptera, Volume 3. Science Herald, Budapest: 571-616.

Szábo, J., 1983. Family Psychodidae. - Fauna Hungariae 156: 1-88.

Szadziewski, R., Krzywinski, J. & Gilka, W., 1997. Diptera Ceratopogonidae, Biting midges. - In: Nilsson, A. (ed.), Aquatic insects of North Europe. A taxonomic handbook. Volume 2. Odonata - Diptera. Apollo Books, Stenstrup: 243-263.

Tanasijtshuk, V.N., 1984a. Family Chamaemyiidae. - In: Soós, Á. & Papp, L. (eds), Catalogue of Palaearctic Diptera 9: 220-232.

Tanasijtshuk, V.N., 1984b. Family Cremifaniidae. - Same as 1984a: 232.

Tanasijtshuk, V.N., 1986. Fam. Chamaemyiidae. Part 1. - Fauna USSR 14(7): 1-335.

Tanasijtshuk, V.N., 1988. Family Pipunculidae. - In: Soós, Á. & Papp, L. (eds), Catalogue of Palaearctic Diptera 8: 230-245.

Tanasijtshuk, V.N., 1989a. Family Psychodidae. - In: Bei-Bienko, G.Y. & Steyskal, G.C. (eds), Keys to the insects of the European Part of the USSR. Volume V. Diptera and Siphonaptera, Part 1. Brill, Leiden, etc.: 167-198 (originally published in Russian in 1969).

Tanasijtshuk, V.N., 1989b. Family Platypezidae (Clythiidae). - In: Bei-Bienko, G.Y. & Steyskal, G.C. (eds), Keys to the insects of the European Part of the USSR. Volume V. Diptera and Siphonaptera, Part 2. Brill, Leiden, etc.: 1-9 (originally published in Russian in 1970).

Tanasijtshuk, V.N., 1989c. Family Pipunculidae (Dorylaidae). - Same as 1989b: 149-161.

Tanasijtshuk, V.N., 1989d. Family Chamaemyiidae (Ochthiphiidae). - Same as 1989b: 331-345.

Tanasijtshuk, V.N., 1992. Morphological differences and phylogenetic relationships of genera of Chamaemyiidae (Diptera). - Entomologicheskoe Obozrenie 71: 199-230 (in Russian; English translation in: Entomological Review, Washington 72(2) (1993): 66-100).

Tanasijtshuk, V.N. & Beschovski, V.L., 1991. A contribution of the study of Chamaemyia species (Diptera, Chamaemyiidae) from Bulgaria and some East European countries. - Acta Zoologica Bulgarica 41: 18-25.

Theodor, O., 1958. 9c. Psychodidae-Phlebotominae. - In: Lindner, E. (ed.), Die Fliegen der paläarktischen Region. Schweizerbart, Stuttgart 3(1)9c: 1-55.

Theodor, O., 1967. An illustrated catalogue of the Rothschild collection of Nycterybiidae (Diptera) in the British Museum (Natural History). With keys and short descriptions for the identification of subfamilies, genera, species and subspecies. - British Museum (Natural History) Publication 655: 1-506.

Theodor, O., 1968. A revision of the Streblidae (Diptera) of the Ethiopean region. - Transactions of the Royal Entomological Society of London 120: 313-373.

Theodor, O., 1975. Diptera Pupipara. - Fauna Palestina 1: 1-70

Theodor, O. & Oldroyd, H., 1964. 65. Hippoboscidae. - In: Lindner, E. (ed.), Die Fliegen der paläarktischen Region. Schweizerbart, Stuttgart 12 (1-3)65: 1-70.

Theowald, Br., 1973-1980. 15. Tipulidae. - In: Lindner, E. (ed.), Die Fliegen der paläarktischen Region. Schweizerbart, Stuttgart 3(5)1: 321-538.

Thomas, A.G.B., 1997. Rhagionidae and Athericidae, Snipe flies. - In: Nilsson, A. (ed.), Aquatic insects of North Europe. A taxonomic handbook. Volume 2. Odonata - Diptera. Apollo Books, Stenstrup: 311-320.

Thompson, F.C. & Rotheray, G., 1998. Family Syrphidae. - In: Papp, L. & Darvas, B. (eds), Contributions to a Manual of Palaearctic Diptera, Volume 3. Science Herald, Budapest: 81-139.

Timmer, J., 1990. De dazen (Diptera, Tabanidae) van de Beneluxlanden. - Wetenschappelijke Mededelingen Koninklijke Nederlandse Natuurhistorische Vereniging 138: 1-38.

Tjeder, B., 1948. The Swedish Prionocera. - Opuscula Entomologica 13: 75-99.

Tjeder, B., 1949. The first Swedish representative of the family Thaumaleidae (Dipt. Nemat.). - Opuscula Entomologica 14: 106-109.

Tjeder, B., 1953. Elephantomyia edwardsi Lack. and Pedicia littoralis Mg. in Sweden. - Opuscula Entomologica 18: 231-232.

Tjeder, B., 1963. Three subapterous crane-flies from Alaska (Dipt. Tipulidae). - Opuscula Entomologica 28: 229-241.

Tomasovic, G., 2000. Connaissances actuelles sur les Bibionidae de Belgique (Diptera, Nematocera). - Notes

Faunistiques de Gembloux 38: 21-42.

Tóth, S., 1977. Bombyliidae-Scenopinidae. - Fauna Hungariae 127: 1-87.

Tóth, S., 2004. The fauna of mosquitoes in Hungary (Diptera: Culicidae). - Natura Somogyiensis 6: 1-327 (in Hungarian).

Trojan, P., 1957. Diptera. - Klucze do Oznaczania owadow Polski 20: 1-145.

Trojan, P., 1967. Czesc XXVIII. Muchowki - Diptera. Zeszyt 24. Bujanki - Bombyliidae. - Klucze Towarzystwo Entomologiczne 55: 1-44.

Trojan, P., 1979. Tabanidae. - Fauna Polski 8: 1-309.

Tschorsnig, H.-P. & Herting, B., 1994. Die Raupenfliegen (Diptera: Tachinidae) Mitteleuropas: Bestimmungstabellen und Angaben zur Verbreitung und Ökologie der einzelnen Arten. - Stuttgarter Beiträge zur Naturkunde (A): 506: 1-170.

Tschorsnig, H.-P. & Richter, V.A., 1998. Family Tachinidae. - In: Papp, L. & Darvas, B. (eds), Contributions to a Manual of Palaearctic Diptera, Volume 3. Science Herald, Budapest: 691-827.

Upton, M.S., 1991. Methods for collecting, preserving and studying insects & allied forms. - Miscellaneous Publications of the Australian Entomological Society 3: 1-88.

Vaillant, F., 1971-1983. 9d. Psychodidae-Psychodinae. - In: Lindner, E. (ed.), Die Fliegen der paläarktischen Region. Schweizerbart, Stuttgart 3(1)9d: 1-358.

Vaillant, F., 1989. Contribution à l'etude des Diptères Lonchopteridae d'Europe et d'Afrique du Nord. - Bulletin de la Societé Vaudoise des Sciences Naturelles 79: 209-229.

Vaillant, F., 2002. Insecta: Diptera: Lonchopteridae. - Süßwasserfauna von Mitteleuropa 21(22): 1-14.

Vala, J.C., 1989. Diptères Sciomyzidae euro-méditerranéens. - Faune de France 72: 1-300.

Vanin, S., 2003. Xenasteiidae: A dipterous family new to Italy. - Bollettino del Museo Civico di Storia Naturale di Venezia 54: 91-93.

Veen, M. van, 1984. De blaaskopvliegen en roofvliegen van Nederland en België. Fifth edition. Jeugdbondsuitgeverij, 's-Graveland: 1-52.

Veen, M.P. van, 1996. De roofvliegen van Nederland. - Wetenschappelijke Mededelingen Koninklijke Nederlandse Natuurhistorische Vereniging 216: 1-120.

Veen, M.P. van, 2004. Hoverflies of Northwest Europe. Identification keys to Syrphidae. - KNNV Publishing, Utrecht: 1-254.

Ventura, D. & Carles-Tolrá, M., 2002. A new species of Tunisimyia Papp from Spain (Diptera: Xenasteiidae). - Studia Dipterologica 9: 673-677.

Venturi, F., 1947. Notulae dipterologicae I. Miltogrammini e Metopini (Dip. Sarcophagidae) dell' Italia centrale. - Redia (2) 32: 119-139.

Verlinden, L., 1991. Fauna van België. Zweefvliegen (Syrphidae). - Koninklijk Belgische Instituut voor Natuurwetenschappen, Brussel: 1-298.

Verrall, G.H., 1909. Stratiomyidae and succeeding families of the Diptera Brachycera of Great Britain. - British Flies 5: 1-780.

Verves, Y.G., 1982-1993. 64h. Sarcophaginae. - In: Lindner, E. (ed.), Die Fliegen der paläarktischen Region. Schweizerbart, Stuttgart 11(64h): 235-504.

Verves, Y.G., 1986. Family Sarcophagidae. - In: Soós, Á. & Papp, L. (eds), Catalogue of Palaearctic Diptera 12: 58-193.

Verves, Y.G., 1994. A key to genera and subgenera of Palaearctic Miltogramminae (Diptera, Sarcophagidae)

with a description of a new genus. - Dipterological Research 5: 239-247.

Vorst, O. & Hijbregts, J., 2001. Overzicht van de wijzigingen in de lijst van de Nederlandse kevers (1987-1999) (Coleoptera). - Entomologische Berichten, Amsterdam 61: 80-81.

Wagner, R., 1982. Palaearctic moth-flies: a review of the Trichomyiinae (Psychodidae). - Systematic Entomology 7: 357-365.

Wagner, R., 1985. A revision of the genus Heleodromia (Dipterra, Empididae) in Europe. - Aquatic Insects 7: 33-43.

Wagner, R., 1990a. Family Psychodidae. - In: Soós, Á. & Papp, L. (eds), Catalogue of Palaearctic Diptera 2: 11-65.

Wagner, R., 1990b. Family Chaoboridae. - Same as 1990a: 71-74.

Wagner, R., 1997a. Family Psychodidae. - In: Papp, L. & Darvas, B. (eds), Contributions to a Manual of Palaearctic Diptera, Volume 2. Science Herald, Budapest: 205-226.

Wagner, R., 1997b. Family Dixidae. - Same as 1997a: 299-303.

Wagner, R., 1997c. Family Thaumaleidae. - Same as 1997a: 325-329.

Wagner, R., 1997d. Diptera Psychodidae, Moth flies. - In: Nilsson, A. (ed.), Aquatic insects of North Europe. A taxonomic handbook. Volume 2. Odonata - Diptera. Apollo Books, Stenstrup: 133-144.

Wagner, R., 1997e. Diptera Dixidae, Meniscus Midges. - Same as 1997d: 145-148.

Wagner, R., 1997f. Diptera Thaumaleidae. - Same as 1997d: 187-190.

Wagner, R., 1997g. Diptera Empididae, Dance flies. - Same as 1997d: 333-344.

Wagner, R., 2002. Insecta: Diptera: Thaumaleidae. - Süßwasserfauna von Mitteleuropa 21(11): 39-110.

Webb, D.W., 1974. A revision of the genus Hilarimorpa (Diptera: Hilarimorphidae). - Journal of the Kansas Entomological Society 47: 172-222.

Webb, D.W., 1981a. Athericidae. - In: McAlpine, J.F. et al. (eds), Manual of Nearctic Diptera, Volume 1. - Research Branch, Agriculture Canada, Monograph 27: 479-482.

Webb, D.W., 1981b. Hilarimorphidae. - In: McAlpine, J.F. et al. (eds), Manual of Nearctic Diptera, Volume 1. - Research Branch, Agriculture Canada, Monograph 27: 603-605.

Weber, M., 1989. Dolichopodidae. - Fauna Hungariae 164: 1-243.

Weinberg, M. & Bächli, G., 1995. Diptera Asilidae. - Insecta Helvetica Fauna 11: 1-124.

Weinberg, M. & Bächli, G., 1997. Faunistik und Taxonomie der Acroceriden (Diptera) der Schweiz. - Mitteilungen der Schweizerischen Entomologischen Gesellschaft 70: 209-224.

Wheeler, T.A., 1987. Drosophilidae. - In: McAlpine, J.F. et al. (eds), Manual of Nearctic Diptera, Volume 2. - Research Branch, Agriculture Canada, Monograph 28: 1011-1018.

Wheeler, T.A., 1998. Family Chyromyidae. - In: Papp, L. & Darvas, B. (eds), Contributions to a Manual of Palaearctic Diptera, Volume 3. Science Herald, Budapest: 457-461.

Wheeler, W.M., 1931. Demons of the dust: a study in insect behavior. London: i-xviii, 1-378.

White, I.M., 1988. Tephritid flies; Diptera: Tephritidae. - Handbooks for the Identification of British Insects 10(5a): 1-134.

White, I.M. & Elson-Harris, M.M., 1994. Fruitflies of economic significance: Their identification and bionomics. - International Institute of Entomology, London. CD-Rom.

Wiederholm, T., (ed.), 1983. Chironomidae of the Holarctic region. Keys and diagnosis. Part 1. Larvae. - Entomologica Scandinavica, Supplement 19: 1-457.

Wiederholm, T., (ed.), 1986. Same as 1983, Part 2. Pupae - Entomologica Scandinavica, Supplement 28: 1-482.

Wiederholm, T., (ed.), 1989. Same as 1983, Part 3. Adult males. - Entomologica Scandinavica, Supplement 34: 1-532.

Williams, F.X., 1939. Biological studies in Hawaiian water-loving insects. III. Diptera or flies. - Proceedings of the Hawaiian Entomological Society 10: 281-315.

Wilson, R.S. & Ruse, L.P., 2005. A guide to the identification of genera of chironomid pupal exuviae occurring in Britain and Ireland (including common genera from Northern Europe) and their use in monitoring lotic and lentic fresh waters. - Freshwater Biological Association Special Publication 13: 1-176.

Wirth, W.W. & Grogan, W.L., 1988. The predacious midges of the world (Diptera: Ceratopogonidae; tribe Ceratopogonini). - Flora and Fauna Handbooks 4: i-xv, 1-160.

Withers, P., 1978. The British species of the genus Suillia (Diptera, Heleomyzidae), including a species new to science. - Proceedings and Transactions of the British Entomological and Natural History Society 20: 91-104.

Withers, P., 1989. Moth flies (Diptera: Psychodidae). - Dipterists Digest 4: 1-83.

Wood, D.M., 1981a. Axymyiidae. - In: McAlpine, J.F. et al. (eds), Manual of Nearctic Diptera, Volume 1. - Research Branch, Agriculture Canada, Monograph 27: 209-212.

Wood, D.M., 1981b. Asilidae. - In: McAlpine, J.F. et al. (eds), Manual of Nearctic Diptera, Volume 1. - Research Branch, Agriculture Canada, Monograph 27: 549-573.

Wood, D.M., 1987a. Rhinophoridae. - In: McAlpine, J.F. et al. (eds), Manual of Nearctic Diptera, Volume 2. - Research Branch, Agriculture Canada, Monograph 28: 1187-1291.

Wood, D.M., 1987b. Tachinidae. - In: McAlpine, J.F. et al. (eds), Manual of Nearctic Diptera, Volume 2. - Research Branch, Agriculture Canada, Monograph 28: 1193-1269.

Woodley, N.E., 1995. The genera of Beridinae (Diptera: Stratiomyidae). - Memoirs of the Entomological Society of Washington 16: 1-231.

Yankovskii, A.V., 2003. A key to black flies (Diptera: Simuliidae) of Russia and adjacent lands (of the former USSR). Zoological Institute, Russian Academy of Sciences, St.Petersburg: 1-570 (In Russian).

Yeates, D.K. & Wiegmann, B.M., 1999. Congruence and Controversy: Toward a higher-level phylogeny of the Diptera. - Annual Review of Entomology 44: 397-428.

Yeates, D.K. & Wiegmann, B.M. (eds), 2005. The Evolutionary Biology of Flies. Columbia University Press, New York: i-ix, 1-430.

Zaitzev, A.I., 1994. Fungus gnats of the fauna of Russia and adjacent regions. Part I. The families Ditomyiidae, Bolitophilidae, Diadocidiidae, Keroplatidae, Mycetophilidae (subfamilies Mycomyiinae, Sciophilinae, Gnoristinae, Allactoneurinae, Leiinae). Nauka, Moscow: 1-288 (In Russian).

Zaitzev, A.I., 2003. Fungus gnats of the fauna of Russia and adjacent regions. Part II. - International Journal of Dipterological Research 14: 77-386.

Zaitzev, V.F., 1984. Family Ulidiidae. - In: Soós, Á. & Papp, L. (eds), Catalogue of Palaearctic Diptera 9: 59-66.

Zaitzev, V.F., 1989a. Family Therevidae. - In: Bei-Bienko,

G.Y. & Steyskal, G.C. (eds), Keys to the insects of the European Part of the USSR. Volume V. Diptera and Siphonaptera, Part 1. Brill, Leiden, etc.: 821-837 (originally published in Russian in 1969).

Zaitzev, V.F., 1989b. Family Bombyliidae. - Same as 1989a: 843-885.

Zaitzev, V.F., 1989c. Family Phoridae. - Same as 1989a: 1156-1172.

Zaitzev, V.F., 1989d. Family Bombyliidae. - In: Soós, Á. & Papp, L. (eds), Catalogue of Palaearctic Diptera 6: 43-169.

Zatwarnicki, T., 1997. Diptera Ephydridae, Shore flies. - In: Nilsson, A. (ed.), Aquatic insects of North Europe. A taxonomic handbook. Volume 2. Odonata - Diptera. Apollo Books, Stenstrup: 383-399.

Zeegers, Th., 1992. Tabel voor de grotere sluipvliegen en horzels van Nederland. Jeugdbondsuitgeverij, Utrecht: 1-85.

Zeegers, Th., 1997. Een veldtabel voor de inheemse Zwarte Vliegen (Bibionidae). - Vliegenmepper 6(2): 4-9.

Zeegers, Th., 1998. Correctie en aanvulling op de Zwarte vliegentabel. - Vliegenmepper 7(1): 5.

Zeegers, Th. & Haaren, T. van, 2000. Dazen en dazenlarven. - Wetenschappelijke Mededelingen Koninklijke Nederlandse Natuurhistorische Vereniging 225: 1-114 (Additions: Vliegenmepper 9(2) (dec. 2000): 2-3).

Zeegers, Th. & van Veen, M., 1993. Pissebedvliegen (Rhinophoridae) in Nederland: een voorlopig overzicht. - Vliegenmepper 2(2): 1-10.

Ziegler, J., 2003. Ordnung Diptera, Zweiflügler (Fliegen und Mücken). - In: Dathe H.H. (ed.), Lehrbuch der Speziellen Zoologie, Band 1: Wirbellose Tiere, 5.Teil: Insecta (Zweite Auflage): 756-860,

Zilahi-Sebess, G., 1960. Nematocera 1. - Fauna Hungariae 55: 1-70.

Zimin, L.S. & Elberg, K.Y., 1989. Family Muscidae. - In: Bei-Bienko, G.Y. & Steyskal, G.C. (eds), Keys to the insects of the European Part of the USSR. Volume V. Diptera and Siphonaptera, Part 2. Brill, Leiden, etc.: 839-974 (originally published in Russian in 1970).

Zimin, L.S., Zinovyeva, K.B. & Stackelberg, A.A., 1989. Family Tachinidae. - In: Bei-Bienko, G.Y. & Steyskal, G.C. (eds), Keys to the insects of the European Part of the USSR. Volume V. Diptera and Siphonaptera, Part 2. Brill, Leiden, etc.: 1111-1310 (originally published in Russian in 1970).

Zimina, L.V., 1989. Family Conopidae. - In: Bei-Bienko, G.Y. & Steyskal, G.C. (eds), Keys to the insects of the European Part of the USSR. Volume V. Diptera and Siphonaptera, Part 2. Brill, Leiden, etc.: 162-175 (originally published in Russian in 1970).

Zuijlen, J.W.A. van, 1999. Notes on the Fallén collection of Opomyzidae (Diptera) in the Naturhistoriska Riksmuseet, Stockholm. - Studia Dipterologica 6: 129-134.

Zuska, J., 1984. Family Piophilidae. - In: Soós, Á. & Papp, L. (eds), Catalogue of Palaearctic Diptera 9: 234-240.

Zuska, J. & Pont, A.C., 1984. Family Sepsidae. - In: Soós, Á. & Papp, L. (eds), Catalogue of Palaearctic Diptera 9: 154-167.

Zwick, P., 1992. Family Blephariceridae. - In: Soós, Á., Papp, L. & Oosterbroek, P. (eds), Catalogue of Palaearctic Diptera 1: 39-54.

FIGURE DETAILS

2a habitus Tabanus americanus (Tabanidae); McAlpine 1981a

2b habitus Brachystoma vesiculosum (Brachystomatidae), male; Niesiolowski 1992

3 habitus brachycerous fly; Lane & Crosskey 1993

4 head calyptrate fly; Lane & Crosskey 1993

5 head calyptrate fly; Lane & Crosskey 1993

6a antenna Aedes alternans (Culicidae), female; Colless & McAlpine 1991

6b antenna Tabanidae; Merz & Haenni 2000 (drawing by B. Merz)

6c antenna calyptrate fly; Lane & Crosskey 1993

7 head calyptrate fly; Lane & Crosskey 1993

8 head calyptrate fly; Colless & McAlpine 1991

9 head Paracantha gentilis (Tephritidae); Hennig 1973

10 head Anorostoma currani (Heleomyzidae); Hennig 1973

11 thorax Tachinidae; Colless & McAlpine 1991

12 thorax Calliphoridae; Lane & Crosskey 1993

13 hypothetical ground plan of the Diptera wing; Lane & Crosskey 1993

14 wing Calliphora vicina (Calliphoridae); Lane & Crosskey 1993

15 elements of the Diptera leg; Oosterbroek 1981

16 bristle (macrotrichium) with socket (alveolus); Merz & Haenni 2000 (drawing by B. Merz)

17 antenna Aedes alternans (Culicidae), female; Colless & McAlpine 1991

18 antenna Aedes alternans (Culicidae), male; Colless & McAlpine 1991

19 antenna Dictenidia bimaculata (Tipulidae), male; Mannheims 1951

20 antenna Tabanidae; Merz & Haenni 2000 (drawing by B. Merz)

21 head calyptrate fly; Lane & Crosskey 1993

22 head Tipula trivittata (Tipulidae); McAlpine 1981a

23 habitus Tipula luna (Tipulidae); Lindner 1948

24 wing Mycetobia pallipes (Mycetobiidae); Krivosheina 1997d (drawing by N.P. Krivosheina)

25 thorax Tipulidae; Hendel 1928

26 thorax Cylindrotoma distinctissima (Cylindrotomidae); Peus 1952

27 wing Trichocera saltator (Trichoceridae); Dahl & Krzeminska 1997

28 wing Diazosma tibeticum (Trichoceridae); Dahl 1957

29 head Prionocera turcica (Tipulidae); Tjeder 1948

30 wing Tricyphona protea (Pediciidae); Alexander & Byers 1981

31 habitus Elephantomyia edwardsi (Limoniidae); Tjeder 1953

32 wing Elliptera tennessa (Limoniidae); Alexander & Byers 1981

33 wing Tipula tricolor (Tipulidae); Alexander & Byers 1981

34 habitus Cylindrotoma distinctissima (Cylindrotomidae); Lindner 1948

35 head Phalacrocera replicata (Cylindrotomidae); Alexander & Byers 1981

36 hypopygium Triogma trisulcata (Cylindrotomidae); Peus 1952

37 ovipositor Triogma trisulcata (Cylindrotomidae); Peus 1952

38 wing Sylvicola fenestralis (Anisopodidae); Peterson 1981

39 wing Mochlonyx velutinus (Chaoboridae); Saether 2002

40 wing Ptychoptera lacustris (Ptychopteridae); Peus 1958

41 haltere Ptychoptera contaminata (Ptychopteridae); Oosterbroek 1981

42 habitus Tinearia alternata (Psychodidae); Wagner 1997a (drawing by A. Szappanos)

42a wing Trichomyia urbica (Psychodidae); Wagner 1997a

43 habitus Dixa maculata (Dixidae), female; Wagner 1997e (drawing by A. Szappanos)

44 antenna Dixella nova (Dixidae); Peters 1981

45 habitus Aedes vexans (Culicidae), female; Minář 2000a (drawing by A. Szappanos)

46 head Chaoborus plumicornis (Chaoboridae), male; Oldroyd 1964

47 wing Thaumalea tricuspis (Thaumaleidae); Tjeder 1949

48 head Thaumalea brevidens (Thaumaleidae); Wagner 2002

49 habitus Cecidomyia pini (Cecidomyiidae), male; Skuhravá 1997 (drawing by M. Skuhravá)

50 wing Greniera spec. (Simuliidae); Peterson 1981

51 head Twinnia spec. (Simuliidae); Peterson 1981

52 wing Chironomidae; Oosterbroek 1998

53 wing Culicoides antennalis (Ceratopogonidae); Colless & McAlpine 1991

54 wing Liponeura bilobata (Blephariceridae); Hennig 1973

55 head Blepharicera tetrophthalma (Blephariceridae), female; Hennig 1973

56 wing Protaxymyia melanoptera (Axymyiidae); Krivosheina 2000 (drawing by N.P. Krivosheina)

57 wing Groveriella carpathica (Cecidomyiidae); Skuhravá 1997 (drawing by M. Skuhravá)

58 head Swammerdamella acuta (Scatopsidae); Haenni 1997a (drawing by A. Szappanos)

59 habitus Scatopse notata (Scatopsidae), male; Colless & McAlpine 1991

60 wing Synneuron decipiens (Synneuridae); Peterson & Cook 1981

61 wing Hyperoscelis eximia (Canthyloscelidae); Haenni 1997b (drawing by S. Podenas)

62 hind leg Hyperoscelis eximia (Canthyloscelidae); Haenni 1997b (drawing by S. Podenas)

63 habitus Bibio marci (Bibionidae), male; Skartveit 1997 (drawing by A. Szappanos)

64 tibia fore leg Dilophus femoratus (Bibionidae); Zilahi-Sebess 1960

65 head Hesperinus brevifrons (Hesperinidae), male; Hardy 1981

66 wing Hesperinus brevifrons (Hesperinidae), male; Hardy 1981

67 wing Plecia americana (Pleciidae), male; Hardy 1981

68 habitus Penthetria funebris (Pleciidae), male; Séguy 1940

69 head Pnyxia scabiei (Sciaridae); Steffan 1981

70 wing Pnyxia scabiei (Sciaridae); Menzel & Mohrig 1997 (drawing by F. Menzel)

71 wing Pachyneura fasciata (Pachyneuridae); Krivosheina 1997a (drawing by N.P. Krivosheina)

72 wing Mycetobia pallipes (Mycetobiidae); Krivosheina 1997d (drawing by N.P. Krivosheina)

73 head Sciara spec. (Sciaridae); Colless & McAlpine 1991

74 wing Zygoneura sciarina (Sciaridae); Menzel & Mohrig 1997 (drawing by F. Menzel)

75 habitus Mycetophila propria (Mycetophilidae), female; Colless & McAlpine 1991

76 wing Ditomyia fasciata (Ditomyiidae); Hendel 1928

77 wing Keroplatus spec. (Keroplatidae); Hendel 1928

78 habitus Keroplatus tipuloides (Keroplatidae), male; Séguy 1940

79 wing Macrocera spec. (Keroplatidae); Hendel 1928

80 habitus Macrocera centralis (Keroplatidae), female; Séguy 1940

81 wing Bolitophila spec (Bolitophilidae); Hendel 1928

82 wing Diadocidia spec. (Diadocidiidae); Hendel 1928

83 wing Sciarosoma borealis (Sciaroidea); Chandler 2002

84 wing Sylvicola fenestralis (Anisopodidae); Peterson 1981

85 head Rachicerus obscuripennis (Rachiceridae), female; James 1981

86 habitus Xylomya maculata (Xylomyidae), female; Lindner 1938

87 wing Vermitigris orientalis (Vermileonidae); Edwards 1932

88 habitus Beris clavipes (Stratiomyidae), female; Rozkošný 1973

89 habitus Coenomyia ferruginea (Coenomyiidae), male; Rozkošný & Nagatomi 1997a (drawing by J. Pál)

89a spined scutellum Coenomyia ferruginea (Coenomyiidae), male; Oosterbroek 1981

90 habitus Xylophagus ater (Xylophagidae), female; Séguy 1926

91 head calyptrate fly; Séguy 1923

92 head Haematopota bigoti (Tabanidae), female; Séguy 1926

93 head Stichopogon trifasciatus (Asilidae), female; Wood 1981b

94 head Arthroceras leptis (Rhagionidae), female; James & Turner 1981

95 head Atherix variegata (Athericidae), female; Webb 1981a

96 head Platypalpus infectus (Hybotidae); Collin 1961

97 head Hercostomus nobilitatus (Dolichopodidae); Oldroyd 1970

98 habitus Ogcodes gibbosus (Acroceridae), male; Sack 1936

99 habitus Cyrtus pusillus (Acroceridae), female; Séguy 1926

100 wing Thereva nobilitata (Therevidae), female; Goot 1985

101 wing Empidideicus hungaricus (Mythicomyiidae); Greathead & Evenhuis 1997 (drawing by L. Papp)

102 wing Oestrus ovis (Oestridae); Minář 1980

103 fifth tarsal segment Inopus rubriceps (Stratiomyidae); Colless & McAlpine 1991

104 fifth tarsal segment Musca domestica (Muscidae); Colless & McAlpine 1991

105 wing Ibisia marginata (Athericidae); Thomas 1997

106 antenna Atherix marginata (Athericidae); Verrall 1909

107 wing Fallenia fasciata (Nemestrinidae); Richter 1997a (drawing by V.A. Richter)

108 habitus Nemestrinus obscuripennis (Nemestrinidae), female; Richter 1997a (drawing by Florinskaya)

109 wing Trichopsidea costata (Nemestrinidae), female; Sack 1933

110a-d antenna: a: Heptatoma pellucens; b: Haematopota pluvialis; c: H. italica; d: Tabanus autumnalis (Tabanidae); Séguy 1926

111 habitus Tabanus autumnalis (Tabanidae), female; Oldroyd 1970

112 habitus Xylophagus ater (Xylophagidae), female; Séguy 1926

112a antenna Xylophagus ater (Xylophagidae), male; Séguy 1926

113 habitus Vermileo degeeri (Vermileonidae), male; Séguy 1926

114 habitus Lampromyia cylindrica (Vermileonidae), male; Séguy 1926

115 habitus Xylomya maculata (Xylomyidae), female; Lindner 1938

116 habitus Oxycera rara (Stratiomyidae), female; Séguy 1950

117 habitus Coenomyia ferruginea (Coenomyiidae), male; Rozkošný & Nagatomi 1997a (drawing by J. Pál)

117a spined scutellum Coenomyia ferruginea (Coenomyiidae), male; Oosterbroek 1981

118 habitus Rhagio scolopaceus (Rhagionidae), female; Verrall 1909

119 habitus Chrysopilus splendidus (Rhagionidae), male; Séguy 1926

120 wing Hispanomydas hispicus (Mydidae); Sack 1933

121 head Leptomydas padischach (Mydidae), male; Richter 1997b (drawing by V.A. Richter)

122 wing Hilarimorpha ditissa (Hilarimorphidae); Webb 1981b

123 head Dioctria engeli (Asilidae), male; Goot 1985

124 head Proagonistes athletes (Asilidae); Oldroyd 1964

125 head Scenopinus fenestralis (Scenopinidae), female; Kelsey 1981

126 habitus Scenopinus fenestralis (Scenopinidae), female; Tóth 1977

127 habitus Thereva nobilitata (Therevidae), female; Goot 1985

128 habitus Bombylella atra (Bombyliidae), male; Tóth 1977

129 wing Amphicosmus elegans (Bombyliidae); Hall 1981

130 habitus Apolysis andalusiaca (Bombyliidae), female; Engel 1932

131 habitus Brachystoma vesiculosum (Brachystomatidae), male; Niesiolowski 1992

132 wing Scenopinus unifasciatus (Scenopinidae); Engel 1932

133 habitus Chelifera diversicauda (Empididae), male; Niesiolowski 1992

134 habitus Clinocera dimidiata (Empididae), male; Niesiolowski 1992

135 habitus Episyrphus balteatus (Syrphidae), male; Thompson & Rotheray 1998 (drawing by J. Pál)

206 wing Trixoscelis marginella (Trixoscelididae); Soós 1981
207 wing Camilla glabra (Camillidae); Hennig 1958
208 wing Euthychaeta spectabilis (Campichoetidae); Chandler 1987
209 head Camilla nigrifrons (Camillidae); Mihályi 1972
210 femur fore leg Camilla flavicauda (Camillidae); Beuk & de Jong 1994
211 habitus Coenia palustris (Ephydridae), male; Papp 1975
212 head Limnellia quadrata (Ephydridae); Zatwarnicki 1997
213 head Thinoscatella spec. (Ephydridae); Zatwarnicki 1997
214 head Odinia boletina (Odiniidae); Oldroyd 1970
215a habitus Prosopantrum flavifrons (Cnemospathidae); Ismay & Smith 1994
215b detail wing Prosopantrum flavifrons (Cnemospathidae); Ismay & Smith 1994
215c thorax Prosopantrum flavifrons (Cnemospathidae); Ismay & Smith 1994
216a head Chiropteromyza wegelii (Chiropteromyzidae); original, drawing by D. Langerak
216b habitus Chiropteromyza wegelii (Chiropteromyzidae); original, drawing by D. Langerak
217 habitus Drosophila spec. (Drosophilidae); Oldroyd 1970
218 head Chymomyza amoena (Drosophilidae); Wheeler 1987
219 wing Drosophila deflexa (Drosophilidae); Shorrocks 1972
220 frons Drosophilidae a: Leucophenga maculata, b: Microdrosophila congesta; Bächli 1998 (drawings by G. Bächli)
221 head Diastata adusta (Diastatidae); Oldroyd 1970
222 habitus Diastata fuscula (Diastatidae); Chandler 1986
223 head Euthychaeta spectabilis (Campichoetidae); Chandler 1987
224 head Campichoeta zernyi (Campichoetidae); Chandler 1987
225 wing Euthychaeta spectabilis (Campichoetidae); Chandler 1987
226 antenna Tunisimyia shalam (Xenasteiidae); Freidberg 1994
227 wing Tunisimyia shalam (Xenasteiidae); Freidberg 1994
228 wing Asteia amoena (Asteiidae); Oldroyd 1970
229 head Asteia elegantula (Asteiidae); Papp 1973
230 wing Stenomicra cogani (Stenomicridae); Irwin 1982
231 head Stenomicra delicata (Stenomicridae); Collin 1944
232 head Santhomyza inermis (Anthomyzidae); Roháček 1998b (drawing by J. Roháček)
233 femur fore leg Anthomyza gracilis (Anthomyzidae); Mihályi 1972
234 wing Paranthomyza nitida (Anthomyzidae); Hennig 1958
235 wing Cercagnota collini (Anthomyzidae); Roháček 1998b (drawing by J. Roháček)
236 wing base Geomyza combinata (Opomyzidae); Hennig 1958
237 habitus Geomyza combinata (Opomyzidae), male; Soós 1981

238a head Canace salonitana (Canacidae); Mathis 1998 (drawing by W.N. Mathis)
238b wing Xanthocanace ranula (Canacidae); Hendel 1928
239a wing Chaetostomella cylindrica (Tephritidae); Mihályi 1960
239b head Chaetostomella cylindrica (Tephritidae); Merz 1994
240 wing Rhodesiella plumiger (Chloropidae); Hennig 1958
241 head Ophiomyia maura (Agromyzidae); Spencer 1987
242 wing Melanagromyza laetifica (Agromyzidae); Spencer 1987
243 habitus Acartophthalmus nigrinus (Acartophthalmidae), female; McAlpine 1987a
244 habitus Themira putris (Sepsidae), male; Séguy 1934
245 head Heteromeringia nigrimana (Clusiidae); Soós 1959
246 habitus Heteromeringia nigrimana (Clusiidae), female; Soós 1981 (drawing by J. Pál)
247 head Thyreophora cynophila (Piophilidae); Hendel 1928
248 habitus Piophila casei (Piophilidae), male; Soós 1959
249 head Aulacigaster leucopeza (Aulacigastridae); Oldroyd 1970
250 wing Aulacigaster leucopeza (Aulacigastridae); Hennig 1958
251 wing Pelomyiella hungarica (Tethinidae); Munari 1998
252 head Pelomyia steyskali (Tethinidae); Munari 1998 (drawing by A. Szappanos)
253 head and thorax Borboropsis puberula (Heleomyzidae); Gill & Peterson 1987
254 habitus Nidomyia cana (Heleomyzidae), male; Papp 1998l
255 head Tethina strobliana (Tethinidae); Soós 1981
256 wing Chyromya flava (Chyromyidae); Hendel 1928
257a wing Parochthiphila coronata (Chamaemyiidae); Hendel 1928
257b head Leucopis psyllidiphaga (Chamaemyiidae); McLean 1998b
258 wing Leiomyza laevigata (Asteiidae); Hennig 1958
259 head Periscelis annulata (Periscelididae); McAlpine 1987b
260 wing Periscelis annulata (Periscelididae); McAlpine 1987b
261 wing Odinia maculata (Odiniidae); Papp 1998c
262 wing Turanodinia tisciae (Odiniidae); Papp 1998c
263 wing Liriomyza alpicola (Agromyzidae); Hendel 1931
264 wing Phytomyza affinis (Agromyzidae); Hendel 1928
265 head Napomyza elegans (Agromyzidae); Oldroyd 1970
266 habitus Agromyza albipennis (Agromyzidae), female; Spencer 1987
267 wing Paranthomyza nitida (Anthomyzidae); Hennig 1958
268 femur fore leg Anthomyza gracilis (Anthomyzidae); Mihályi 1972
269 head Amygdalops thomasseti (Anthomyzidae); Roháček 1998b (drawing by J. Roháček)
270 head Aphaniosoma quadrivittatum

(Chyromyidae); McAlpine 1987c
271 head Cacoxenus indagator Loew (Drosophilidae);
Papp 1973
272a head Hemeromyia spec. (Carnidae); Papp 1998b
(drawing by A. Szappanos)
272b wing Hemeromyia spec. (Carnidae); Sabrosky
1987
273 head Desmometopa m-nigrum (Milichiidae);
Hennig 1958
274 head Phyllomyza securicornis (Milichiidae);
Hennig 1937b
275 head Phyllomyza longipalpis (Milichiidae);
Hennig 1937b
276 head Madiza britannica (Milichiidae); Hennig
1937b
277 head Madiza glabra (Milichiidae); Hennig 1937b
278 wing Milichia speciosa (Milichiidae); Hennig
1937b
279 wing Leiomyza laevigata (Asteiidae); Hennig
1958
280 wing Asteia amoena (Asteiidae); Oldroyd 1970
281 head Asteia elegantula (Asteiidae); Papp 1973
282 head Chloropidae: a: Trachysiphonella scutellata;
b: Gampsocera numerata; c: Eurina calva; d:
Chlorops scutellaris; Dely-Draskovits 1978
283 thorax Chlorops certimus (Chloropidae);
McAlpine 1981b
284 habitus Meromyza pratorum (Chloropidae),
female; Balachowski & Mesnil 1935
285 wing Rhodesiella plumiger (Chloropidae);
Hennig 1958
286 head Carnus hemapterus (Carnidae); Sabrosky
1987
287 head Meoneura obscurella (Carnidae); Hennig
1937b
288 wing Carnus hemapterus (Carnidae); Sabrosky
1987
289 wing Meoneura obscurella (Carnidae); Hennig
1937b
290 habitus Coenia palustris (Ephydridae), male;
Papp 1975
291 wing Paralimna cinerella (Ephydridae); Hennig
1958
292 head Limnellia quadrata (Ephydridae);
Zatwarnicki 1997
293 head Thinoscatella spec. (Ephydridae);
Zatwarnicki 1997
294 head Pseudopomyza atrimana
(Pseudopomyzidae); Chandler 1983
295 head Pseudopomyza atrimana
(Pseudopomyzidae); Hennig 1958
296 wing Pseudopomyza atrimana
(Pseudopomyzidae); Chandler 1983
297 head Chymomyza amoena (Drosophilidae);
Wheeler 1987
298 wing Drosophila deflexa (Drosophilidae);
Shorrocks 1972
299 habitus Myopa stigma (Conopidae), male;
Bankowska 1979
300 habitus Leopoldius calceatus (Conopidae), male;
Bankowska 1979
301 wing Ulidia albidipennis (Ulidiidae); Hennig 1940
302 wing Colobaea bifasciella (Sciomyzidae);
Stackelberg 1989ac
303 wing Eusapromyza poeciloptera (Lauxaniidae);
Papp 1978
304 head Sciosapromyza advena (Lauxaniidae);
Hennig 1958

305 head Saltella sphondylii (Sepsidae); Hennig 1958
306 wing Platystoma seminationis (Platystomatidae);
Oldroyd 1970
307 wing Salticella fasciata (Sciomyzidae); Rozkošný
1998a
308 head Coremacera marginata (Sciomyzidae);
Hendel 1928
309 habitus Psacadina verbekei (Sciomyzidae);
Knutson & Lyneborg 1965
310 habitus Pelidnoptera nigripennis (Phaeomyiidae),
male; Rozkošný 1998b (drawing by J. Pál)
311 head Eurygnathomyia bicolor (Pallopteridae);
Merz 1998b (drawing by B. Merz)
312 wing Eurygnathomyia bicolor (Pallopteridae);
Merz 1998b (drawing by B. Merz)
313 katepisternum Eurygnathomyia bicolor
(Pallopteridae); Merz 1998b (drawing by B. Merz)
314 head Neuroctena anilis (Dryomyzidae); Czerny
1930
315 habitus Dryomyza flaveola (Dryomyzidae);
Lindner 1948
316 ventral view thorax Dryomyza flaveola
(Dryomyzidae); Hennig 1958
317 ventral view thorax Helcomyza ustulata
(Helcomyzidae); Hennig 1958
318 ventral view thorax Helcomyza mirabilis
(Helcomyzidae); Steyskal 1987
319 head Helcomyza ustulata (Helcomyzidae);
McAlpine 1998c (drawing by A. Szappanos)
320 wing Helcomyza ustulata (Helcomyzidae);
Czerny 1930
321 head Heterocheila buccata (Heterocheilidae);
Czerny 1930
322 habitus Cryptochetum grandicorne
(Cryptochetidae), female; Papp 1978
323 head Adapsilia coarctata (Pyrgotidae); Hendel 1928
324 wing Adapsilia coarctata (Pyrgotidae); Hendel
1928
325 head Loxocera aristata (Psilidae); Hendel 1928
326 wing base Psilidae; Oldroyd 1970
327 habitus Chyliza annulipes (Psilidae): Lyneborg
1964
328 head Orellia falcata (Tephritidae); White 1988
329 wing base Tephritidae; Oldroyd 1970
330 wing Stemonocera cornuta (Tephritidae); Merz
1994
331 habitus Urophora cardui (Tephritidae); Lindner
1948
332 wing base Strongylophthalmyia ustulata
(Strongylophthalmyiidae); Hennig 1958
333 habitus Dalmannia punctata (Conopidae), male;
Bankowska 1979
334 habitus Sepedon spinipes (Sciomyzidae), male;
Séguy 1934
335a femur hind leg Megamerina dolium
(Megamerinidae); Oldroyd 1970
335b head Megamerina dolium (Megamerinidae);
Hennig 1942
336 habitus Megamerina dolium (Megamerinidae);
Soós 1959
337 habitus Sphyracephala europaea (Diopsidae),
male; Simova-Tosic & Stojanovic 1999
338 habitus Themira putris (Sepsidae), male; Séguy
1934
339 wing Micropeza corrigiolata (Micropezidae);
Hendel 1928
340 head Micropeza corrigiolata (Micropezidae);
Hendel 1928

341 head Calobata petronella (Micropezidae); Oosterbroek 1981
342 wing Tanypeza longimana (Tanypezidae); Hennig 1937a
343 habitus Tanypeza longimana (Tanypezidae); Chandler 1975
344 wing Lipoleucopis praecox (Chamaemyiidae); McLean 1998a (drawing by I. McLean)
345 wing base Parochthiphila spectabilis (Chamaemyiidae); Hennig 1958
346 habitus Parochthiphila spectabilis (Chamaemyiidae); McLean 1998a (drawing by I. McLean)
347 wing Cremifania nigrocellulata (Cremifaniidae); Czerny 1936
348 head Cremifania lanceolata (Cremifaniidae); Papp 1994
349 wing Ulidia albidipennis (Ulidiidae); Hennig 1940
350 habitus Seioptera vibrans (Ulidiidae), female; Lyneborg 1964
351 habitus Myennis octopunctata (Ulidiidae), female; Lindner 1948
352 wing Palloptera ustulata (Pallopteridae); Morge 1974
353 habitus Palloptera modesta (Pallopteridae), female; Morge 1974
354 wing Platystoma seminationis (Platystomatidae); Oldroyd 1970
355 head Platystoma lugubre (Platystomatidae); Mihályi 1972
356 head Ulidia erythrophthalma (Ulidiidae); Mihályi 1972
357 habitus Dorycera graminum (Ulidiidae), female; Mihályi 1972
358 wing Geomyza hackmani (Opomyzidae); Drake 1993
359 wing Opomyza punctata (Opomyzidae); Drake 1993
360 head Anomalochaeta guttipennis (Opomyzidae); Trojan 1957
361 head Calamoncosis sasae (Chloropidae); Nartshuk 1987
362 wing Rhodesiella plumiger (Chloropidae); Hennig 1958
363 habitus Azorastia mediterranea (Nannodastiidae), male; Papp 1980
364 wing Azorastia mediterranea (Nannodastiidae); Papp & Mathis 2001
365 wing Paralimna cinerella (Ephydridae); Hennig 1958
366 habitus Ochthera mantis (Ephydridae); Séguy 1926
367 habitus Pelomyia occidentalis (Tethinidae); Merz et al. 2001
368 wing base Strongylophthalmyia ustulata (Strongylophthalmyiidae); Hennig 1958
369 habitus Strongylophthalmyia ustulata (Strongylophthalmyiidae); Papp 1978
370 tarsal claw Ornithomya (Hippoboscidae); Falcoz 1926
371 habitus Ornithomya rupes (Hippoboscidae), female; Büttiker 1994 (drawing by W. Büttiker)
372 wing Brachytarsina diversa (Streblidae); Theodor 1968
373 head and fore legs Brachytarsina flavipennis (Streblidae), male; Hůrka 1998b (drawing by A. Szappanos)
374 thorax Tachinidae; Colless & McAlpine 1991

375 head Trixa caerulescens (Tachinidae); Tschorsnig & Richter 1998 (drawing by H.-P. Tschorsnig)
376 habitus Gasterophilus intestinalis (Gasterophilidae), female; Grunin 1969
377 habitus Oedemagena tarandi (Hypodermatidae), female; Grunin 1964-1969
378 head Crivellia corinnae (Hypodermatidae), female; Grunin 1964-1969
379 head Oestrus ovis (Oestridae); Colless & McAlpine 1991
380 habitus Pharyngomyia picta (Oestridae), female; Grunin 1966-1969
381 habitus Rhinoestrus purpureus (Oestridae), female; Grunin 1966-1969
382 head Cephalopina titillator (Oestridae); Grunin 1966-1969
383 habitus Cephalopina titillator (Oestridae); Grunin 1966-1969
384 thorax Pegomyia quadrivittata (Anthomyiidae); Suwa & Darvas 1998 (drawing by M. Suwa)
385 habitus Graphomya maculata (Muscidae), female; Séguy 1923
386 meron Chrysotachina alcedo (Tachinidae); Wood 1987b
387 meron Winthemia fumiferanae (Tachinidae); Wood 1987b
388 wing Phyllomya volvulus (Tachinidae); Tschorsnig & Richter 1998 (drawing by H.-P. Tschorsnig)
389 wing Steleoneura czernyi (Tachinidae); Tschorsnig & Richter 1998 (drawing by H.-P. Tschorsnig)
390 scutellum Hylemya alcathoe (Anthomyiidae); McAlpine 1981b
391 habitus Anthomyia spec. (Anthomyiidae), male; Papp & Schumann 2000
392 head Craspedochoeta pullata (Anthomyiidae); Hennig 1976
393 head Scathophaga stercoraria (Scathophagidae); Séguy 1934
394 habitus Spaziphora hydromyzina (Scathophagidae); Séguy 1934
395 wing Fannia canicularis (Fanniidae); Huckett & Vockeroth 1987
396 habitus Musca domestica (Muscidae), female; Lane & Crosskey 1993
397 wing Muscina levida (Muscidae); Huckett & Vockeroth 1987
398 head Azelia cilipes (Muscidae), female; Hennig 1964
399 thorax Exorista larvarum (Tachinidae); McAlpine 1981a
400 scutellum and subscutellum Melanophora roralis (Rhinophoridae); Wood 1987a
401 scutellum and subscutellum Calliphoridae; Mihályi 1979
402 posterior spiracle Peribaea orbata (Tachinidae); Draber-Monko 1989 after Crosskey 1977
403 posterior spiracle Phyto discrepans (Rhinophoridae); Draber-Monko 1989 after Crosskey 1977
404 posterior spiracle Melanophora roralis (Rhinophoridae); Draber-Monko 1989 after Crosskey 1977
405 habitus Ectophasia crassipennis (Tachinidae); Mihályi 1986
406 habitus Cylindromyia brassicaria (Tachinidae); Mihályi 1986

407 habitus Eginia ocypterata (Muscidae), male;
Séguy 1923
408 meron bristles Eginia ocypterata (Muscidae);
Gregor et al. 2002
409 wing Lucilia caesar (Calliphoridae); Rognes 1991
410 wing Stevenia deceptoria (Rhinophoridae); Pape
1998b (drawing by P. Lidmark)
411 habitus Melanomya nana (Calliphoridae); Mihályi
1986
412 posterior spiracle Melanophora roralis
(Rhinophoridae); Draber-Monko 1989 after
Crosskey 1977
413 posterior spiracle Phumosia lutescens
(Calliphoridae); Draber-Monko 1989 after
Crosskey 1977
414 lower calypter and scutellum Rhinophoridae;
Mihályi 1986
415 scutellum and subscutellum Melanophora roralis
(Rhinophoridae); Wood 1987a
416 head Rhinomorinia sarcophagina
(Rhinophoridae); Mihályi 1986
417 scutellum and subscutellum Calliphoridae;
Mihályi 1979
418 wing Angioneura cyrtoneurina (Calliphoridae);
Rognes 1991
419 thorax Sarcophagidae (detail); Mihályi 1979
420a habitus Sarcophaga carnaria (Sarcophagidae);
Pape 1987
420b wing Sphenometopa fastuosa (Sarcophagidae);
Venturi 1947
421 thorax Calliphoridae; Mihályi 1979
422 thorax Protophormia terraenovae (detail)
(Calliphoridae); Rognes 1991
423 thorax Protophormia atriceps (detail)
(Calliphoridae); Rognes 1991
424 leg Frirenia tenella (Cecidomyiidae); Skuhravá
1997 (drawing by M. Skuhravá)
425 thorax Tipulidae; Hendel 1928
426 head Tipula carinifrons (Tipulidae); Tjeder 1963
427 habitus Tipula carinifrons (Tipulidae), female;
Lantsov & Chernov 1987
428 head Tricyphona hannai antennata (Pediciidae);
Tjeder 1963
429 habitus Tricyphona hannai antennata
(Pediciidae); Tjeder 1963

430 habitus Dicranomyia lindrothi (Limoniidae);
Tjeder 1963
431 copula Chionea scita (Limoniidae); Byers 1983
432 thorax Tricyphona hannai antennata (Pediciidae);
Tjeder 1963
433 copula Clunio pacificus (Chironomidae); Oka
1930
434 wing Telmatogeton pectinata (Chironomidae);
Goetghebuer & Lenz 1950
435 habitus Epidapus gracilis (Sciaridae), female;
Menzel & Mohrig 1997 (drawing by F. Menzel)
436 head Swammerdamella acuta (Scatopsidae);
Haenni 1997a
437 habitus Aprionus miki (Cecidomyiidae);
Skuhravá 1997 (drawing by M. Skuhravá)
438 underside Melophagus ovinus
(Hippoboscidae); Theodor & Oldroyd 1964
439 tarsal claw Braula coeca (Braulidae); Hennig 1938
440 habitus Braula coeca (Braulidae); Hendel 1928
441 habitus Penicillidia dufouri (Nycteribiidae); Soós
1955
442 habitus Crataerina pallida (Hippoboscidae),
female; Büttiker 1994 (drawing by R. Heinertz
after original by Schneider-Orelli)
443 habitus Melophagus ovinus (Hippoboscidae),
female; Büttiker 1994 (drawing by R. Heinertz)
444 head and fore legs Brachytarsina flavipennis
(Streblidae), male; Hůrka 1998b (drawing by A.
Szappanos)
445 habitus Aenigmatias lubbocki (Phoridae),
female; Oldroyd 1970
446 habitus Ariasella pandellei (Hybotidae), male;
Séguy 1926
447 habitus Aptilotus paradoxus (Sphaeroceridae),
female; Séguy 1926
448 habitus Tricimba brachyptera (Chloropidae),
female; Soós 1980
449 habitus Carnus hemapterus (Carnidae), female;
De Meijere 1913
450 head Rhynchopsilopa nitidissima (Ephydridae);
Séguy 1934
451 habitus Stiphrosoma sabulosum
(Anthomyzidae); Roháček 2006

GURE CREDITS

Permission to use figures was gratefully received from:

Minister of Public Works and Government Services, Canada, 2005. Agriculture and Agri-Food Canada (from McAlpine et al. (eds) 1981, 1987, Manual of Nearctic Diptera: 2a, 22, 30, 32, 33, 35, 38, 44, 50, 51, 60, 65, 66, 67, 69, 84, 85, 93, 94, 95, 122, 125, 129, 146, 147, 194, 195, 204, 218, 241, 242, 243, 253, 259, 260, 266, 270, 272b, 283, 286, 288, 297, 318, 386, 387, 390, 395, 397, 399, 400, 415, 463, 567)

Apollo Books, Denmark (from Nilsson (ed.), 1997. Aquatic Insects of North Europe, Vol. 2: 105, 212, 213, 292, 293, 498, 503)

Brill Academic Publishers, the Netherlands (from Fauna Entomologica Scandinavica: 88, 156, 159, 160, 161, 166a, 166b, 167, 168, 170, 171, 172, 173, 199, 409, 418, 420, 422, 423, 512)

British Entomological Society, United Kingdom (from British Journal of Entomology and Natural History: 358, 359)

Peter Chandler, United Kingdom (83, 157, 158, 159, 160, 165, 208, 222, 223, 224, 225, 294, 296, 343, 533, 546)

Chapman & Hall, United Kingdom (From Lane & Crosskey, 1993, Medical Insects and Arachnids: 3, 4, 5, 6c, 7, 12, 13, 14, 21, 153, 176, 177, 179, 396)

Milan Chvála, Czech Republic (156, 161, 166a, 166b, 167, 168, 170, 171, 172, 173, 512)

CSIRO Australia, Australia (from Colless & McAlpine, 1991, Insects of Australia: 6a, 8, 11, 17, 18, 53, 59, 73, 75, 103, 104, 162, 374, 379)

Deutsches Entomologisches Institut des Leibniz-Zentrums für Agrarlandschafts- und Landnutzungsforschung, Germany (from Beiträge zur Entomologie): 191, 205, 207, 234, 236, 240, 250, 258, 267, 273, 279, 285, 291, 295, 304, 305, 316, 317, 332, 345, 362, 365, 368)

Entomological Society of Israel and Amnon Freidberg, Israel (226, 227)

Yves Gonseth, Switzerland (174b, 371, 442, 443)

Frantisec Gregor, Czech Republic (178, 408)

Jean-Paul Haenni, Switzerland (58, 61, 62, 436, 556)

John Ismay, United Kingdom (215, 216, 485)

Leif Lyneborg, Denmark 1964 (327, 350)

I.F.G. McLean, United Kingdom (257b)

Bernhard Merz, Switzerland (6b, 16, 20, 239b, 311, 312, 313, 330, 367, 573)

Henk Meuffels, the Netherlands (164)

NEST Foundation, Hungary (from Papp & Darvas (eds), 1997 - 2000: Contributions to a Manual of Palaearctic Diptera: 6b, 16, 20, 24, 27, 42, 42a, 43, 45, 49, 56, 57, 58, 61, 62, 63, 70, 71, 72, 74, 89, 101, 107, 108, 117, 121, 135, 139, 141, 158, 159, 160, 175, 190, 203, 220, 232, 235, 238a, 251, 252, 261, 269, 272a, 307, 310, 311, 312, 313, 319, 344, 346, 373, 375, 384, 388, 389, 391, 410, 424, 435, 436, 437, 444, 474, 479, 489, 493, 494, 506, 509, 516, 524, 533, 535, 537, 565, 578)

Natural History Museum, Basel, Switzer-land (from Mitteilungen der Entomol-ogischen Gesellschaft Basel: 367)

Stefan Niesiolowski, Poland (2b, 131, 133, 134)

Thomas Pape, Denmark (410, 420)

Laszlo Papp, Hungary (183, 186, 192, 193, 198, 200, 201, 206, 211, 229, 237, 239, 244, 245, 246, 248, 254, 255, 261, 262, 271, 281, 290, 303, 322, 336, 338, 348, 363, 364, 369, 441, 448, 452, 459, 462, 467, 471, 472, 478, 483, 484, 492, 495, 515, 520, 525, 528, 531, 538, 549, 553, 557, 562, 566, 570, 575, 577, 581)

Adrian Pont, United Kingdom (199)

Knut Rognes, Norway (409, 418, 422, 423)

Jindrich Rohácek, Czech Republic (185, 451, 457)

Royal Entomological Society, United Kingdom (from Handbooks for the Identification of British Insects: 97, 111, 138, 155, 163, 180, 181, 214, 217, 221, 228, 249, 265, 280, 306, 326, 329, 335a, 354, 445, 464, 466, 497, 511, 540, 552, 554, 558, 560, 568)

Rudolf Rozkosny, Czech Republic (88, 89, 93, 117, 150, 307, 310, 564)

E. Schweizerbart'sche Verlagsbuchhandlung, Germany (from Lindner (ed.), 1930-1974, Die Fliegen der Paläarktischen Region: 19, 23, 26, 34, 36, 37, 40, 86, 98, 109, 115, 120, 130, 132, 142, 174a, 263, 274, 275, 276, 277, 278, 287, 289, 301, 314, 315, 320, 321, 331, 335b, 342, 347, 349, 351, 376, 377, 378, 380, 381, 382, 383, 392, 398, 434, 438, 439, 453, 490, 505, 513, 532)

Bryan Shorrocks, United Kingdom (219)

Duska Simova-Tosic, Serbia (337, 496)

The author has taken much care to ask permission from the copyright owners for the figures in this book. If you think that your copyright is affected, please contact the publisher.

ACKNOWLEDGEMENTS

The author would like to thank the following persons and institutions: My good friend Nico Schonewille for putting me on this project. The co-authors of the Dutch precursor, Herman de Jong and Liekele Sijstermans, who because of lack of time could not participate in preparing the present book. Willem Hurkmans for translating the Dutch version. The publisher (KNNV Publishing, Utrecht) for the generous financial support with the translation and with the distribution of the test copy, and especially Wilma Seijbel and Rijnvis van Wirdum, for advice and guidance during the development stage of the book.

Special thanks are due to all the people who tested the key and/or critically reviewed the text. The generous help received with the Dutch version is acknowledged therein. A test copy of the English version was mailed to some 40 Dipterists throughout Europe and even further. In return, many valuable corrections, comments and additions were received. This substantially improved the book and it is therefore a great pleasure to say thanks to Daniele Avesani (Italy), Rudolf Bährmann (Germany), Miroslav Barták (Czech Republic), Miguel Carlos-Tolrá (Spain), Peter Chandler (UK), Pierfilippo Cerretti (Italy), Graham Collins (UK), Amnon Freidberg (Israel), Paul Gatt (Malta), David Gibbs (UK), David Greathead (UK), Lita Greve Jensen (Norway), Jean-Paul Haenni (Switzerland), Willem Hurkmans (Netherlands), Ladislav Jedlička (Slovakia), Paul Kramer (UK), Nina Krivosheina (Russia), Franco Mason (Italy), Bernard Merz (Switzerland), Lorenzo Munari (Italy), Emilia Nartshuk (Russia), Andrey Przhiboro (Russia), Vera Richter (Russia), Jindrich Rohacek (Czech Republic), Rudolf Rozkošný (Czech Republic), Jukka Salmela (Finland), Wolfgang Schacht (Germany), Vasily Sidorenko (Russia), Bradley Sinclair (Germany), Daniel Whitmore (Italy), Kaj Winqvist (Finland) and Joachim Ziegler (Germany).

Finally, with respect to *Chiropteromyza wegelii* Frey, I would like to thank Pekka Vilkamaa (Finland) for checking the typematerial, Jean-Paul Haenni (Switzerland) and Laszlo Papp (Hungary) for the loan of specimens, and Dick Langerak (Netherlands) for the drawings he made so that illustrations for the family Chiropteromyzidae could be included as well.

BLISHING DATA:

AUTHOR:
Pjotr Oosterbroek
Dept of Entomology, Zoological Museum
University of Amsterdam

TRANSLATION:
Willem Hurkmans,
Bureau Hurkmans, Zwolle

GRAPHIC LAYOUT AND DESIGN:
Erik de Bruin, Varwig Design, Hengelo

PRINTED BY:
DZS d.d.

COVER ILLUSTRATIONS:
Front cover illustration:
Habitus Anthrax anthrax
(Bombyliidae), male
After Tóth 1977
Used by courtesy of Lázló Papp

Back cover illustration:
Brachycerous fly
After Lane & Crosskey, 1993
Reproduced with permission
from Chapman & Hall.

© KNNV Publishing, Utrecht, 2006.
ISBN 90-5011-245-5 / 978-90-5011-245-1
www.knnvpublishing.nl

OTHER AVAILABLE TITLES ON INSECTS BY KNNV PUBLISHING

HOVERFLIES OF NORTHWEST EUROPE
Identification keys to the Syrphidae
M.P. van Veen

This book identifies the hoverflies of Northwest Europe and incorporates the knowledge that has been published in many articles over past decades. The book covers 500 species – nearly twothirds of the known European syrphid fauna. The book is a rich source of reference for entomologists and specialists working on nature conservation, as well as for amateurs who are fascinated by hoverflies!

Hardcover | 2005 | 16,5 x 24 cm | 256 pp | € 34,95 | English | 978-90-5011-199-8

WATERBEETLES OF THE NETHERLANDS
M.B.P. Drost, H.P.J.J. Cuppen, E.J. van Nieukerken

The present book is an excellent and still up-to-date key to all the 350 species of waterbeetles occurring in The Netherlands. With 824 detailed illustrations, contributing to a reliable identification.

Paperback | Second edition | 2006 | 17,5 x 25,5 cm | 280 pp | € 34,95 | Dutch | 978-90-5011-053-2

THE DUTCH SPECIES OF CADDIS LARVAE
Identification and ecology
Bert Higler

The main body of the book consists of keys to all Dutch species (about 150) of Caddis larvae. The author did research on the subject for about 40 years, and is the Dutch expert on the subject. He made all of the drawings himself. To complete the book it contains 24 wonderful full-colour pictures.

Paperback | 2005 | 16,5 x 24 cm | 144 pp | € 29,95 | Dutch | 978-90-5011-212-3

HOW TO ORDER THESE BOOKS?
KNNV Publishing, Publishing Foundation of the Royal Dutch Natural Historic Society. Please contact KNNV Publishing for orders or additional information.
www.knnvpublishing.nl

Printed in the United States
By Bookmasters

Printed in the United States
By Bookmasters